Volume

1 gallon = 4 quarts = 231 cubic inches = 3.79 liters = 8.34 pounds (weight)

1 quart = 2 pints = 946.33 ml = 0.946 liters

1 pint = 2 cups

1 cup = 16 tablespoons

1 tablespoon = 3 teaspoons

1 teaspoon = 4.734 milliliters(ml) or cubic centimeters (cm^3 or c.c.)

1 cubic yard = 27 cubic feet = 46,656 cubic inches = 0.765 cubic meters

1 cubic foot = 1728 cubic inches = 7.48 gallons = 0.0078 cords = 0.028 cubic meters = 28.32 liters

1 board foot = 144 cubic inches

1 cubic inch = 0.00058 cubic feet = 16.387 cubic centimeters

1 liter = 1000 cubic centimeters = 1.06 quarts

1 cubic foot of air = 0.0817 pounds

1 cubic foot of ice = 57.5 pounds

Densities of Various Common Substances

Material	gm/cm^3	lb/ft^3
alcohol	0.79	49.3
benzine	0.90	56.2
blood	1.04	65.0
gasoline	0.69	41.2
mercury	13.6	849
milk	1.03	64.3
sea water	1.025	63.96
gold	19.3	1200
lead	11.3	708
copper	8.89	555
iron	7.85	490
aluminum	2.70	168
wood (maple)	0.55	34.1
cork	0.25	15.6
dry air at sea level	0.00129	0.081
humid air at sea level	0.00119	0.074
carbon dioxide	0.00198	0.123
carbon monoxide	0.00125	0.077
hydrogen	0.00009	0.0056
oxygen	0.00143	0.088

Pressure

Air pressure at sea level = 14.7 pounds/square inch

Water pressure = 0.433 pounds/square inch/foot of depth

Power

1 horsepower = 550 foot-pounds/second = 746 watts

1 watt = one joule/second = 10^7 ergs/second

Activity

1 curie = 3.70×10^{10} disintegrations/second

Metric Prefixes

giga = $\times 10^9$ deci = $\times 10^{-1}$

mega = $\times 10^6$ centi = $\times 10^{-2}$

kilo = $\times 10^3$ milli = $\times 10^{-3}$

Modern
Physical
Science

Modern Physical Science

Charles A. Payne
Professor of Chemistry
Dean, School of Sciences and Mathematics
Morehead State University

William R. Falls
Professor of Science Education
Head, Department of Science Education
Morehead State University

WM. C. BROWN COMPANY PUBLISHERS
Dubuque, Iowa

contents

contents

19

the birth of earth 393

20

the chemical origin of life 417

24

chemical and physical pollution 493

appendices 509

glossary 519

index 535

To
Essie and Bea

preface

The permeation of science into the lives of modern society is a matter of fact. One needs only to study his immediate surroundings briefly to become aware of science's importance to our advancement and well-being. Almost each day yields a major breakthrough into the unknown that will influence society for eons to come. Our social foundation is built around the continuance of investigation through the application of scientific principles and the development of new theories. The reading of any daily newspaper furnishes proof that for a general citizen to find his place in society he must develop a basic understanding of science. Colleges have realized this need and have included courses in science as part of the general education requirements for all degree candidates.

The accomplishments of science scan the spectrum of nature from the microscopic world to the macroscopic world. Yet, the attainment of knowledge well beyond the present ends of available knowledge is an absolute necessity if society is to continue to better itself. Not until we find solutions to numerous problems, including cures for cancer and heart disease, new sources of available energy, a replacement for our diminishing natural resources, and the relief from want for the underdeveloped nations and for the multitudes within the more fortunate nations will man's search for knowledge slacken. And then, more problems will arise and man will change direction to encounter them.

In their search for scientific knowledge, many students desire to discover the manner in which scientists secure their information and understanding of the universe. Our aim, as authors of this textbook, is to help the students obtain an understanding of science and the scientist. The fields of astronomy, chemistry, earth science, and physics are presented through an interrelated approach. The book contains numerous original illustrations and photographs which we, the authors, conceived to strengthen various concepts discussed. Many other illustrations and photographs were furnished by leading scientists, scientific equipment companies, manufacturing companies, and outstanding research organizations. Professor Allen L. Lake has displayed his considerable talents in the art of photography by contributing many unique and meaningful illustrations.

We have included topics in this text that are of special interest to today's students which are not found in other physical science texts. These special topics are the science of sound and music, the chemical origin of life, and chemical and physical pollution.

In writing this text, we have tried to write in a simple and direct manner so that the text is easy to read. In addition, we have emphasized concepts of science instead of simply presenting statements to be memorized and regurgitated at examination time. Finally, we have attempted to capture the interest of the non-science majors who, after all, are the audience for whom the text is intended.

We are indebted to a number of our colleagues at Morehead State University for many stimulating conversations concerning various scientific concepts and philosophies that we wish to convey to the students in this text. Some of the colleagues who have helped us are James R. Chaplin, David R. Cutts, David K. Hylbert, Madison E. Pryor, Charles J. Whidden, and John C. Philley.

We are further indebted to the following reviewers whose many helpful suggestions were incorporated into the final manuscript:

Professor C. W. Dowse, chairman, Department of Physics, University of Wisconsin-Whitewater; Professor Whit Marks, chairman, Department of Physics, Central State University, Edmond, Oklahoma; Professor John P. Schellenberg, Associate Professor of Physical Science, Kutztown State College, Kutztown, Pennsylvania; and Professor Paul R. Wignall, chairman, Math-Science Division, Brevard Community College, Cocoa, Florida.

Finally, and with special thanks, we are indebted to our long-suffering wives who read and advised us on the student readability of the manuscript, who kept us from becoming too pedantic in many places, and who typed the manuscript.

WILLIAM R. FALLS
CHARLES A. PAYNE

to the student

In the writing of this text, careful attention was given to making it clear, concise, and easily readable by the student. It is neither possible nor desirable, however, to eliminate all scientific or literary terms that a student may be unfamiliar with. Therefore, at the beginning of each chapter there is an important listing called *Vocabulary* the mastery of which is essential to the understanding of the scientific concepts contained in that chapter. *Before you begin to read any chapter, you should look up the Introductory Vocabulary words in the Glossary at the back of the text.* Once you are familiar with the definitions of the words in the Introductory Vocabulary, the chapter will be much easier to understand.

The Glossary is very extensive and covers the technical words with which you would normally be unfamiliar.

At the end of each chapter is a summary section that highlights for you the main concepts and facts that are central to an understanding of the material taken up in the chapter.

Following the chapter summary section are questions and problems relating to the important concepts discussed in the chapter. Each question and problem is followed by a page number in parentheses that indicates where you should turn in the event that you are not able to answer the question.

A *mathematics refresher* is included in the back of the text for those of you who need a brief review of mathematics.

on understanding science

1

WHAT IS SCIENCE
SCIENCE AND THE NON-SCIENCE MAJOR
UNITS AND SYSTEMS OF MEASUREMENT
VECTOR AND SCALAR QUANTITIES

vocabulary

Magnitude	Metric System
Scalar Quantity	Science
Vector Quantity	

There are countless writings of what science has done or is expected to do in the immediate future. Each generation is aware of the role that science plays in helping man control his surroundings in a manner beneficial to him. Man is also aware of the duty of science to see that knowledge is used in a manner that will preserve, control, and benefit society.

Science and its techniques have had more influence on man in the last century than in all of previous civilized time combined. Science has overcome many natural forces and has permitted man partially to change his environment to one that is more productive and beneficial. Science has led to means of replenishing the soil and increasing its growing capacity several fold, of controlling population growth, of finding and controlling new forms of energy, and of enhancing man's ability to solve the increasingly complex problems of our technological society through the use of new generation computers. Some of the age-old dreams of man such as space exploration, vital organ transplants, and the unraveling of the genetic code of life have in the last decade or so been realized through the relentless curiosity and brilliant investigations of scientists throughout the world.

What scientific breakthroughs lie just around the corner in the future? The control of nuclear fusion reactions similar to the reactions used by our sun to produce energy promises man the almost limitless supply of energy that his future society will need. It now appears that controlled fusion reactions will become a reality in the next decade. It also seems likely that cancer, the disease that has stubbornly resisted scientists' efforts to unravel its secrets for so long, will finally fall prey to scientists' persistent scrutiny and research. Neither knowledgeable and intelligent extrapolation nor extravagant and fanciful guesses may be sufficient to foretell what benefits science may produce for man in the distant future.

The scientific resources of our country have often been mobilized in an effort to aid in the solution of various national emergencies. Man can attribute to these efforts the control of malaria and yellow fever and the development of atomic energy, synthetic materials, computers, and insecticides.

Science is not free of errors, because science is an enterprise managed by human beings. Each of us has seen the products of numerous scientific achievements suddenly become an integral part of our everyday lives and then promptly disappear because further research indicates the product is hazardous to our health or environment.

Often new scientific discoveries may be used foolishly by man to undermine his environment, waste our planet's natural resources, or even develop more efficient methods of exterminating each other through warfare. The discoverer of new knowledge, however, is not to blame for the misuse of that knowledge by ignorant, misguided, or selfish individuals. For instance, society cannot blame the inventor of the internal combustion engine for the atmospheric pollution that results today from use of untold millions of these engines without adequate pollution control devices. Nor can the discoverers of various insecticides be blamed because society has so often used insecticides without sufficient thought about the possible ecological imbalances that may result.

WHAT IS SCIENCE

The term "science" is derived from the Latin *scientia* (*scire,* to learn or to know) and from the German *wissenschaft* (an organized body of knowledge). Science, then, includes history, philosophy, and other areas that are based on systematized knowledge, the accepted truths of which can be studied or learned. For the purposes of this book, however, *science* will refer to the general areas of nature that incorporate matter and energy. Even with this stipulation, the meaning of the term is extremely complex.

Science has been defined as what scientists do. A difficulty immediately arises regarding this definition in that any two scientists do not necessarily do the same thing. Any two scientists will, however, attack problems with some degree of logic and with some means of making measures to establish limits. Perhaps a better attempt to define science would be to define science not in terms of people, but in terms of methods of knowledge acquisition. In this light, science is a branch of knowledge which deals with information systematically arranged so as to exhibit the natural laws governing the many processes of nature. Science is similar to other branches of knowledge in this respect. Where science quite often differs from other branches of knowledge is in the use by scientists of highly complex means to collect data. The data are then critically studied and become the basis for prediction. Further intense study of data leads to the development and application of the conceptual model. In other words, science is a process by which one observes and interprets information that leads to the development of new or corrected concepts, the application of which extends our overall understanding of the world. Additionally, these observations may suggest possible new routes and areas for exploration by other interested people.

One can summarize the meaning of science as being both a process and a product. Science is a process in that it makes use of scientific procedures to determine what experiments to perform and to interpret logically the experimental data obtained. Then use is made of prior observations to determine how existing theories should be modified. Science is a product in that man then adds the information gained to his vast storehouse of knowledge.

SCIENCE AND THE NON-SCIENCE MAJOR

One of the most important goals of a liberal education is to develop in society an understanding of science. Our society has advanced to the point that scientific literacy is imperative. One needs only to look at the general reading media to realize the importance of some degree of understanding of science and to see the immediate concern of educators.

Each member of society is a potential parent, and this fact in itself shows the need for an understanding of science and its nature. No longer is science only for adults, as can readily be verified by a brief look into the elementary school curriculum. Unless each adult has available some understanding of the concepts involved in science, he is unable to advise his children with some degree of satisfaction. It can be said that the individual who understands science also realizes that science is an inseparable part of the framework of human thought. He realizes too that it is not the responsibility of the scientist alone to interpret the consequences of national scientific decisions, but that all mankind must be involved. For this reason alone, science is for everybody.

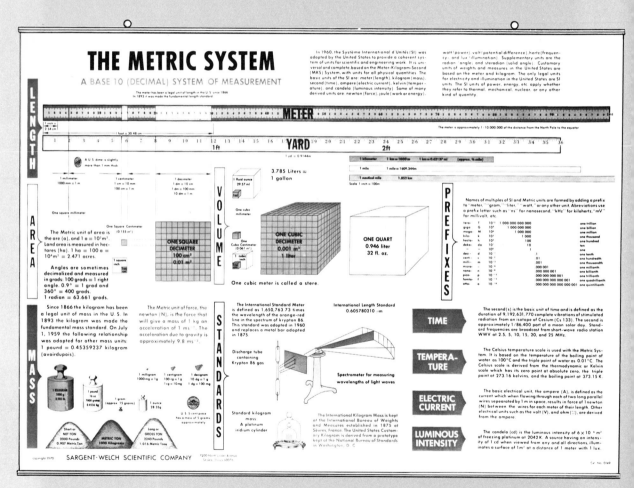

(Courtesy Sargent-Welch Scientific Company, Skokie, Illinois)

understanding science

UNITS AND SYSTEMS OF MEASUREMENT

For a full understanding of science, one must be aware of the units used to make quantitative measurements in science. The fundamental concepts of length, mass, and time serve to define most other concepts used in science. For instance, density is mass divided by volume, and speed is distance divided by time.

Standardization of measurements is as important in the fields of commerce and trade as it is in the field of measurement. The first known civilization on Earth recognized the need for standard units of measurement. Therefore, about 4,000 years ago the Sumerian civilization used certain stones, designated by the high priest, as legal weights to be used as standard weights throughout the Sumerian kingdom. Standards of length and volume were also established. Later the Sumerian standards of weight, length, and volume were adopted by the Babylonians, who, in turn, passed a modification of these standards on to the Greek civilization.

The ancient Egyptians used the weight of a certain number of grains of wheat as their standard of weight, and the cubit, the length from a man's elbow to the tip of his outstretched middle finger, as their unit of length.

It is not so important what arbitrary objects are declared the standard units of measurements as what the availability is of exact copies of the object that can be used for general measurements throughout society. Thus, the distance from the elbow to outstretched middle finger made an acceptable unit of length for the ancient Egyptian except for the fact that the distance varied with each individual. Likewise, the early use by the British of the length of three barley seeds as representing an inch had the disadvantage that the length of barley seeds varied from one barley plant to another.

It was obvious to scientists, government officials, and traders that there existed a need for a standard set of weights and measurements that could be used by all people and that could be easily copied for use in all lands.

In 1799, the French adopted what is today called the *metric system*. In the metric system, the first fundamental unit is that of length—the meter. The meter is defined as one ten millionth the distance from the equator to the north pole. Eventually, the standard meter became the distance between two lines drawn on a platinum bar which is kept at 0°C by the International Bureau of Weights and Measures located at Sevres, France. In order to permit accurate measures of this length by various people throughout the world, the standard was redefined in 1961 as being 1,650,763.73 times the wave length of the orange line seen in a spectroscope emitted by krypton-86 when the element is heated to incandescence. The standard yard is thus defined as 3600/3937 of the standard meter.

The second fundamental unit is that of the second. As was true of the concept of length, various attempts at standardizing this unit of time were made, and as man increased his need for greater accuracy, better standards were developed. Until the year 1960, various standards were used, the most prominent being based on the period of Earth's rotation. The second was defined as 1/86,400 of a mean solar day, a measure which reflected the supposed constancy of Earth's rotation on its axis. Evidence collected in later years pointed out the slight irregularities in the period of Earth's rotation and the more nearly constant orbital motion of Earth about the sun. The exactness that was needed for this measurement of time was realized, so in 1960 the Bureau redefined the standard second as 1/31,556,925.747 of the year 1900. The year 1900 was chosen because man established that the length of time for Earth to rotate was increasing about 0.001 seconds per rotation each century, due to tidal friction and other uncontrollable factors.

The third fundamental unit, mass, is a quantitative measure of the concept known as inertia, a property of all matter. The weight of a body, though mistakenly taken as identical to mass, is quite different in many respects from mass. The two concepts are discussed at length later in the text. The standard of mass is the kilogram, which is defined as the mass of a certain block of platinum-iridium alloy preserved at Sevres, France. For many years the gram was defined as the mass of one cubic centimeter (cm^3 or cc) of water at $4°C$, but this standard was dropped due to the difficulty in measuring a cm^3 of water and in controlling other variables involved in the precise measurements. The accuracy of measure, however, was improved by making the kilogram (1000 grams) equivalent to the mass of 1000 cm^3 of water at $4°C$, probably the reason that the kilogram is the accepted unit of mass rather than the gram.

Three systems of units are in common use today:

1. Scientists worldwide use the meter-kilogram-second system (*mks* for the initial letter in each unit) as the preferred system of units. The basic unit of length in this system is the meter, a measure which is equal to 39.37 inches. The meter is divided into submultiples such as centimeters (1/100 meters) and millimeters (1/1000 meters). Multiples are also used such as kilometer (1000 meters) and megameter (1,000,000 meters). Some important prefixes used in the metric system are shown in Table 1.1.

TABLE 1.1

Common prefixes used in the Metric System of Measurement
The same prefixes apply for units of mass, time, and volume.

kilo = 1000 (10^3)	kilometer = 1000 meters
hecto = 100 (10^2)	hectometer = 100 meters
deka = 10 (10^1)	dekameters = 10 meters
unit = 1 (10^0)	meter = 1 meter
dici = 1/10 (10^{-1})	decimeter = 0.1 meter
centi = 1/100 (10^{-2})	centimeter = 0.01 meter
milli = 1/1000 (10^{-3})	millimeter = 0.001 meter
micro = 1/1000000 (10^{-6})	micrometer = 0.000001 meter
nano = 1/1000000000 (10^{-9})	nanometer = 0.000000001 meter

The mass of an object which is large enough to permit the measure of its length in meters would naturally have its mass measured in a large unit; thus the correct unit is the kilogram. If one were to measure the length, width, and height of an object, he would express each in meters in this system. The product of the three measures would yield the volume of the object in cubic meters. If the volume, then, were divided into the mass of the object, a related concept known as *density* would result.

Example 1.1

A block of granite 3m x 2m x 1m was found to have a mass of 16,320 kgm. What is the density of granite?

$$D = m/V$$

$$= \frac{16{,}320 \text{ kgm.}}{3\text{m x } 2\text{m x } 1\text{m}}$$

$$= \frac{16{,}320 \text{ kgm.}}{6\text{m}^3}$$

$$= 2720 \text{ kgm/m}^3$$

The *mks* system unit of time is the second, a unit very familiar to everyone. One can safely assume that the standard second, as previously defined in this chapter, will have to be more accurately defined. In fact, a current attempt in redefining the second is through the use of radioactive decay, an extremely reliable measure, since the rate of decay of a given isotope is unaffected by any physical change. Another method under study makes use of the vibrations of the nitrogen atom in the ammonia molecule as well as atoms in other molecules. This method is generally known as the "atomic clock" method and is extremely constant.

2. The centimeter-gram-second (*cgs*) system is the smaller of two systems within the metric system of units. The basic unit of mass in this system is the gram, about the mass of a cubic centimeter of water at 4°C, as previously discussed. The density of water in this system is 1 gm/cm³, since $D = m/v = 1$ gm/cm³. There are about 28.35 grams in one ounce, thus some 453.6 grams to the pound. The length of time within the system is the second.

In this system there is a second unit of volume known as the milliliter which is equal in magnitude to the cubic centimeter. This unit is very useful to the chemist or to others who deal primarily with liquid measures. It is defined as the amount of liquid which can be held in a container one cm in each dimension; thus the milliliter (ml) is equivalent to the cm³ except that it deals with liquid measure entirely. The druggist and the physician still find the cubic centimeter, or cc, as the convenient dimension and term for liquid measure, however. The liter, then, is the amount of liquid volume held in a container 10 cm in all dimensions and thus has a volume of 10 cm $\times$ 10 cm $\times$ 10 cm = 1000 cm³ = one liter.

3. The English system of units is used in the United States for most non-scientific purposes. The foot is described as the fundamental unit of length; but

the mile, the yard, and the inch are in common use. Areas are generally measured in square inches, square feet, and square yards within this system, while volumes are measured in cubic inches, cubic feet, and cubic yards. One also finds volumes of various liquids measured in such units as teaspoons, tablespoons, cups, pints, quarts, and gallons. How many teaspoonfuls would be equivalent to one gallon? Imagine all of the conversion factors one must know and all the calculations one must do in order to determine the answer.

The origin of the inch has been discussed, but nothing has been said about the origin of the mile. The mile originates from the Latin *mille passuum,* meaning one thousand paces. The Roman soldier marched with a pace which was two steps of 2½ feet each; thus the mille passuum was 5000 feet; the modern mile is now 5280 feet. The foot is now defined as exactly 1200/3937 of the standard meter, which makes the inch 2.54000508 cm, rounded off to 2.54 cm for useful purposes.

In the English system, common use is made of a unit of weight rather than of a unit of mass. Since 1893, the pound has been defined as being exactly 1/2.2046 the weight of an object whose mass is one kilogram, but the comparison must be made under specific conditions. According to this observation the English system does not have an absolute unit of measure as does the metric system. Further discussion on the concepts of mass and weight is provided in Chapter 3.

The origin of the pound has been traced to Rome also. In ancient Rome, a *libra* was a unit of weight which equals 7,680 grains of wheat and was the forebearer of our pound, hence the abbreviation of the pound today, *lb.*

The United States is the only major nation not using or not eminently undergoing a total conversion to the metric system. The rapid growth of international trade in terms of its volume and its importance to our economy seems to make it imperative that the United States eventually adopt the metric system entirely.

Almost two centuries from the time that Thomas Jefferson suggested that the United States adopt the metric system, as was done by France and other leading countries at that time, the first of our Federal agencies has initiated proceedings to change its system of measures completely. The National Aeronautics and Space Administration (NASA) directed that all measurement values used in sponsored publications be expressed in metric system units, known as the International System of Units (ISU). Imagine the degree of difficulty in converting gauges, weights, screws, linear measuring devices, and the like into equivalent values.

Great Britain, since 1965, has been in the process of converting entirely to the metric system and will have completed the conversion by 1975. The Commonwealth nations have expressed intentions to change also. The United States realizes that she must accomplish the changeover to the metric system at the earliest and in the shortest feasible amount of time. There is evidence that the conversion is underway. Motorists often find the distance between cities listed on highway markers in miles and in kilometers. In addition, canneries are listing on the labels of their goods the weight and volume of the item in both English and metric units of measure. The cost of the conversion will be great, but if the United States is to retain her place among the leaders in world trade competition, there is no other recourse.

VECTOR AND SCALAR QUANTITIES

Many physical measurements that include area, mass, volume, and time can be totally specified by their sizes, or *magnitudes*. These quantities are called *scalars*. Other examples of scalar quantities are temperature, speed, and energy. Scalars cause no difficulty as to mathematical manipulation since two or more like scalar quantities can be added, subtracted, multiplied, or divided by ordinary mathematical procedures. For example, the sum of five gallons and three gallons is eight gallons. Also, an automobile which is moving at seventy mph and decreases its speed by forty mph would be traveling at thirty mph.

Other quantities, however, must be specified by direction as well as magnitude. *Vector* quantities, as these are called, are more complex to manipulate than scalars. Some examples of vectors are distance, velocity, acceleration, and force. The vector quantity is commonly represented graphically by an arrow. The length of the arrow is drawn to some suitable scale. The direction of the vector is indicated by the arrowhead, and the tail of the arrow represents the point of application. The method of adding velocities, a vector quantity, is presented in Chapter 3.

CHAPTER SUMMARY

Science consists of an extremely large body of knowledge that has been accumulated primarily through observation and experimentation. Investigators critically analyze data that they and others have accumulated with which they make predictions and establish conceptual models. Generally, *science* refers to the study of nature that is centered around matter and energy.

Scientific knowledge governs our actual existence and well-being. Science and its applications in the practical field have had a tremendous influence on our society. In order to receive a well-rounded education, a student must achieve a basic understanding of science so that he can take his rightful place in our society.

Units of measure should be carefully standardized among all countries because of world trade and other international activities. The units of *length, weight* (mass), *volume,* and *time* are basic to all systems of measurement. The most acceptable system of measurement is the metric system. All major countries will have soon adopted this system as their standard system of measurement.

All physical measurements are readily placed into one of two categories. *Scalar* quantities, such as time, volume, and speed, are totally specified by their magnitudes. *Vector* quantities must be specified by both magnitude and direction. Examples of vector quantities are weight, force, and velocity.

QUESTIONS AND PROBLEMS

1. Science is both self-correcting and self-destroying. How would you interpret this assertion? Cite various examples of your interpretation. (p. 2)

2. Science is considered both a process and a product. Can you provide an analogy with this supposition as it would apply to industry? (p. 3)

3. Contrast the relative need for a universally accepted system of measurement as required of a major nation of today with the same nation's need in the eighteenth century. (p. 5)

4. At the present time, the United States is making preparations for the complete conversion to the metric system of measurement. What evidence of this effort have you observed recently? (p. 8)

5. What units in the metric system (cgs or mks) are comparable in magnitude to the following units from the English system: (See inside cover.)

ounce	pint	foot
pound	quart	yard
ton	square yard	mile
teaspoon	inch	

6. The wave length of a given orange light is now recognized as the primary standard of length. For what reasons do you suppose the standard meter bar was replaced? (p. 5)

7. A block of material was sold as pure aluminum. The block was found to have a mass of 6,250 gm and measured 40.0 cm long, 10.0 cm wide, and 5.00 cm high. (a) What is the density of the block in gm/cm^3? (b) If the density of pure aluminum is 2.70 gm/cm^3, was the block pure aluminum? (p. 7)

8. Give two examples of quantities in everyday life that vary directly and two examples of quantities that vary inversely. (See Mathematics Refresher, Appendix I.)

9. What are the advantages of the metric system over the English system of measurement? For what reasons does the United States retain the English system? (p. 5)

10. Cite examples of objects too small to be measured directly. What objects can you mention that are too large to measure directly?

11. A platinum alloy has a density of twenty-five grams/cubic centimeter. Determine the volume of a nation's standard made from this alloy whose mass is: (a) one gram; (b) one kilogram. (p. 7)

12. Compare the mass of an object with the object's volume and weight. (p. 6)

13. Science has been said to be a Prometheus bringing gifts and also a Frankenstein monster. Can you clarify both beliefs? (p. 2)

14. Should man cease his continued investigations of the macroscopic and the microscopic world? Discuss your belief. (p. 2)

15. Accuracy in measurement has become increasingly important in all phases of the scientific world. Cite examples which illustrate this importance in the areas of space ventures, sports, electronics, and automobile design. (p. 5)

early astronomy

2

PRIEST — ASTRONOMERS
THE GREEK PHILOSOPHER — ASTRONOMERS
MEASURING THE SIZE OF EARTH
MEASURING THE DISTANCE TO THE MOON
MEASURING THE DIAMETER OF THE MOON
MEASURING THE SPEED OF LIGHT

vocabulary

Constellations	Eclipse (of Sun)
Geocentric	Heliocentric
Period (of Satellite)	Zenith

It would be incorrect to say that astronomy had a definite beginning at a specific time during the history of man. Surely Cro-Magnon man, Neanderthal man, and probably even *Homo habalis* man (by some dating methods estimated to have lived at least 1,750,000 years ago) studied with awe the myriad points of cold but soft light that dotted the black void of night. Because of the vastness and breathtaking beauty of the night sky, it is little wonder that early man associated the stars with gods of great power. Certainly, when man started domesticating animals, it was necessary for shepherds to stay with their flocks at night in order to guard them against attack by predators. As a result of these nightly vigils, the shepherds became very familiar with the night skies and noticed the changes in the stars that were visible at different seasons. Certain groupings of stars, called constellations, assumed fanciful shapes in the minds of the shepherds. These constellations became associated with certain mythological gods and goddesses and were eventually named in their honor.

Fig. 2.1 Ancient observers fancied that the collection of stars shown above represented a bull. Today we call this constellation Taurus, the bull.

In general, the simple agricultural life led by the ancients made it possible for the average person to quite frequently view the starry skies of night and to develop a very deep appreciation for the beauty of the night heavens. Our present society, on the other hand, with its complex and extensive technology, does not lend itself very readily to the pleasant and leisurely pastime of stargazing. Indeed, in towns and cities where nightfall is greeted by thousands and tens of thousands of lights being turned on, one cannot see the stars even if one tries.

PRIEST–ASTRONOMERS

In ancient Egypt the astronomers were the pagan temple priests. These priest-astronomers found it very advantageous to their prestige and power to be able to predict when eclipses of the sun and moon would occur since this ability was interpreted by their king as an example of their understanding of and communication with various gods.

It was also of great economic importance to be able to predict when the Nile River would annually overflow its banks and renew the fertility of Egypt's farm lands with its silt-laden waters, since much preparation was needed in order for the people to reap maximum advantages from the event.

About 2700 B.C., predating the building of the great pyramid of Cheops at Giza, the Egyptian priest-astronomers noticed that when the bright star Sotkis (called Sirius today) peeped above the horizon just a little south of and simultaneous with the rising of the sun, the Nile River would overflow its banks within a few days. They further observed that after this twin rising of Sirius and the sun, Sirius rose slightly earlier than the sun each day. After approximately 91 days Sirius was high in the night sky, halfway between rising and setting, when the sun was just beginning to rise. After 182 days Sirius would be setting at the same time that the sun was rising. Then after 365 days, Sirius and the sun would once again rise simultaneously, followed in a few days by the overflowing of the Nile River. The whole performance would be repeated in another 365 days. Thus, the establishment of a fairly accurate year came into being. Actually, the Egyptians found that it took a period of four years plus one day for the sun and Sirius to rise together four times and, therefore, the length of the year was in reality 365¼ days. Thus, the concept of leap year was born in which a day is added every fourth year to give 366 days in a leap year. Besides the economic value of Egyptian astronomy, the priest-astronomers actively pursued the study of the stars in order to add to the information that they thought necessary to the prediction of human events through astrology.

This superstitious belief in astrology, a sort of astronomical magic, was generally held by the ancients to be based on self-evident truths. It is interesting that the development of the pure science astronomy owes such a great debt to the charlatanry non-science, astrology. It is perhaps even more interesting that in our supposedly enlightened present civilization untold numbers of people still believe that astrologers can predict a person's destiny by consulting the positions of the planets!

By the year 2000 B.C. the Babylonians were using a calendar divided into twelve months of thirty days each. The number of days in a month was based on the time necessary for the moon to go through all its phases: about twenty-nine and one-half days. The Babylonians also devised the week as a unit of time and named the seven days of the week after the sun, the moon, and the five planets known to them. Thus, Sunday is the day of the sun, Monday is the day of the moon, and Saturday is the day of Saturn. Since the names of the remaining days of the week are taken from the Anglo-Saxon or Middle English language, the names of the remaining days of the week do not sound like the names of the planets; therefore, it is not obvious which planets correspond to which days of the week. However, Tuesday, Wednesday, Thursday, and Friday are named after Mars, Mercury, Jupiter, and Venus respectively.

The Babylonian priest-astronomers also compiled lists indicating the positions of many of their constellations. In fact, many of the gods, goddesses, and animals that the Babylonians assigned to various constellations are the same ones we use today. Thus, we still associate a bull with the constellation Taurus and the Centaur-Archer with the constellation Sagittarius. By 250 B.C. the Babylonians had produced a compilation of lunar eclipses, both past and future, and described the calculations necessary to predict other future eclipses.

Many ancient as well as more modern but equally non-technical societies have relied upon the stars and the sun for long distance navigation. Thus, in Homer's epic poem, the *Odyssey,* (eighth century B.C.) the goddess Calypso gave Odysseus directions on how to sail from what must have been the Madeira Islands to the southwest coast of Spain. These directions included using three constellations, the Pleiades, Boötes, and the Bear (the Big Dipper), as navigational steering aids. The Polynesians voyaged many hundreds of miles across portions of the uncharted Pacific Ocean by noting at night which stars and constellations were rising or setting on the horizon at a point near the island to which they were headed. During the day they closely observed the position of the sun in order to estimate their approximate latitude.

THE GREEK PHILOSOPHER–ASTRONOMERS

The Greek philosopher Aristotle (384–322 B.C.) reasoned that Earth must be spherical since the shape of Earth's shadow on the moon was part of a circle. Only spherical objects produce a round shadow.

Aristarchus (310–230 B.C.), the noted Greek astronomer at Alexandria, was the first man to teach the *heliocentric* concept in which it is assumed that Earth and all the other planets revolve around the sun (see Fig. 2.2). He also believed that Earth rotated on its axis daily. The heliocentric concept (the word *heliocentric* comes from two Greek words, *helio* meaning sun and *kentron* meaning center), though ably defended by Aristarchus, was not destined for general acceptance. Aristotle had considered the heliocentric concept as a possibility but had rejected it in favor of the *geocentric* concept (from the Greek *geo* meaning *earth* and *kentron* meaning *center*) in which it was assumed that the sun, the moon, and all other planets revolved around Earth (see Fig. 2.2). Because of Aristotle's great reputation, the heliocentric concept was accepted by only a moderate number of philosophers and astronomers during Aristarchus' life. Furthermore, another great Greek-Egyptian astronomer who studied and taught at Alexandria, Claudius Ptolemy (85–165 A.D.), sided with Aristotle in the heliocentric versus geocentric argument and managed to deal a death blow to the heliocentric concept. It should be pointed out that Ptolemy's geocentric representation of the motion of the sun, the moon, and the planets did rather accurately represent the observable facts known during Ptolemy's life. And, indeed, Ptolemy could make reasonably accurate predictions about the future position of the planets based on his ingenious geocentric concept.

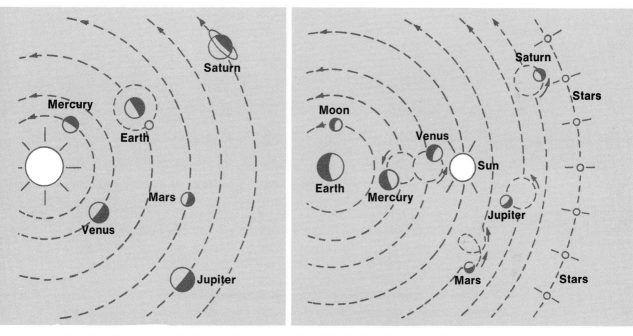

Fig. 2.2a Copernicus' heliocentric view of the solar system.

Fig. 2.2b Ptolemy's geocentric view of the solar system.

The heliocentric concept was not seriously reintroduced into science until 1530 A.D. when Nicolas Copernicus, a Polish astronomer published a book in which he pointed out the logical reasons why the heliocentric concept must be the correct interpretation of planetary motion. Galileo (1564–1642) vigorously defended the heliocentric concept as outlined by Copernicus and, as a result and in spite of his world renowned reputation, was forced during the Roman Inquisition to publicly renounce his teachings. Furthermore, Galileo was confined to his house and grounds under continual surveillance for the last nine years of his life for publicizing heretical scientific doctrine that was contrary to the teachings of the Church.

MEASURING THE SIZE OF EARTH

The first reasonable measurement of the size of our planet was accomplished by Eratosthenes (284–195 B.C.) who was the Greek librarian for the museum at Alexandria, Egypt. In order to understand the principles that Eratosthenes used in his measurements, we must realize that light reaching us from a source very, very far away travels approximately in parallel lines. As we see in Figure 2.3, as an observer is farther and farther away from a source of light, the source's light rays become more and more nearly parallel. Eratosthenes made the essentially correct assumption that the sun is sufficiently far from Earth for light from the sun to reach Earth in essentially parallel lines.

To the Sun ➡

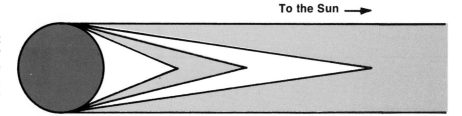

Fig. 2.3 If a light source is far enough away, the rays of light from the source are essentially parallel.

From Erastosthenes' many travels throughout Egypt, he had noticed that at Syene (Syene was a town on the Nile River approximately where the Aswan Dam stands today) at noon, the sun shone directly down a well so that the bottom of the well was completely illuminated. This observation could only mean that the sun was directly overhead at the time. In the following year, at the same day and time, Eratosthenes was at Alexandria, which is approximately due north of Syene, and he measured the angle between his zenith, a point directly overhead, and the center of the sun as shown in Figure 2.4. The angle measured about 7°. The light ray shining down the well at Syene and the light ray shining 7° from the vertical at Alexandria are parallel. Since these two parallel lines are cut by the straight line CZ, angle B must equal angle A. Since the unit of Greek length at that time was the length of the Greek sports stadium, Eratosthenes had a slave determine the number of stadium units or stadia between Syene and Alexandria by actually walking with measured paces between the two cities. The slave reported a distance of 5000 stadia. Since 7° is about 1/50 (7/360°) of a circle, the distance from Syene to Alexandria must be 1/50 of a great circle around Earth or of Earth's circumference. Therefore, Earth's circumference must be fifty times the distance between Syene and Alexandria or 50 x 5000 stadia, which gives a circumference of 250,000 stadia. If the length of the stadium Eratosthenes used is taken as one-tenth of a mile (this is the generally accepted value), it means that Eratosthenes' polar circumference of Earth was 25,000 miles, which is only 96 miles more than its modern value! The diameter of Earth is determined by dividing the circumference by π (3.14):

$$\frac{25{,}000 \text{ miles}}{3.14} = 7962 \text{ miles}$$

MEASURING THE DISTANCE TO THE MOON

The last famous astronomer of antiquity was Claudius Ptolemy (85–165 A.D.), who, like Eratosthenes did his main astronomical work at the Museum of Alexandria (founded by Alexander the Great) which by now had a staff of over one hundred professors and a library of over half a million scrolls.

One of Ptolemy's greatest accomplishments was measuring the distance from the earth to the moon as shown in Figure 2.5. Ptolemy knew that on a certain day of the year the moon would be directly overhead (that is, at his zenith) at Alexandria (point B in Figure 2.5). At that same time he had a fellow astronomer stationed at point D on the earth determine the angle between his zenith at point D and the center of the moon; that is, he found the angle EDC. Since the angle of a straight line, for instance line EA, is 180°, then angle CDA equals 180° − angle EDC. Ptolemy had the distance on Earth between points B and D carefully measured. Knowing the distance between B and D and knowing the circumference of Earth, Ptolemy determined what fraction of the circumference of Earth was represented by the distance BD. It follows then that

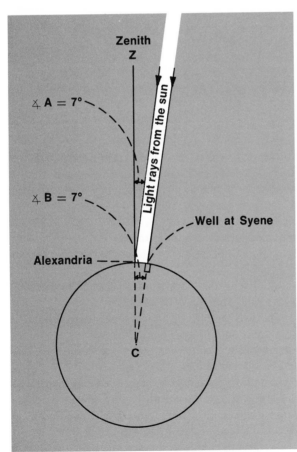

Fig. 2.4 Eratosthenes' method of determining the circumference of the Earth.

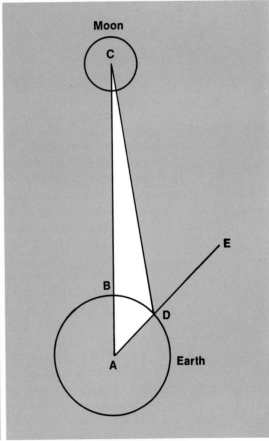

Fig. 2.5 Ptolemy's geometrical method of determining the distance between the Earth and the moon.

the angle BAD must be the same fraction of the total degrees of a circle (360°). For instance, if the distance BD were 25,000 stadia, BD would be one-tenth of the 250,000 stadia circumference of Earth and the angle BAD would be one-tenth of 360°, or 36°.

From Eratosthenes' measurement of the circumference of Earth, Ptolemy knew the radius of Earth, line AD $\left(\text{radius} = \frac{\text{circumference}}{2\pi} \right)$. Now Ptolemy knew the angle CDA, the angle BAD, and the distance AD (thus he knew two angles and the included side of the triangle CAD). With this information Ptolemy solved the triangle for the distance CA, the distance from the center of Earth to the center of the moon. Ptolemy obtained a value for CA of 233,050 miles, which is only two and one-half percent less than the modern value of 239,000 miles! Ptolemy's result was an outstanding accomplishment, considering the instruments available during this period of history. The determination of the distance to the moon also allowed Ptolemy to calculate the distance that the moon travels to complete one orbit around Earth. The circumference of the moon's orbit would be equal to the distance of the moon from Earth times 2π since the circumference of any circle equals its radius times 2π.

Today, the distance to the moon is measured to within inches by use of laser beams that are sent from Earth and bounced off special reflectors placed on the moon by astronauts.

MEASURING THE DIAMETER OF THE MOON

From Eratosthenes' measurement of the circumference of Earth, Earth's diameter and radius could be calculated. Knowing the radius of Earth, Ptolemy was able to determine the distance to the moon. From this information the circumference of the moon's orbit could be calculated. From a knowledge of the circumference of the moon's orbit, the diameter of the moon can be determined and so it is with science. Each new discovery stands upon the shoulders of previous discoveries.

The moon has an angular diameter of approximately 31′ of an arc or approximately ½° since there are 60′ in 1° of an angle (the exact value of the moon's angular diameter is thirty-one minutes and five seconds, written 31′5″). Because the moon is relatively small compared to the circle made by the moon's orbit around Earth, the small arc of the moon's orbit passing through the moon (see Fig. 2.6) is essentially the moon's diameter. Even though this arc is curved, a small portion of a large circle is essentially a straight line. The solution to the problem of the diameter of the moon is basically the reverse of Eratosthenes' method of determining the circumference of Earth. Thus, if we determine the fraction of the total degrees in a circle represented by 31′, then the diameter of the moon should be the same fraction of the total circumference of the moon's orbit. There are 360° in a circle or 360 × 60 = 21,600′. The angular diameter of the moon then is 31′/21,600′ or 0.001435 of the total degrees in a circle. Therefore, the moon's diameter should be 0.001435 of the total circumference of the moon's orbit around Earth. The circumference of the moon's orbit is equal to the radius of its orbit (239,000 miles) times 2π, which gives a value of 1,500,920 miles. The diameter of the moon then is $0.001435 \times 1,500,920$ miles or 2146 miles. Very accurate measurements of the moon's diameter give a value of 2160 miles.

MEASURING THE SPEED OF LIGHT

Light travels so very fast, 186,000 miles per second, that it was impossible to measure the speed of light by methods and instruments available to early scientists. Although some scientists had attempted to measure the speed of light before 1676, all had failed. In 1676, however, the Danish astronomer, Roemer, was able to measure the speed of light as a result of his observations of the length of time it took a satellite of Jupiter to make one revolution around Jupiter; that is, he measured the *period* of the satellite. As shown in Figure 2.7, Roemer determined the period of a Jovian satellite by observing the instant that it came out of Jupiter's shadow and determining how long it took for the satellite to circle the planet and once again start to emerge from Jupiter's shadow.

In order to keep the arithmetic as simple as possible, we will assume that Roemer was determining the period of one of Jupiter's satellites that circles Jupiter in exactly six days. Actually, Jupiter does not have a satellite with a period of exactly six days, but its largest satellite (3162 miles in diameter), Ganymede, has a period closer to six days than any other Jovian satellite since its period is $7^\mathrm{d}3^\mathrm{h}42^\mathrm{m}33^\mathrm{s}$. Assume that Roemer started timing the six-day satellite when Earth was at position one and the satellite was just emerging from Jupiter's shadow. Six days later the satellite would have made one complete revolution around Jupiter and would again just be emerging from Jupiter's shadow. During that six days, however, Earth would have traveled in its orbit from position one to position two. It follows, therefore, that light from the satellite just emerging from Jupiter's shadow when Jupiter is in position B

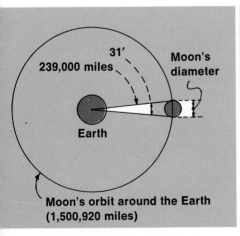

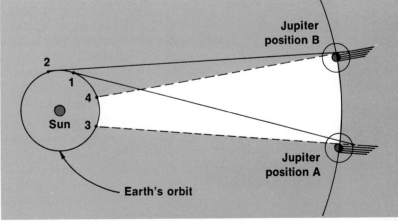

Fig. 2.6 **A method of determining the diameter of the moon.**

Fig. 2.7 **Roemer's method of determining the speed of light by measuring the period of one of Jupiter's satellites at different times of the year.**

would have to travel a greater distance to reach Earth in position two than would light from the emerging satellite when it travels from Jupiter in position A to Earth in position one. The increased distance that the light would have to travel would be the distance between Earth in position one and Earth in position two. Consequently, the period of the satellite that Roemer measured as Earth traveled from position one to position two would be longer than a similar measurement of the satellite's period as Earth traveled from position three to position four. Although Earth would be moving during the six days required to go from position three to position four, Earth would be traveling approximately parallel to Jupiter's orbit. Consequently, the distances between Jupiter's satellite and Earth would remain essentially the same.

Roemer knew that Earth traveled in its orbit at a speed of 18½ miles per second or 1,598,400 miles per day. In the six days it took Earth to go from position one to position two, Earth would have traveled 6 × 1,598,400 or 9,590,400 miles. It turned out that the measurement that Roemer made of the satellite's period as Earth went from position one to position two was 51.56 sec. longer than when he measured the period as Earth went from position three to position four. This increase in the satellite's revolutionary period means then that it took light 51.56 sec. to travel the 9,590,400 miles extra distance between Earth position one and Earth position two. It follows then that the velocity of light is $\frac{9,590,400 \text{ miles}}{51.56 \text{ sec.}}$ or 186,000 miles per second.

CHAPTER SUMMARY

Astronomy had its beginning in the astrology practiced by the ancient civilizations of Egypt and Mesopotamia.

Aristotle believed the solar system to be geocentric (Earth-centered) in nature. Aristarchus, a great Greek astronomer at Alexandria, Egypt, introduced the idea that the solar system was heliocentric (sun-centered) in design. Aristarchus' heliocentric concept failed to gain widespread recognition. The heliocentric concept was not seriously reintroduced until Nicolas Copernicus published a book in 1530 A.D. supporting the heliocentric concept. The concept was finally established beyond serious question by Galileo in the seventeenth century.

Earth was first measured with reasonable accuracy about 200 B.C. by the Greek astronomer Eratosthenes while he was at Alexandria, Egypt.

Claudius Ptolemy, the last famous astronomer of Alexandria, Egypt, measured the distance from Earth to the moon in approximately 150 A.D.

In 1676, Ole Roemer, for the first time in human history, measured the speed of light. Roemer had set out to determine the period of one of Jupiter's moons. His inability to attain consistent values for the period of one of Jupiter's moons at different times of the year caused Roemer to consider why he was obtaining erratic results. He eventually realized that the increase in the values of revolution which he was obtaining was due to the length of time it took light to travel through space. His calculations gave a value for the speed of light of approximately 186,000 miles per second.

QUESTIONS AND PROBLEMS

1. How did the economy of Egypt influence the growth of astronomy? (p. 15)
2. Why are there twelve months in a year, seven days in a week, and 365¼ days in a year? (p. 15)
3. Even though Aristarchus presented the heliocentric concept about 250 B.C. the concept was not generally accepted until about 1530 A.D. Explain. (p. 17)
4. Suggest a method for determining the distance from Earth to the sun. (p. 19)
5. Could Ptolemy have made his determination of the distance to the moon without knowledge of Eratosthenes' determination of the circumference of Earth? (p. 20)

newton's laws applied 3

INERTIA
THE FORCE ITSELF
FALLING OBJECTS
THE THIRD LAW
CIRCULAR MOTION
COMMON COMPARISONS OF VELOCITY
NEWTON'S LAW OF UNIVERSAL GRAVITATION
NEWTON'S LAWS AND AIR TRAVEL

vocabulary

Acceleration	Centripetal Force
Force	Gravity
Inertia	Mass
Weight	

The middle of the sixteenth century was an era of turmoil in the field of scientific knowledge. Italy, the center of the cultural world at that time, was the scene of the widespread practice of witchcraft along with numerous other superstitious beliefs and teachings. Also, during this period man was deprived of his right to speak freely against the dogma of Aristotle (384–322 B.C.), the Greek philosopher whose word was irrefutable, or against the Church, whose doctrine conflicted with certain scientific beliefs. Matter, according to Aristotle, could be maintained in motion only as long as the force which caused movement remained in direct contact with the matter. Should the force cease or lose contact, the object would stop abruptly. For example, when a stone or other projectile left a catapult, the medieval equivalent to a cannon, the propelled object was maintained in motion by the air which streamed in behind it and thus maintained a continuous and physical contact with the catapult. In addition, Aristotle contended that the weight of a falling object directly affected its rate of free fall. A fifty pound object, for example, fell decidedly faster than a five pound object. Both of his contentions were beyond reproach and were universally accepted well into the seventeenth century.

The university of the sixteenth century taught other narrow-minded and incorrect concepts also which included the premise that Earth stood still, that it was the center of the universe, and that the sun revolved about Earth as did other heavenly bodies. Such also were the beliefs of the Church. The Church widely influenced much official thinking and would not tolerate alternate interpretation of Holy Scripture from its own established dogma.

At the close of the sixteenth century, Galileo, a scholar of physics at the University of Pisa, dared question many of the teachings of Aristotle. He was the first investigator to use a telescope to study the heavens. He studied the moons of Jupiter as they revolved about the planet and concluded that the moon also orbited Earth. As his studies continued, he expressed his belief that the model of the universe proposed by Copernicus (1473–1543) was more nearly correct than the model presented by Aristotle. As Copernicus contended, Earth and the other planets revolved about the sun. In addition, Galileo (1564–1642) is reputed to have taken his students to the Leaning Tower of Pisa where they dropped various objects from different heights. As a result of the investigation, Galileo concluded that all objects, disregarding air friction, fall with the same rate of change of velocity.

Numerous times Galileo tried to illustrate to his instructors the fallacies in Aristotle's teachings, but, to his dismay, he was rewarded with abuse and the threat of poor marks in his classes. Galileo's writings brought disfavor also from the Church, since many concepts he discussed were in defiance of the contemporary interpretations of the Holy Scriptures. Though he had become one of the outstanding men of science in Italy by this time, his persistence could not go unheeded by the Church, and eventually he was brought before the Inquisition in Rome. Because of his age and fame though, his punishment consisted only of reciting the seven penitential psalms at least once per week for three years. It is said that after each psalm was recited, Galileo would add in a soft manner "And still Earth moves," in obvious disagreement with Aristotle's teachings of a static Earth.

Importantly enough, Galileo's numerous authentic experiments did much to disprove the misconceptions attributed to Aristotle and others. Of equal value was the contribution to the scientific world of the manner in which he attacked the fallacies—through experimentation. Many of his experiments were readily duplicated by others, a feature which added to the flavor of verification in the conception of theories. One of his major objectives was to develop an adequate description of motion from which he could conceive an accurate theory. In fact, Galileo, along with Newton, Descartes (1596–1650), and Huygens (1629–1695) should be credited collectively with the formulation of the accepted theory of motion, since all made significant contributions.

Although Galileo and others introduced the concept of inertia, Isaac Newton, a seventeenth century genius, realized its significance. Newton is credited with conceiving three of the most outstanding discoveries in the entire realm of human thought. These contributions include calculus, the field of mathematics which deals with variations in functions and areas bounded by curves; the nature and properties of light; and the laws of universal gravitation. His works were published in a three-volume series called *Principia* or *Mathematical Principles of Natural Philosophy,* in 1686 and 1687. The first two volumes were entitled *The Motion of Bodies* and explained the conditions and principles of force and motion as well as discussed such concepts as light, space void, density, and body resistance. The third volume he published was called *Of the System of the World* and summarized the laws of gravitation and the writings of the first two volumes. Newton's writings which involved force and motion have been condensed into three related and useful principles which have come to be known as "Newton's Three Laws of Motion."

INERTIA

The "First Law" of Newton was concerned with the concept we know as *inertia.* The law maintains that an object which exists in a motionless state remains motionless until an external force of some size acts upon the object. For example, an automobile must have some external force applied to it to cause it to move. This force is produced by the expanding gases which result from the combustion of gasoline. Similarly, the steam engine operates by virtue of an external force furnished by water and coal or fuel oil. The human body receives its ability to move from an internal force created by the muscles and controlled by the brain. The force, however, can be traced back to our intake of food and water. Without these substances as fuel, our ability to move would decrease to zero in less than a month's time.

The force which creates motion is composed of the wind, gravity, or the conversion of energy from one form to another. The action which produces motion is commonly classed as a push or pull that acts on a body and sets the object in motion. In contrast, an object which exists in a state of motion remains in motion until an external force is applied to stop it. The force may be physical, such as an auto's collision with another, or it may be an invisible force which offers resistance to the object's motion. For instance, an auto which is coasting comes to rest because of the force of gravity if it rolls up an incline. The auto comes to rest on a horizontal surface through rolling friction offered between the tires and the highway. Inertia, the First law of Newton, can be correctly stated: "A body of itself is not capable of changing its condition of rest or motion." The law applies to matter whether solid, liquid, or gas. However, another concept enters into the full interpretation of the concept of inertia. The greater the amount of mass, the more inertia an object contains. (The concept of mass will be discussed in detail later.) A truck with attached trailer has more inertia than an auto by virtue of the greater amount of mass the truck and trailer have. For this reason, the auto may not be able to furnish the force required to bring the large truck with its trailer to rest. The persistence with which the truck and trailer remain in motion equals the persistence it required to start it in motion. The external force necessary to stop the truck equals the external force which was required to start its motion. The auto, though it is moving at a relatively high rate of speed, may collide with the truck without the truck's being set in motion because of the vast difference in mass and, therefore, inertia.

Another example of inertia is apparent when a passenger who is standing in a moving bus is considered. The bus can be stopped by an external force which exists between the braking wheels and the pavement. The same force, however, is not applied to the standing passenger so he continues forward and must apply a resistive force with his feet or grab a fixed object on the bus to control his forward motion.

Objects in motion also have a decided tendency to maintain the direction in which they are moving. A cue ball in the game of pool moves in a straight line from the force applied to it by the cue stick. As the ball collides with the cushion of the pool table, the ball rebounds in a straight line path unless a rotational force was applied to the ball by the player or the collision. Similarly, a falling object falls downward in a straight line path unless some external force other than gravity acts on it such as the wind or air friction.

Newton's First Law, then, consists of two parts, an observation which is commonly overlooked because of the lack of clarity with which the Law is often stated. A literal translation of Newton's writings points out that "Every body continues in a state of rest or uniform motion in a straight line except insofar as it may be compelled by force to change that state." Therefore, the Law actually includes two distinct observations: (1) A body at rest tends to remain at rest, and (2) A body in uniform motion in a straight line tends to remain in that motion.

THE FORCE ITSELF

The final velocity which a force causes an object to attain is directly related to the length of time the force is in contact with the object. A greater force is necessary to cause a stalled auto to move than is required to keep the auto in motion. If the same amount of force which started the auto to move is maintained in contact with it, the auto will continuously increase in velocity. If the accelerator in a stationary auto is pushed to its maximum and maintained so, the auto moves forward very abruptly and continues to increase in velocity within limits as long as the accelerator is held down. The action of the accelerator provides the force necessary for *acceleration*—a rate of change of velocity. Acceleration, then, is the comparison of the amount the velocity changes to that of the time which lapses. The student should keep in mind that a change in a quantity is equal to the final measure of the quantity minus the initial measure of the quantity. Therefore, the change in velocity equals the final velocity minus the initial velocity. If the time to change velocity from its initial measure to its final measure is considered, the concept of acceleration follows:

$$\text{acceleration} = \frac{\text{final velocity} - \text{initial velocity}}{\text{lapsed time between measures}}$$

or algebraically:

$$a = \frac{v_f - v_i}{t}.$$

An auto which starts from rest (v_i) and in ten seconds attains a velocity of fifty miles per hour (v_f) has obviously increased its velocity during each second. If the increase is considered uniform, that is, the velocity changes by equal quantities in equal time intervals, the change in velocity must amount to five miles per hour for each of the ten seconds, and the acceleration is five miles per hour per second (five mi/hr/sec). At the end of fifteen seconds, the auto reaches the velocity of seventy-five miles per hour (see Fig. 3.1). If the

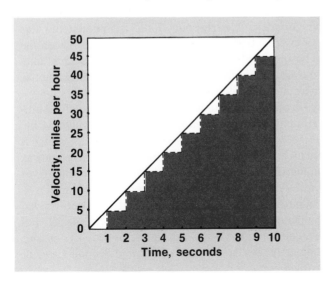

Fig. 3.1 **The time variation of the velocity of an auto undergoing uniform acceleration of five ft/sec/sec is represented by the diagonal line.**

brakes are applied and the auto slows down so that it stops in ten seconds, the rate of change of velocity or acceleration is considered negative and amounts to −7.5 miles per hour per second (−7.5 mph/sec). The final velocity of an accelerating object which starts from rest always equals the product of the acceleration and the time the object is moving or:

$$v_f = a \times t = at.$$

The average velocity of the moving object, assuming it starts from rest and is uniformly accelerated by a constant force, is equal to one-half the final velocity or:

$$\bar{v} = \frac{v_f}{2} = \frac{a \times t}{2} = \frac{1}{2}\, at.$$

If the distance an object travels is directly determined by the rate and time of travel, then:

$$d = \bar{v} \times t \text{ or } \frac{a \times t}{2} \times t$$

$$= \frac{1}{2} a \times t^2 = \frac{1}{2}\, at^2.$$

As the concepts involved are considered, the distance that a uniformly accelerated object travels, starting from rest, equals one half the product of its acceleration and the square of the time the force acts on it.

Newton's Second Law may be interpreted as: "The acceleration of an object is directly related and in the same direction as the force which produces the acceleration." A force, according to the First Law, produced motion in the body and the same amount of force is required to stop the body, assuming that the time during which the forces are applied is identical. If the body is stopped more abruptly than it was started, the magnitude of the stopping force must be greater than the starting force. If, however, the body is brought to a stop in a period of time which is longer than the time required for the body to reach its maximum velocity, the magnitude of the stopping force is undoubtedly less than the starting force. Suppose the body reaches a uniform velocity. Then the rate of acceleration would be zero, and the magnitude of the force is equal only to that which is required to overcome frictional, gravitational, and other counter forces; thus the object is in a state of equilibrium.

Acceleration is always in the direction in which the force is applied, assuming the object acquires a straight line path. If an auto increases in velocity, the force and acceleration are forward. If the auto is slowing down, the force is opposite the direction in which the auto is moving and so is the acceleration.

According to Newton, mass is the characteristic a body possesses that gives it an attractive force toward another mass. (The attraction between masses is governed by the inverse square law, an established rule that has been proved numerous times by experimentation.) Mass is also considered a measure of inertia or of the resistance an object has to being moved or stopped. The greater the mass of a body, the greater its inertia. Newton's First and Second Laws of

Motion are inseparable, since according to the First, the applied force over-comes this resistance of a mass to being moved. Then the same force acting on an object, according to the Second Law, causes the object to accelerate. In mathematical terms, force is the product of mass and acceleration, or

$$F = m \times a = ma.$$

If the same force is applied to two objects in which m_1 is twice the mass m_2, m_1 is accelerated at only half the rate of m_2. As suggested in this discussion, acceleration is inversely related to mass. If two forces, the second force twice the magnitude of the first, are individually applied to mass m_1, the second force causes mass m_1 to be accelerated twice the value produced by the first force. The acceleration, then, is directly related (proportional) to the magnitude of the force applied to the object. The relation of acceleration to both mass and force may be combined and expressed as:

$$a = \frac{F}{m}$$

Force and weight are interchangeable terms in some instances since weight is a measure of the attractive force of Earth or other heavenly body on an object. The expression of Newton's Second Law, $F = ma$, in terms of weight may be written: $w = mg$. Weight should be represented by force (F), and mass (m) would be, as previously discussed, a measure of the object's inertia. But how does an object, presumably at rest, accelerate? An object on Earth's surface moves in a circular path as Earth rotates on its axis. Earth's circumference obviously varies with distance from the equator, but let's assume the circumference of Earth at our location is 24,000 miles. The period of rotation of Earth, approximately twenty-four hours, causes an object on the surface of Earth at the point described to assume a velocity of 1000 miles per hour. We are not aware of the object's constant motion because we are in motion also.

Acceleration for objects revolving on Earth's surface is toward the center of Earth. A body of mass m is acted on by one significant force, the force Earth exerts on the object. The downward acceleration of the object is determined from $F = ma$, F representing the gravitational pull of Earth on the object. The application of Newton's Second Law permits scientists to measure the acceleration of an object because of the gravitational force known as *gravity*. The acceleration due to gravity (g), about 32 ft/sec^2 (9.8 m/sec^2), varies slightly with location and with changes in atmospheric conditions (friction). Newton's Second Law is commonly expressed as weight, $w = mg$. The mass of the object can be calculated by the expression of Newton's Second Law in terms of mass. That is, $m = \frac{w}{g}$. Since the mass of an object is constant, the relationship of w to g is such that if Earth's attraction for an object is less at one location than at another, the acceleration due to gravity is proportionally less. Experimentally, the weight of an object is measured by a scale that uses a tension spring. Since the tension of the spring remains constant at any location, the scale reads less the farther above sea level the object is measured. Mass, on the other hand, is not affected by distance above sea level. The measure of mass may be accomplished by a device as is illustrated in Figure 3.2.

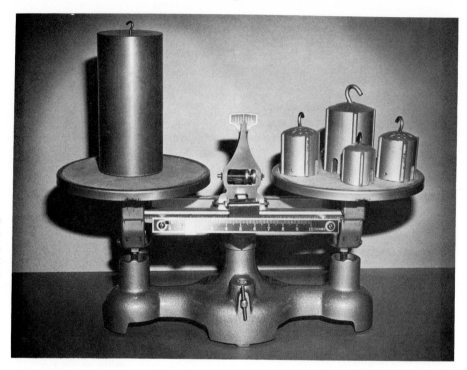

Fig. 3.2 Masses on each side of the scale are equal. A loss of weight would be proportional on both sides.

The inertia of an automobile remains constant independent of location. If frictional forces and their differences are disregarded, an automobile at rest requires the same force to cause it to move with a given acceleration whether the automobile is located on the surface of Earth, the moon, or suspended in outer space. Just as the inertia of the automobile remains constant, so does the automobile's mass, since mass is a measure of inertia. The weight of the automobile differs at each location because the force due to gravity varies at each location. Mass, a scalar quantity, is a separate entity from weight, a vector quantity or one that has both magnitude and direction.

As has been discussed, since a body has mass, it obeys Newton's First Law of Motion. From Newton's Second Law of Motion, the mass of a body determines the acceleration caused by a given force. On the other hand, because a body has mass, that body can exert a gravitational force on another body that has mass. The magnitude of the force depends on the mass of each body as well as on the distance between them. (Gravitational attraction is discussed in detail later in the chapter, but is introduced at this time for points of clarification.) The concept of mass, then, can be viewed according to different characteristics of matter. There are two distinct kinds of mass, *inertial mass* and *gravitational mass*. The amount of inertial mass a body possesses is exactly equal to the amount of gravitational mass it possesses, a phenomenon known as the *principle of equivalence*. The equivalency of the two measures of mass accounts for the reason all bodies, regardless of weight, accelerate toward Earth at the same rate, if frictional forces are disregarded.

Convention plays an important and integral part in our society. As was presented in Chapter 1, there are three common systems of measurement in use throughout the world today. The meter-kilogram-second (*mks*) system is an absolute system of units because the meter, kilogram, and second are defined in a manner that does not involve local variations in Earth's gravitational field. Mass, a measure of inertia, rather than weight is used. The centimeter-gram-second (*cgs*) system differs from the *mks* system only in that the centimeter and gram, measures which are submultiples of the meter and kilogram, are used. The foot-pound-second (*fps*) system is often used in English speaking countries, including the United States. The pound is legally defined as exactly 1/2.2046 the weight of an object whose mass is exactly one kilogram. The foot is legally defined as exactly 1200/3937 meters. Recently, the American Standards Association has defined the inch as equal to 2.54 centimeters, though legal adoption has not been completed.

The *mks* and *cgs* systems are referred to as *metric absolute* systems, the *fps* system as the *British gravitational* system. Since the three systems are typically presented simultaneously as measurement is discussed, confusion often results. The reference of the pound both as a unit of weight and as a unit of mass (pound-mass) has been discontinued in contemporary textbooks; therefore, the pound represents weight in all cases as used in this book.

In the United States the *fps* system uses the *slug* as its unit of mass. The unit of force or weight in this system is, of course, the *pound* (slug-ft/sec^2 from $F = ma$). In the *cgs* system, the unit of mass is the *gram* and the unit of force or weight is the *dyne* (gm-cm/sec^2). In the *mks* system, the unit of mass is the *kilogram* and the unit of force or weight is the *newton* (kgm-m/sec^2). (Note that "gm" is used to abbreviate the mass unit gram rather than g to eliminate the confusion which develops between g representing both a unit of a mass and a measure of acceleration.) From $w = mg$, the weight of a ten kgm mass is ninety-eight newtons (n) and of a five gm mass is 4900 dynes (d).

Examples which point out the application of the units of force follow.

Example 3.1

Neglecting friction, determine the value of a constant force that gives a mass of 90 grams an acceleration of 5 cm/sec/sec (cm/sec^2).

$$Solution: \quad F = ma$$
$$force = mass \times acceleration$$
$$F = 90 \, gms \times 5 \, cm/sec^2$$
$$= 450 \, gm\text{-}cm/sec^2$$
$$= 450 \, dynes$$

newton's laws applied   **33**

Example 3.2

A small wagon which weighs 64 pounds (lbs) is at rest on a level sidewalk. What force is required to give the wagon an acceleration of 3 ft/sec²?

Solution: From $F = ma$

$$\text{force} = \text{mass} \times \text{acceleration}$$

$$\text{mass} = \frac{\text{force}}{\text{acceleration}}$$

or $$\text{mass} = \frac{\text{weight}}{\text{acceleration due to gravity}}$$

$$m = \frac{w}{g}$$

$$= \frac{64 \text{ lbs}}{32 \text{ ft/sec}^2}$$

$$= 2 \text{ lbs/ft/sec}^2$$

$$= 2 \text{ slugs (the unit of mass)}$$

Then the solution becomes:

$$F = ma$$

$$= 2 \text{ slugs} \times 3 \text{ ft/sec}^2$$

$$= 6 \text{ slug-ft/sec}^2$$

$$= 6 \text{ pounds}$$

Often, more than one force is in action against the same object, a feature which results in the object's moving at a velocity commensurate with the sum of the forces. A typical illustration appears in Figure 3.3 in which two forces cause the object to move in a path between the direction of the two forces. Force

Fig. 3.3 The combined effort of the two forces causes the boat to move onto the trailer. The resultant force, then, is in the direction the boat moves.

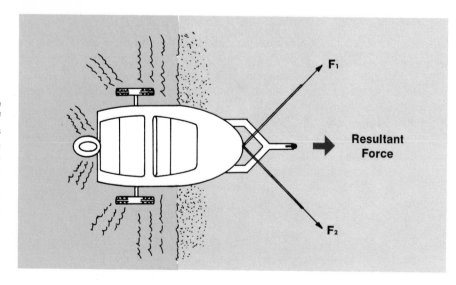

newton's laws applied

and velocity include both size and direction; therefore both are vector quantities and can be illustrated by means of a graphical representation. Typical examples of velocities which are "additive" or capable of being combined are illustrated in Figure 3.4.

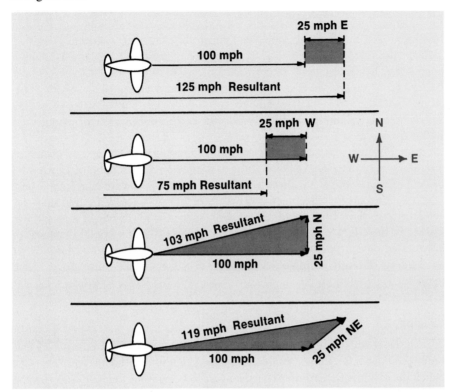

Fig. 3.4 **The resultant value of the vector quantities, wind velocity and plane velocity, depends on the direction of each velocity.**

FALLING OBJECTS

The decline of ancient Greek civilization marked the temporary end of Aristotle's impact on the area of science, an effect which remained until the thirteenth century A.D. Eventually, however, his writings became part of the scholastic teachings of Christianity in this period of the Middle Ages and included areas of astronomy, biology, literature, logic, philosophy, physics, and psychology.

Sadly enough, many of Aristotle's observations and projections were not sound in theory, though his numerous valuable contributions outweighed his misconceptions. According to Aristotle, for example, an object which fell toward Earth quickly reached a velocity commensurate with its innate characteristics. The factors which affected this maximum velocity included weight, the medium in which the object fell freely (objects fall faster in air than in water), and, conceivably, even color and temperature of the object.

The study of freely falling bodies was undertaken by several investigators, but not in detail for several centuries after Aristotle's death. In the fifth century A.D., John Philoponus of Alexandria concluded from his observations that the velocity of a freely falling object is determined by subtracting the resistance of the medium from the weight of the object instead of dividing the weight by the resistance as stated by Aristotle. Few people accepted Philoponus' findings because of Aristotle's great influence. Galileo differed from Aristotle in that he realized the role of mathematics in terrestrial motion and explained in these coherent terms how falling objects and objects rolling down inclines actually performed. He was correct in his assumptions that a falling object moves with a uniform change in velocity at any given time interval. The velocity, according to his way of reasoning, is directly related to the time in which an object is falling freely.

Investigators who tested the reasoning of Galileo were soon to confirm that a familiar force in existence is the force of gravitational attraction of Earth on a body. This force causes an object to fall faster in its path toward Earth, an aspect which is constant for all objects regardless of mass. A freely falling body accelerates in its rate of fall about 32 ft/sec for each second the object is in motion. Earlier in the chapter, the relation which displayed the horizontal distance an accelerated object would traverse in a given time t was shown to be $d = \bar{v}t$ or $d = \frac{1}{2}at^2$. For freely falling objects, the letter g is often substituted for a in the mathematical relation, so $d = \frac{1}{2}gt^2$ and $g = 32$ ft/sec/sec (ft/sec^2) or 9.8 m/sec^2.

Example 3.3

A steel ball is dropped from the top of the Empire State Building. The ball requires about 8.9 seconds to hit the ground. Determine (a) the velocity with which the ball hits the ground, (b) the average velocity of the ball, and (c) the distance the ball fell (which represents the height of the building).

Solution: (a) The final velocity, v_f

$$= gt$$
$$= 32 \text{ ft/sec}^2 \times 8.9 \text{ sec}$$
$$= 284.8 \text{ ft/sec}$$

(b) The average velocity, $\bar{v} = \dfrac{v_f}{2}$

$$= \dfrac{284.8 \text{ ft/sec}}{2}$$
$$= 142.4 \text{ ft/sec}$$

(c) The distance the ball fell, (two possible solutions)

$d = \bar{v}t$ $\qquad\qquad\qquad\qquad$ $d = \frac{1}{2}at^2$

$= 142.4 \text{ ft/sec} \times 8.9 \text{ sec}$ $\quad$ or $\quad$ $= \frac{1}{2} \times \dfrac{32 \text{ ft}}{\text{sec}^2} \times 8.9 \text{ sec} \times 8.9 \text{ sec}$

$= 1267 \text{ ft.}$ $\qquad\qquad\qquad\qquad$ $= 1267 \text{ ft.}$

A mathematical relationship between the horizontal velocity that a horizontally applied force impresses on an object and the vertical velocity that an object attains by virtue of gravitational forces reveals that the maximum range of the object exposed to both forces is reached when the path the object makes is a 45° angle with the horizontal. An angle of 45° is one which an imaginary line would make drawn halfway between vertical and horizontal from the point from which the object is given its velocity. Regardless of how fast an object is moving horizontally, gravity is unaffected, as is displayed in Figure 3.5. The conclusion holds true with all velocities less than the escape velocity of an object from the pull of Earth's gravity, about seven miles per second. For example, a rifle bullet fired horizontally from four feet above the ground will hit the ground in 0.5 seconds as will a rifle bullet dropped vertically. The greater the velocity that the bullet fired from the rifle attains, the farther from the rifle it will strike the ground.

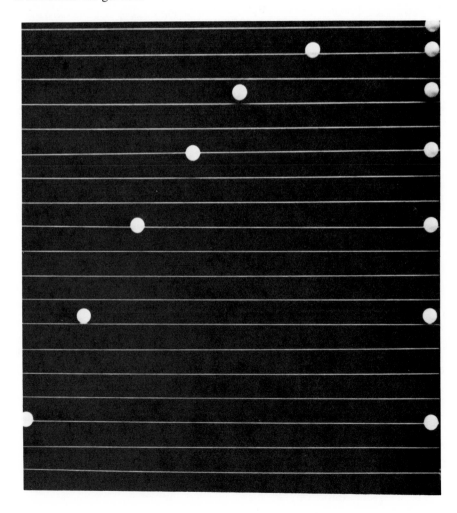

Fig. 3.5 Two objects in various conditions of free fall. One object has velocity in a vertical direction only, the other in both horizontal and vertical directions. Note that each object crosses the time lines in identical intervals.

Another related condition involves objects which are projected upward and then return to Earth (see Fig. 3.6).

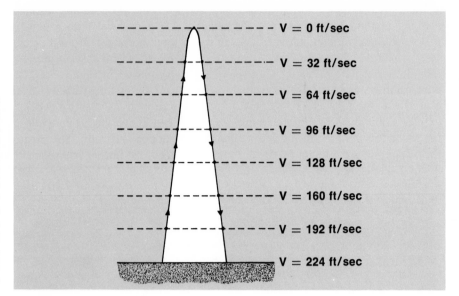

Fig. 3.6 The motion of an object projected upward is the same as the downward motion, but in reverse. An object projected upward returns to the ground at the same rate as the object's initial velocity, disregarding frictional forces.

$V = 0$ ft/sec

$V = 32$ ft/sec

$V = 64$ ft/sec

$V = 96$ ft/sec

$V = 128$ ft/sec

$V = 160$ ft/sec

$V = 192$ ft/sec

$V = 224$ ft/sec

Example 3.4

A rifle bullet whose initial velocity is 1200 ft/sec is fired directly upward. (a) How long does the bullet require to reach its maximum height? (b) How long until it returns to Earth? (c) How far does the bullet travel upward?

Solution:

(a)   From $v = gt$

$$1200 \text{ ft/sec} = 32 \text{ ft/sec}^2$$

$$t = \frac{1200 \text{ ft/sec}}{32 \text{ ft/sec}^2}$$

$$= 37.5 \text{ sec}$$

(b) The bullet required 37.5 seconds to reach its maximum height where its velocity upward is zero. The bullet then becomes a freely falling object and is pulled back to Earth by gravity. As the bullet falls, its velocity increases because of the acceleration due to gravity, g. The bullet reaches the velocity in its first second of the fall downward as it had at the start of its last second upward. In the next second of its downward flight the bullet reaches the velocity it had at the start of its next to last second upward and so on. In other words, the bullet reaches the level of its starting point where its initial velocity was achieved (the end of the rifle barrel) at the same velocity it had when projected upward. So 37.5 seconds added

newton's laws applied

to the 37.5 seconds the bullet required to reach its maximum height equals 75 seconds, assuming that frictional resistance from the air is neglected.

(c) If the initial velocity of the bullet is 1200 ft/sec and the final velocity, $v_f = 0$ ft/sec, then the average velocity, $\bar{v} = 600$ ft/sec. From the formula $d = \bar{v}t$, $d = 600$ ft/sec $\times$ 37.5 sec or 22,500 ft. The bullet, then, reaches a height of four miles above Earth's surface before it starts the return journey.

One of the most outstanding breakthroughs in science occurred when Russia launched its first successful orbiting satellite, called Sputnik, in 1957. All mankind was amazed that objects could remain in space with no apparent external forces acting on them. The facts that the moon orbits Earth and the planets the sun were accepted, but were never a topic of popular conversation by the general public. The explanation was simple; scientists had learned to apply Newton's Laws of Motion in yet another dimension.

The satellite was obeying all laws of motion, since in reality it was continually falling back to Earth. The horizontal velocity the vehicle was initially given was exactly equal to that required to cause the satellite to follow the curvature of Earth.

The principles of motion involved are fundamental. Imagine a cannon placed on a tall mountain, as is illustrated in Figure 3.7. If the cannon could hurl its projectile at different velocities, the length of the projectile's path downrange would increase as the horizontal velocity was increased. At a given velocity the projectile would fail to return to Earth. Instead, the projectile would fall around Earth until its velocity appreciably dropped because of air friction and other external forces which might act upon it.

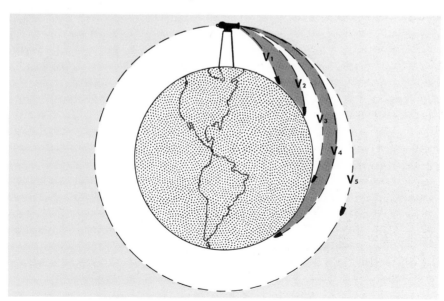

Fig. 3.7 **As the horizontal velocity of the projectile is increased, the range of the projectile is extended until gravity causes the projectile to "fall" around Earth.**

THE THIRD LAW

The Third Law of Motion discussed in Newton's writings is applicable to all motion. The concept involves the realization that forces always occur in pairs. The Third Law may be stated as follows: "When one object exerts a force on another, the second object exerts an equal but opposite force on the first." This relationship is often referred to as that of action-reaction. The illustrations in Figure 3.8 (a), (b), (c), (d), and (e) display the concept from five slightly different points of view. In Figure 3.8(a), the oarsmen advance the canoe through the application of a force on the water opposite the direction they wish to move and in Figure 3.8(b), the tennis ball and the racket are moving in opposite directions before collision. In Figure 3.8(c), the football is set in motion from a state of rest by the kicker's foot. In Figure 3.8(d), both the bullet and the rifle react to each other from the state of rest, and in Figure 3.8(e), no reaction to equal but opposite force is obvious; however, the chair pushes downward, Earth upward, and, therefore, a state of equilibrium exists.

(a)

(b)

(c)

(d)

Force

Reactive force

(e)

Fig. 3.8 Various conditions which involve action and reaction.

The relation of equal but opposite forces may be expressed by: $F_1 - F_2$ where F_1 is acting on mass m_1 and F_2 is acting on m_2. If the established equation of Newton's Second Law, $F = ma$ is applied, the expression $F_1 = F_2$ can be written $m_1a_1 = m_2a_2$.

If the time that the force is applied to the object is noted, the acceleration, hence the final velocity of each, can be calculated from $a = \dfrac{v_f}{t}$. In the event the object is in motion prior to contact with the force, the amount of acceleration is related as:

$$\text{acceleration} = \frac{\text{final velocity} - \text{initial velocity}}{\text{time}}$$

$$\text{or } a = \frac{v_f - v_i}{t}$$

When two objects at rest react and move in opposite directions, $a_1 = \dfrac{v_f}{t_1}$ and $a_2 = \dfrac{v_f}{t_2}$ so $m_1a_1 = m_2a_2$ can be substituted with $\dfrac{m_1v_1}{t_1} = \dfrac{m_2v_2}{t_2}$. Since $t_1 = t_2$, the equation is simply $m_1v_{f_1} = m_2v_{f_2}$. The product of mass and velocity is in reality a concept known as *momentum*, a familiar term indeed. To express Newton's Third Law in this light, "The momentum of the acting object is equal to the momentum of the opposing object."

Example 3.5

The bullet from a hunter's rifle weighs one ounce ($\frac{1}{16}$ lb). When the rifle is fired, the bullet leaves the rifle with a velocity of 1600 ft/sec. If the weight of the rifle is 5 lbs, what is the velocity with which the rifle recoils?

Solution:

From $m_1v_1 = m_2v_2$ and recalling that $m = \dfrac{w}{g}$, then

$$\frac{w_1}{g} \times v_1 = \frac{w_2}{g} \times v_2 \text{ or } \frac{w_1v_1}{g} = \frac{w_2v_2}{g}$$

Since both sides of the equation are divided by "g", the equation can be simply written:

$$w_1v_1 = w_2v_2, \text{ since the "g's" cancel.}$$

This condition points out that in such a relation mass and weight can be interchanged. The solution concludes:

$$\tfrac{1}{16} \text{ lb} \times 1600 \text{ ft/sec} = 5 \text{ lbs} \times v_2$$

$$v_2 = \frac{\tfrac{1}{16} \text{ lb} \times 1600 \text{ ft/sec}}{5 \text{ lbs}}$$

$$= 20 \text{ ft/sec.}$$

Similarly, a 300 pound football player who moves 3 mph has no more momentum than a 150 pound player who moves 6 mph. In other words, if the masses (or weights) of two objects are in a ratio of two to one (2:1), the velocity of the objects need only be in a ratio of one to two (1:2) to have equal momentum. If the objects collide under these conditions, both are stopped as the momentum of each counteracts the other's momentum. This relationship has helped preserve a place for the small but fast athlete in the areas of contact sports.

CIRCULAR MOTION

When an object to which a string is attached travels in a circular path, an inward force is exerted by the string on the object. This force is known as *centripetal force* (center-seeking), and is the only significant force which acts on the revolving object. Since the force remains constant, the object continually accelerates inward. The revolutionary speed, however, of the object remains constant. In addition the object maintains a constant distance from the center (see Fig. 3.9).

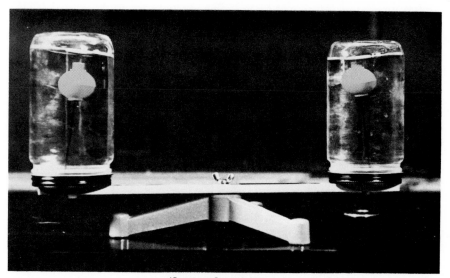

Fig. 3.9 **The revolving central force unit causes the floating bobbers to point inward, an indication of the direction of acceleration.**

(Courtesy Sargent-Welch Scientific Company, Skokie, Illinois)

The discussion may imply to the reader that Newton's Laws of Motion are not applicable to revolving objects since the object's speed remains constant although the object is constantly accelerated. One should keep in mind that velocity is a vector quantity and may consist of a change in direction without a change in speed. Therefore, since the object is constantly changing direction the object must be accelerating.

The acceleration produced by a centripetal force is known as *centripetal acceleration*. In circular motion an object is continually accelerated toward the center of the object's circular path. The velocity of the object is always at a right angle to the direction of acceleration at any given instant. The centripetal acceleration of an object which is revolving obeys Newton's First Law as does an object that is moving along a straight line path. Therefore, the acceleration of the revolving object is equal to the change in velocity during a given time interval. That is, $a = \frac{v_f - v_i}{t}$. For an object that is moving along a circular path, however, acceleration $(a) = \frac{v^2}{r}$. The derivation of this formula is beyond the scope of this book. Newton's Second Law allows us to determine the centripetal force of a revolving object. From $F = ma$, centripetal force (C. F.) $= \frac{mv^2}{r}$.

If the string were to be released or break, the object would move in a straight line path rather than a circular one and would assume a constant velocity according to Newton's First Law, neglecting gravitational forces which would come into play.

Centripetal force necessarily has a reaction force, according to Newton's Third Law of Motion. The reaction force is known as the *centrifugal force* (center-fleeing) and is equal in magnitude but opposite in direction to the centripetal force. The centrifugal force is then an outward force exerted by the object on the string. The string in turn transfers the centrifugal force to the hand or device which holds the string. The centrifugal force, therefore, does not act on the object as does the centripetal force.

COMMON COMPARISONS OF VELOCITY

The velocity with which man can run has increased throughout the years of modern sports as a result of better training techniques and more beneficial diets for athletes. A distance runner of today strives to run a measured mile in approximately four minutes. In order for him to accomplish this feat, his average velocity can be determined by:

$$\bar{v} = d/t$$

$$= \frac{1 \text{ mile}}{4/60 \text{ hour}}$$

$$= 15 \text{ miles per hour.}$$

The athlete who competes in the 100-yard dash endeavors to run this distance in a time which approaches nine seconds. His average velocity thus would be calculated from:

$$\bar{v} = d/t$$

$$= \frac{300 \text{ ft}}{9 \text{ sec}}$$

$$= 33.3 \text{ ft/sec.}$$

The units of velocity are sometimes difficult to conceive in the manner calculated. A conversion to other units of measure is often desirable to enhance the situation. Recall that one mile is 5280 feet in length. A velocity of 1 mph, then, would be equal to $\dfrac{5280 \text{ ft}}{1 \text{ hr } (3600 \text{ sec})} = 1.4667$ ft/sec. A useful comparison to bear in mind is obtained from converting 60 mph to its equivalent in ft/sec: 60 mph $\times$ 1.4667 ft/sec/mph = 88 ft/sec. Then a proportion can be written to convert mph to ft/sec or ft/sec to mph. For example, to convert 33.3 ft/sec to mph, the proportion would be:

$$\frac{60 \text{ mph}}{88 \text{ ft/sec}} = \frac{X \text{ mph}}{33.3 \text{ ft/sec}}$$

Cross-multiplying:

$$88 \text{ ft/sec} \times X \text{ mph} = 33.3 \text{ ft/sec} \times 60 \text{ mph}$$

To ease the method of solution, let's disregard the units for the moment:

$$88 \times X = 33.3 \times 60$$
$$88 X = 1998$$
$$X = 22.7 \text{ mph}$$

How fast has man been able to travel with the aid of his applied technology? An outstanding comparison helps the reader to achieve an appreciation for scientific accomplishments. The velocity of a bullet from a common target rifle, the .22 caliber, is approximately 1200 ft/sec, a velocity which is equivalent to 818 mph. Most of our sleek Air Force planes exceed this velocity as do some of our jet passenger planes. But consider the fantastic velocities at which our brave and adventuresome astronauts have traveled in the last decade as they have journeyed through space. In their conquest of outer space, the space pilots have guided their vehicles to velocities which have exceeded 25,000 mph—over twenty-five times the velocity of a rifle bullet! Imagine the momentum the space vehicle would have at this tremendous velocity!

The variables which affect the velocity of an object have been previously discussed. The velocity an object is capable of reaching is directly controlled by the magnitude of an external force and the length of time the force is applied to the object. Falling objects naturally reach their velocities as a result of the force created by the attraction of Earth upon them. Objects which slide down an inclined plane do so because of the same external force we know as gravity. Each object, sliding or falling freely, meets resistance in varying degrees. This reactive force is generally known as friction and is offered to the sliding object by its surface contact with the incline. Friction is also produced by the air in the case of objects in a state of free fall. In both cases, the frictional forces in action create a condition which limits the maximum velocity Earth's gravitational attraction can impose on a given object. *Terminal velocity,* the utmost velocity an object can attain under defined conditions, varies with shape, size, and mass. This feature is the reason a feather takes so long to fall to Earth when it is dropped and is in free fall. An object of equal mass such as a small steel ball would win the race with the feather consistently. The closer to Earth an object reaches, the more effect air resistance has on the velocity of the object because of the increase in the density of the air that surrounds Earth.

A parachutist, for example, falls with a continuous increase in velocity up to a maximum, a value governed by several variables. For example, his weight and the size of his parachute permit him to fall to Earth at a velocity which may be less than 25 mph. Without a parachute, the man would reach a terminal velocity of about 125 mph regardless of the height from which he jumps. The maximum velocity he would reach is dependent on his weight, his ability to maneuver his body to offer greater air resistance, and the wind currents to which he is exposed.

Raindrops also reach a terminal velocity, a fact that is extremely important. Without the terminal velocity the drops reach because of air resistance, imagine the danger in being exposed to a sudden downpour in which the raindrops approach the velocity of those which comprise a powerful stream from a garden hose. The destruction of plants, insects, and other creatures on Earth would be total as would be the damage done to man and his possessions.

Automobiles and airplanes also reach a terminal velocity because of frictional forces. Streamlining of cars and planes means just what the name implies —choosing a design which will permit streamline flow of air at high velocities and hence minimize frictional forces that react against motion (see Fig. 3.10).

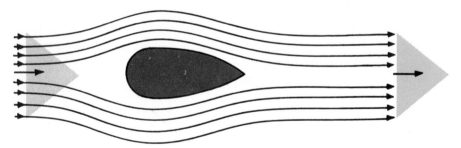

Fig. 3.10 The fluid flow (air or water) around a body offers minimal resistance if the flow is smooth and continuous.

Needless to say, objects which are set in motion by an external force have their actual paths and velocities affected by forces that resist their forward motion. An earlier discussion pointed out the effects of simultaneous forces in action on the same object. Both the speed and the direction of the object are capable of being altered by air and wind resistance.

Man has learned to make use of the change in velocity and direction caused by wind resistance as it acts on a body. With its aid, he can take advantage of this additional force and add its effect to those external forces he applies on an object. The sailboat and the manner in which the skilled helmsman can cause the boat to progress against the wind are prime examples and are illustrated in Figure 3.11. Man has also learned to use the resultant force of two or more

newton's laws applied

ball individually. However, the force of attraction between the two masses is real and measurable.

If an object were to be weighed on the moon by means of a spring scale, the scale reading would be about twenty percent as much as the scale would read if the object were weighed on Earth. This observation would lead the observer to conclude that the force of attraction between the object and the moon is only one-fifth that between the object and Earth. An athlete who can jump vertically seven feet on Earth could leap over a three-story building on the moon with the same effort. An object which weighs 100 pounds on Earth would weigh almost 300 pounds on the planet Jupiter and over 2.5 tons on the surface of the sun. The mass of the object would remain constant regardless of location, disregarding other phenomena beyond the scope of this discussion.

NEWTON'S LAWS AND AIR TRAVEL

The farther from Earth an object is located, the less the acceleration due to gravity created by the tremendously more massive Earth. The attractional forces in action on the object and on Earth are equal, but, due to the object's insignificant mass compared to that of Earth's, the object falls or is attracted toward Earth. If the object is some 4000 miles from Earth, the attraction between Earth and the object is one-fourth the value when the object is on Earth's surface. Other comparisons of weight variations are illustrated in Figure 3.13.

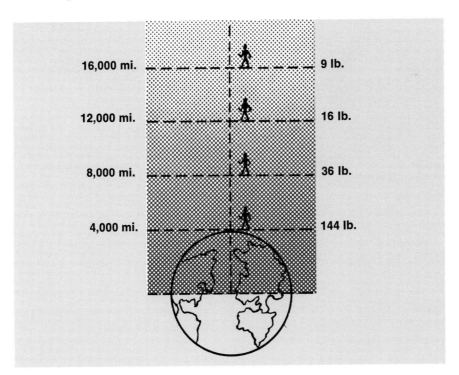

16,000 mi.	9 lb.
12,000 mi.	16 lb.
8,000 mi.	36 lb.
4,000 mi.	144 lb.

Fig. 3.13 The weight of an astronaut decreases as his distance from the center of Earth increases.

In order to propel a space ship into outer space swiftly enough to overcome the effects of gravity, the velocity of the vehicle must reach about 25,000 miles per hour. When this velocity is attained and the rocket engines are extinguished, Newton's First Law becomes evident in that the ship ventures onward with a constant velocity since no forces are present to act on it, including friction and gravity. The vessel continues to coast away from Earth at a constant velocity and remains in such a state of motion until it enters the gravitational influence of some heavenly body other than Earth. As this new force increases, the ship gradually changes direction toward the body and increases its speed as the distance between the two objects diminishes.

If the craft is projected vertically but does not reach the escape velocity of 25,000 miles per hour, however, it falls back to Earth once its fuel is exhausted. If the ship is sent upward at some angle other than vertical, its path is curved as it falls toward Earth. Because of the gravitational attraction of Earth, the ship's velocity increases. If this increase is great enough, the balance of forces acting on the vessel may be such that the vessel is prevented from being pulled back to Earth but is also prevented from escaping into outer space. The object, then, is in a circular orbit around Earth. The number of satellites, manned and unmanned, that technologists have placed in orbit is increasing and will continue to enlarge as attempts are continued by scientists to learn about our weather patterns, worldwide communication systems, and the secrets of outer space.

The method by which space vehicles reach velocities high enough to overcome Earth's pull is much different from other types of air travel. The airplane engine and the jet engine are not powerful enough nor are their modes of operation capable of reaching much above their present limits. Each must have air in order to operate and to move forward. The rocket engine functions without air much more efficiently than when air is present. The engine carries fuel which, when ignited, releases hot gases that are forced out the rear of the ship and that cause the craft to be projected forward at tremendous velocities, as mentioned earlier, many times the speed of a bullet.

The maximum velocity produced by a rocket engine is directly affected by the combination of fuels it uses and the overall mass of the space ship. For this reason, the rocket contains a series of individual rocket engines such as are illustrated in Figure 3.14. The first engine is designed to thrust the rocket upward at a velocity of about two miles per second. Then the second rocket engine ignites, and the spent stage of the rocket is ejected. The second engine increases the ship's velocity an additional two miles per second. This stage of the rocket is ejected, the third engine is fired, and so forth until the rocket reaches a velocity of over seven miles per second (about 25,000 mph) to overcome Earth's gravitational pull.

Both jet and rocket engines employ Newton's Third Law of action and reaction. When a balloon is filled with air and released, the escaping air causes its path to be somewhat irregular as is its velocity. The rocket is so well-balanced that seldom is its path irregular. The rate of fuel consumption by the engine is carefully adjusted to provide proper acceleration. Rockets consume

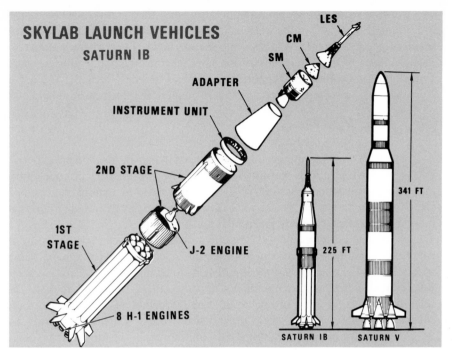

SKYLAB LAUNCH VEHICLES
SATURN IB

LES
CM
SM
ADAPTER
INSTRUMENT UNIT
2ND STAGE
1ST STAGE
J-2 ENGINE
8 H-1 ENGINES

341 FT
225 FT

SATURN IB   SATURN V

Fig. 3.14 **The rocket is constructed of stages which detach at various phases of the flight. The Saturn V and Saturn IB launch vehicles illustrated will be involved in future Earth-orbital missions such as the Skylab Program.**

(Courtesy National Aeronautics and Space Administration)

the fuel they carry, and the exhausted gases produce the action-reaction effect. Jet planes pull air into their engines, expand it by heating, and then expel it, a feature which produces the forward motion. The vertical lift of the rocket is produced by the fuel exhaust system whereas the difference in air pressure on the wings of the jet plane creates the lift which carries the plane upward. The jet changes direction by creating a dragging effect on one wing or the other by raising or lowering the ailerons, part of the wing. The rocket changes direction by expelling gases from its right or left side and slows down by expelling gases from the front of the rocket.

The point is that Newton's Laws of Motion apply anywhere in the universe. The effect of gravitational forces is influenced by the velocity with which an object moves. If an object plummets toward a heavenly body such as Earth or the moon, gravity causes the velocity of the object to increase according to the distance the centers of the two bodies are apart. The closer the bodies become, the faster the object moves toward the larger body. If an object moves away from a heavenly body, the effect of the larger body's gravitational attraction is lessened as a result of an increasingly greater distance between centers of the two bodies. If the forces which propel the object away from the heavenly body remain constant, the object steadily increases in velocity. If the same forces cease, the object continues at its attained velocity because of its inertia except for that amount required to overcome the decreasing gravitational force that remains.

Newton's Laws of Motion are universal since they apply anywhere and to everything. Each law governs various features of space travel and will continue to affect man as he ventures farther into the unknown. Everywhere he looks, he sees examples of the application of the laws whether the application be of his own doing or by nature. Regardless of where he goes, he must use all of Newton's laws to leave Earth and must rely on all of them to return to her.

CHAPTER SUMMARY

The three laws of motion conceived at least in part by Newton are generally known as Newton's Laws. Newton's *First* Law concerns *inertia,* the concept that an object at rest remains stationary unless acted upon by external forces sufficient to cause the object to move. Likewise, an object in motion retains its motion until sufficient external forces cause it to stop.

A force is a vector quantity, that is, a quantity that can be specified only when the magnitude and direction are noted. A constant force applied to any object causes the object to *accelerate,* that is, to change its speed and/or direction. Newton's *Second* Law points out that the rate of change of velocity (acceleration) of an object is always in the same direction as is the applied force. Mass and weight are different in several ways. *Mass* is a scalar quantity and weight is a vector quantity. Mass is a measure of inertia, the resistance an object possesses to being changed from its present state of motion. *Weight* is a measure of the force with which Earth or other heavenly bodies attract an object. The constant force of attraction of Earth on an object produces a rate of change of velocity known as acceleration due to *gravity.*

The investigations of freely falling objects have continued for many centuries. Observations eventually revealed that, within limits, freely falling objects accelerate. The rate of change of velocity has been computed to be about 32 ft/sec^2. Even objects such as projectiles fall toward Earth. The force of gravity is constant on all objects, regardless of the object's horizontal velocity until this horizontal velocity is sufficient to overcome gravitational forces. The escape velocity from Earth is much greater than the velocity of a bullet.

The *Third* Law of Newton is commonly known as the law of action-reaction. For every force applied to an object, there must be a reactive force of equal value. The recoil of a rifle is a prime example. The greater the mass of the rifle, the less is its recoil velocity. The product of the mass and velocity of an object is known as *momentum*. Newton's Third Law reveals that the momentum of the recoiling object is equal to the momentum of the projected object.

For revolving objects, the force of the string on an object is the only significant force in action. *Centripetal* force, as this force is called, remains constant. Therefore, the revolving object undergoes a constant centripetal acceleration, the direction of which is toward the center of revolution. In accordance with Newton's Third Law, there must be a reactive force to centripetal force. The force, equal in magnitude but opposite in direction to the centripetal force, is known as *centrifugal* force.

An athlete at best can run about 23 mph. A rifle bullet attains a horizontal velocity that seldom exceeds 1,000 mph. Man in his travels into space has reached velocities of over 25,000 mph. Objects which fall toward Earth are exposed to reactive forces produced by the surrounding air. The shape of an object, along with other characteristics, governs the maximum velocity the freely-falling object may reach. This maximum velocity is known as *terminal* velocity. Automobiles also reach a terminal velocity because of air resistance and other factors. The presence of air resistance permits a pitched baseball to deviate from a straight-line path, according to Newton's Third Law.

Each object in the universe attracts every other object. The force of *attraction* is dependent on the masses of the two objects and on the distance between them.

Airplanes, rockets, and all other modes of air travel apply Newton's Laws of Motion. External forces must be in action on any space vehicle in order for the vehicle to experience a change in velocity. All of Newton's Laws apply anywhere in the universe.

QUESTIONS AND PROBLEMS

1. Why is Newton's First Law known as the "law of inertia"? Define inertia in your own words. (p. 27)

2. Newton's Second Law is commonly known as the "law of acceleration." How are the First and Second Laws related? (p. 30)

3. Many times, invisible forces act on various objects, a condition which produces or prevents motion. List five examples of such cases. (p. 32)

4. A train is moving with a velocity of 100 mph. A young passenger on the train jumps vertically to touch the ceiling of the passenger car. Where does he land relative to his original position? Illustrate his path as viewed by a fellow passenger. Illustrate his path as viewed through a window by an observer located on a stationary platform outside the train. (p. 37)

5. Why does a carpenter find the chore of driving a nail into a loose board a difficult one? (p. 40)

6. Why does the nozzle of a garden hose push forcibly backward when the hose projects a swift stream of water forward? (p. 40)

7. Give several illustrations of equal and opposite forces which occur in pairs. (p. 40)

8. If an astronaut fires a rifle while he is floating in space, what types of motion might he experience? (p. 50)

9. A 2-gram bullet strikes a wooden plank with a velocity of 400 m/sec and exits on the other side of the board at 300 m/sec. How many such planks would be required to stop an identical bullet? (p. 42)

10. Early investigators of the natural world suspected numerous accepted concepts were erroneous; however, their views often remained secret until long after their death. Have such situations occurred in the 1900's? Support your answer. (p. 26)

11. Define and illustrate a force in terms of Newton's Second Law. (p. 32)

12. How does gravity as a force and gravity as an acceleration differ, according to Newton's Second Law? (p. 31)

13. A ball thrown vertically with the same force from the moon's surface would travel much higher than would the ball thrown from Earth's surface. Discuss the reason why this statement is acceptable. Would a ball dropped from a height of 100 ft above Earth's surface fall faster than a ball dropped from 100 feet above the moon's surface if air resistance in both cases is neglected? Support your answer. (p. 49)

14. A boy swims a river that is one mile wide. If the boy can swim three miles per hour and the current of the river is five miles per hour, how far downstream does the current carry him? (p. 37)

15. Would there be a point between Earth and the moon where an object would be weightless and would not be pulled toward either heavenly body? Discuss your conclusion. (p. 48)

newton's laws applied

the structure of matter

4

THE ATOMIC CONCEPT
THE ANATOMY OF ATOMS
ISOTOPES
THE REALITY OF ATOMS AND MOLECULES
THE STATES OF MATTER
THE KINETIC MOLECULAR THEORY OF GASES
TEMPERATURE-MOLECULES IN MOTION
THE CALORIE AND CHANGES IN STATE

vocabulary

Calorie	Absolute Zero
Heat of Fusion	Heat of Vaporization
Isotopes	Kinetic Energy

Our universe consists of an immense void in which there is scattered a comparatively minute amount of matter. It is, however, the matter in our universe with which we are concerned. It is matter that makes up our galaxies, our planets, and our human bodies. Understandably then, man needs to comprehend the secrets of matter, for unless man understands what matter is and how it responds to various conditions, he has no hope of ever understanding the planet on which he lives, much less the universe.

THE ATOMIC CONCEPT

A desire to understand matter is not unique to modern nuclear physicists or theoretical chemists but is a desire that has been held by men for many millenia.

Man has always been confronted with matter in a bewildering array of sizes, shapes, and colors. The ancients reasoned that there must be some fundamental principle linking together matter in all diverse forms. If one were to split a rock in half and split each half in half again, and so on, would there eventually be an infinitesimal portion of the rock that could not be split and still retain the properties of what is called a rock? Are there pieces of matter too small for the eye to see that combine with other pieces of similar matter to form a rock or a drop of water? Careful thinking about just such questions caused the Greek philosopher Democritus (460–362 B.C.) to declare that all matter was composed of minute particles, too small to be seen with the naked eye, that he called atoms. Democritus conceived, then, that matter is discontinuous on a micro scale even though to the eye it appears to be continuous and unbroken. Unfortunately, both Plato and Aristotle did not accept the atomic theory of matter that Democritus taught, and it was not until about 2000 years later that the atomic concept of matter was reintroduced into scientific thought. In 1808, the English chemist John Dalton expanded the atomic concept to explain the action of gases as well as chemical reactions. Essentially Dalton's atomic theory can be summed up in five statements:

1. Matter consists of indivisible minute particles called atoms.
2. All the atoms of any particular element are exactly alike in terms of shape and mass.
3. The atoms of different elements differ from each other in their masses.
4. Atoms chemically combine in definite whole number ratios to form chemical compounds.
5. Atoms can neither be created nor destroyed in chemical reactions.

Society accepted Dalton's atomic concept, and he is today called the father of the modern atomic theory of matter.

THE ANATOMY OF ATOMS

If uranium or other radioactive substances are placed in a deep well in a block of lead (see Fig. 4.1) and if a photographic film is placed in front of the well, the film is darkened at a spot in line with the well. If electrically charged plates are placed on both sides of the emanations from the well, three separate

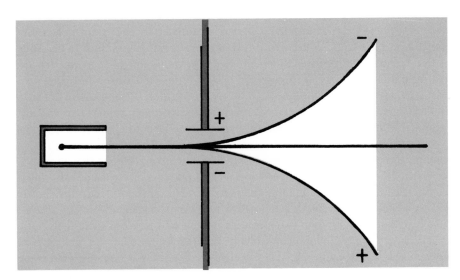

Fig. 4.1 **Electrically charged plates caused positively charged particles to be deflected toward the negative plate and negatively charged particles to be deflected toward the positive plate. Neutral particles and γ–radiation are unaffected by the charged plates.**

spots show up on the photographic film. Notice that one spot has been shifted toward the positive plate indicating that since unlike charges attract, the shifted spot must have been produced by a stream of negatively charged particles. Another spot has been shifted toward the negative plate and must have been produced by a stream of positively charged particles. However, a spot still shows up at the original position. Since the spot was not shifted by the charged plates, the spot must have been caused by either a stream of neutral particles, invisible radiation similar to x-rays, or both particles and radiation.

The negatively charged particles are called *electrons* (or *beta particles*). The emanation which produces the spot in the center that is unaffected by the charged plates is composed of *neutrons* and electromagnetic radiation, called *gamma radiation* (γ-radiation). Actually, it is the gamma radiation that produces the center spot. The film is unaffected by neutrons. The positively charged particles are called *alpha particles*. Alpha particles are composed of two neutrons and two positively charged particles called *protons*. The emanations from some radioactive substances then would lead us to assume that there are three basic types of subatomic particles that are the building blocks of atoms—electrons, protons, and neutrons.

With the exception of certain atoms of the element hydrogen, as will be made clear later on in our discussion, all the atoms in the universe contain all three of the above subatomic particles. Symbols and certain physical properties of the subatomic particles are listed below.

Particle Name	Symbol	Charge	Comparative Weight	Actual Weight
electron	e^-	-1	1	0.00091×10^{-24}g
proton	p^+	$+1$	1836	1.672×10^{-24}g
neutron	n	0	1837	1.675×10^{-24}g

Hydrogen

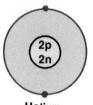

Helium

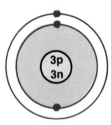

Lithium

All atoms except the atom of the element hydrogen consist of a very small and very dense nucleus of protons and neutrons surrounded by one or more electrons grouped into shells. The atom with the simplest possible structure is an atom of the element hydrogen, designated by the symbol H. The proton in the hydrogen nucleus has been represented by (p), and the electron has been represented by (●). Note that the positively charged proton is electrically balanced by the negatively charged electron; that is, the hydrogen atom is electrically neutral. The next simplest atom is an atom of the element helium (He), which contains two protons, two neutrons, and two electrons. The element lithium (Li) contains three protons, three electrons and four neutrons. Notice that the three electrons of lithium are not placed together but that two electrons are shown in the first electron shell surrounding the nucleus, and the third electron is shown in a second electron shell surrounding the first electron shell. Atoms of all elements contain no more than two electrons in their first electron shell. This limit of two electrons is a fundamental law of nature. As we shall see, the second electron shell can contain from zero to eight electrons—but no more than eight electrons. In fact, there is a maximum number of electrons that an atom can have in any of its shells.

Shell	Maximum number of electrons
1st	2
2nd	8
3rd	18
4th	32
5th	? (no atom in nature or as yet produced by man has this shell filled)

The simplified structure of the atoms of a few common elements are shown below:

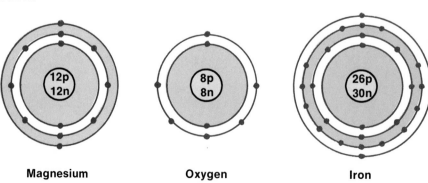

Magnesium   **Oxygen**   **Iron**

The number of protons in the nucleus of an atom is called the *atomic number* of the atom and is referred to as the atom's Z number. Thus, the atomic number of atoms of the element oxygen is 8, and the atomic number of atoms of the element iron is 26. Since all atoms are electrically neutral, the atomic number also designates the number of electrons surrounding the nucleus. The atomic numbers of elements are shown as a left-hand subscript of the symbol of an element ($_8O$ and $_{26}Fe$). Although the exact weight of the atoms of various elements is known with great accuracy, the weight of individual atoms is not generally used. Instead, the comparative weights of the atoms of various elements are employed. Since atoms of helium contain two protons and two neutrons (a total of four *nucleons*) and since atoms of the element hydrogen contain only one proton, atoms of helium should be approximately four times heavier than atoms of hydrogen. Since the weight of an electron is almost negligible in comparison to the weight of a proton or neutron, electrons influence the comparative weights of atoms very little. The relative weight of atoms of different elements is referred to as the *atomic weight* of the element. In a sense, the atomic weight of an atom gives the number of nucleons (protons plus neutrons) in the nucleus of the atom and is shown as a left-hand superscript with the symbol of an element (^{16}O and ^{24}Mg). The number of nucleons in an atom is called the atom's *A* number. Today, the relative weight of atoms is based on the weight of an atom of carbon which is assigned a weight of 12.00000 since the carbon atom contains six protons, six neutrons, and six electrons ($^{12}_6C$). For instance, an oxygen atom ($^{16}_8O$) has a relative weight (atomic weight) of 15.9994 since it is $\frac{15.9994}{12.0000}$ times as heavy as an atom of $^{12}_6C$, and an aluminum atom ($^{27}_{13}Al$) has a relative weight (atomic weight) of 26.9815 since it is $\frac{26.9815}{12.0000}$ times as heavy as an atom of $^{12}_6C$. Since the atomic weight of an atom reflects mainly the number of protons plus neutrons in the nucleus, one can subtract the atomic number (the number of protons in the nucleus) from the atomic weight (using a rounded off value) to obtain the number of neutrons in the nucleus. The number of neutrons in an atom is known as the atom's *N* number.

ISOTOPES

What determines whether an atom is an atom of hydrogen or an atom of oxygen or an atom of uranium? It is the number of protons in the nucleus of the atom. Thus, if an atom has only one proton in its nucleus, the atom must be an atom of hydrogen. If there are eight protons present, the atom must be an atom of oxygen. Even though all the atoms of a certain element must have the same number of protons, the number of neutrons present in the atoms of an element may vary. The most widely known and celebrated example of atoms of an element that has different numbers of neutrons is the atoms of the element uranium. Uranium 235 ($^{235}_{92}U$) is used as a fuel in atomic power plants and is used to make atomic bombs, whereas uranium 238 cannot be used for these purposes. Atoms of uranium 235 contain 235–92 or 143 neutrons whereas atoms of uranium 238 contain 238–92 or 146 neutrons. Atoms of a single element that contain different numbers of neutrons are called *isotopes*. Different isotopes of the element uranium are $\frac{235}{92}U$ and $\frac{238}{92}U$.

In general, the various isotopes of an element do not have separate names but are simply designated as shown above for uranium. The isotopes of the element hydrogen, however, have been given individual names. Quite frequently

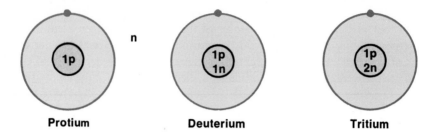

Fig. 4.2 Isotopes of the element hydrogen.

Protium   Deuterium   Tritium

deuterium is represented by the symbol, D, and tritium is represented by the symbol, T. Tritium does not exist naturally on Earth but is produced in the laboratory. The arrangement of two neutrons and one proton in the nucleus of tritium is apparently an unstable grouping of subatomic particles and, consequently, tritium decomposes rather rapidly. Deuterium, on the other hand, is quite stable. Approximately one atom of the isotope deterium exists for every 5000 atoms of the isotope protium. This means then that for every 5000 molecules of regular water (H_2O) there should be a molecule of "heavy water" (D_2O) in which protium atoms are substituted by deuterium atoms. Since there are about 3.2×10^{21} pounds of water in our oceans, there must be 6.4×10^{17} pounds or three hundred and twenty thousand billion tons of heavy water in our oceans!

THE REALITY OF ATOMS AND MOLECULES

Until recently all the evidence for the existence of atoms and molecules was strictly indirect evidence. Do not get the impression, however, that scientists doubted the existence of atoms and molecules because many phenomena observed by scientists could only be explained if they assumed the reality of the particular theory of matter. Still, no one had ever seen a molecule or an atom. Why then have scientists for over a hundred and fifty years been convinced that indeed matter does consist of atoms and molecules? Consider what happens when one places a very small amount of a dye in a glass of water. The color of the dye slowly spreads out, and eventually the color permeates the whole glass of water. How can one explain the diffusion of the dye unless one assumes that the water and the dye are both made up of very minute particles which intermingle to form a completely homogeneous mixture of water particles and dye particles.

Also you have observed, in a room in which the air is still and shafts of light cross the room, the erratic and zigzag motion of smoke particles or dust particles. This random and irregular motion of smoke and dust particles, called *Brownian motion,* can be explained only if one assumes that the dust or smoke particles are being bombarded by minute particles (molecules) of air.

In 1968 two DNA (deoxyribonucleic acid) molecules twined together were photographed with a special electron microscope that magnified the two molecules over seven million fold. In 1970 single atoms of thorium and uranium were seen as shown in Figure 4.3. Man has finally observed that which through logic he knew must exist.

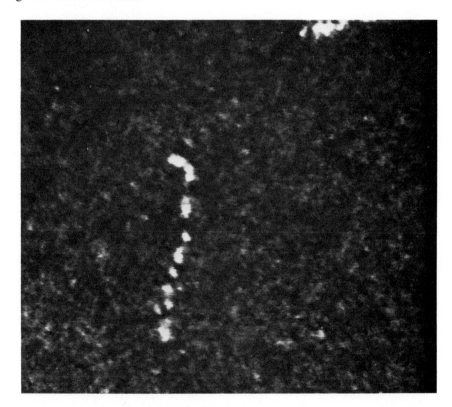

Fig. 4.3 **The first view of atoms (thorium) ever obtained by man. This photograph of a chain of thorium atoms was made by Professor Albert Crewe of the University of Chicago.**

THE STATES OF MATTER

There are three states in which matter can exist—solid, liquid, and gaseous. The substance that we have most often observed existing in all three of these states at various times is water. We have seen ice and snow, the solid state of water, and we have seen "water," the liquid state of water, but we have never seen gaseous water. A gas, unless it has a definite color is invisible just as the mixture of oxygen gas and nitrogen gas that make up the main ingredients of our atmosphere is invisible. Some gases can be seen since they have an intrinsic color, such as the green gas chlorine and the red gas bromine. But water in the gaseous state is colorless and, therefore, invisible. We have, however, on many occasions made observations that prove to us the presence of gaseous water. Many times we have observed water droplets forming on the outside of a cold glass of liquid. The appearance of liquid water droplets on a cold surface, seemingly from nowhere, is the result of gaseous water being cooled to the point where it changes from the gaseous state to the liquid state.

Why should there be such an apparent difference between the solid, liquid, and gaseous state of a substance? Solids are rigid substances and can hold a certain shape even though they are not in a container. This property would suggest then that the atoms or molecules in a solid must be held together by certain attractive forces that produce the rigidity of a solid, that is, a solid's ability to hold a definite shape. The above concept assumes that the atoms or molecules of a solid are restricted in their movements since excessive movement of a solid's atoms or molecules would result in a continual change in a solid's shape with a corresponding reduction in rigidity. A liquid has no rigidity, nor can it hold its shape without being confined in a container. The attractive forces between atoms and molecules of a liquid, then, must be considerably less than the same forces in a solid. Liquids, like a solid, are not very compressible and, therefore, the atoms or molecules in a liquid like those in a solid, must be quite close together so that an applied pressure can not squeeze the atoms or molecules much closer together than they normally are. The atoms or molecules of liquids must be relatively free to move about within the body of the liquid since a colored and colorless liquid placed together will soon assume a single shade of color, which could be the case only if the atoms or molecules of the two liquids were able to mingle and mix freely. Contrary to solids and liquids, gases must consist of atoms and molecules that are very far apart since gases are very easily compressible. Very large volumes of air can be compressed into an extremely small space. This large compression could occur only if a gas is mostly empty space. The atoms or molecules of a gas are certainly free to move about. If you open a bottle of perfume in one corner of a room, it does not take very long before a person on the other side of the room can smell the perfume. The molecules of the perfume in the gaseous state are obviously quite free to move and mingle with the air molecules in the room.

We have, then, built up the following picture of solids, liquids, and gases:

1. Solids consist of atoms or molecules held together in a rigid structure because of rather large attractive forces between the atoms or molecules. The atoms or molecules of a solid are close together and not easily compressible.

2. Liquids are composed of atoms or molecules that are free to move about and mingle with each other, but attractive forces keep the atoms or molecules within the main body of the liquid; that is, they will not uniformly fill a closed container.

3. Gases are made up of atoms or molecules that are completely free to move about at random and that will, therefore, uniformly fill a closed container. A gas is mostly empty spaces, and the distance between atoms or molecules is

very large compared to the size of the atoms or molecules themselves. The attractive forces between the atoms or molecules of a gas are very small.

THE KINETIC MOLECULAR THEORY OF GASES

We found previously that a body in motion has the capacity to do work, and, therefore, it possesses energy. Consequently, all moving bodies have energy, called kinetic energy, because of their motion. The kinetic molecular theory of gases assumes the following:

1. Gases are composed of discrete particles (molecules) that travel in random directions.
2. The molecules of a gas travel at a wide variety of speeds; that is, different molecules possess different amounts of kinetic energy.
3. Colliding molecules conserve their kinetic energy; that is, no kinetic energy of the molecules is converted into heat.
4. The distance between molecules is very large in comparison to the size of the molecules themselves. Thus, the total volume of the molecules of a gas is negligible in comparison to the volume of the gas itself.

The above assumptions of the kinetic molecular theory of gases explain rather readily such observations as the diffusion of perfume throughout a large room, the escape of a gas from only a pinpoint size hole, and the compression of a gas to a very small volume. Furthermore, gases are very light in weight because gases are mostly empty space. The theory also explains the pressure that a gas exerts on the sides of a container. This pressure is due to the incessant bombardment of the container walls by the molecules of the gas, such as in balloons and bubbles.

TEMPERATURE—MOLECULES IN MOTION

It has been mentioned that the molecules of a gas travel at a great variety of speeds. Just how fast do molecules travel? As an example let's assume that we can see the molecules of oxygen gas in a container. What can be observed? We can see some molecules that have practically no motion at all whereas other molecules are hurtling along at fantastically high speeds. The majority of the molecules have speeds in between these extremes. If the oxygen gas is 32°F (0°C), 1.3% will have speeds ranging from 2,237 miles per hour down to zero miles per hour and 7.7% will have speeds in excess of 15,659 miles per hour. Approximately 73.7% of the molecules will have speeds between 4,474 miles per hour and 13,422 miles per hour. The average speed of all the oxygen molecules present will be 10,313 miles per hour.

What will happen to the average speed of the above oxygen molecules if the temperature of the oxygen gas is increased? At 86°F (30°C) the average speed of oxygen molecules is 10,872 miles per hour, and at 212°F (100°C) the average speed of the molecules is 12.057 miles per hour. Obviously, then, the higher the "temperature" of a gas the greater is the average speed of the molecules of the gas. Putting it another way, a thermometer used to measure the temperature of an object is actually a device to measure the average molecular speed of the molecules of that object. If this thing called "temperature" is simply a measure of the average speed of molecules, what is meant when a person says that his skin feels hot or cold? If one is in a room that has a temperature of 78°F, the molecules of air in the room are continually bombarding the molecules of the skin with the consequence that there is attained an equilibrium between the motion of the molecules in the air and the motion of the molecules of the skin. It should be pointed out here that although the molecules of a solid must move within a very restricted volume (they vibrate), nonetheless, the molecules are in constant motion. If the temperature of the room is increased, then the average speed of the molecules of air in the room is increased; therefore, the air molecules have greater kinetic energy at the higher temperature. When the molecules, traveling at a higher speed than previously, come in contact with the molecules of the skin, they impart this extra kinetic energy to the molecules of the skin and make the skin molecules vibrate at a faster rate or move at a higher average speed. It is this increase in the average speed or vibration of the skin molecules that we interpret physiologically as an increase in temperature of our skin. Conversely, if the temperature of the room is decreased, the average kinetic energy of the air molecules becomes less than the average kinetic energy of the molecules of the skin. Now the molecules of the skin, on collision with air molecules, impart some of their extra kinetic energy to the molecules in the room. The exchange of kinetic energy results in a decrease in the speed of motion of the skin molecules with the consequence that we physiologically detect a cooling of our skin.

Can the speed or kinetic energy of molecules be increased or decreased indefinitely? According to the Lorentz contraction and Einstein's equation relating mass to velocity (see Chapter 9), it is apparent that nothing can travel faster than the speed of light. If temperature is simply a measure of the speed of molecular motion, then it would seem, at first thought, that there would be a limit to how hot an object can become. Thus, since molecules cannot move faster than the speed of light, the maximum temperature obtainable would correspond to a temperature at which the molecules were traveling just barely below the speed of light. It turns out, however, that instead of an object's temperatures being the direct result of the speeds of the molecules composing the substance, the temperature of an object is determined by the kinetic energy of the molecules of the substance. The kinetic energy of a molecule depends on the mass of the molecule as well on as the molecule's speed as shown in the following equation. $KE = \frac{1}{2}mv^2$ (see page   ). In this equation KE is the average kinetic energy of a molecule, m is the mass of the molecule, and v is the velocity of the molecule: Even this kinetic energy equation may not make it immediately obvious why there isn't a maximum value for kinetic energy that molecules cannot exceed. From Einstein's special Theory of Relativity, however, it is known that the faster an object moves, the greater becomes its mass. Of course, this relationship is rather negligible for speeds that are not a major fraction of the speed of light. On the other hand, as the speed of an object approaches the speed of light, the mass increases very rapidly as the speed of the object is further increased. Consequently, as a molecule approaches the speed of light, the mass term in the kinetic energy equation increases at such a rapid rate that the total kinetic energy increases very rapidly. As the speed of the molecule approaches the speed of light, its mass and, therefore, its kinetic energy and its temperature approach infinity.

We have seen that there is no limit to how hot an object can become. But is there a limit to how cold an object can be? Can an object become infinitely cold as well as infinitely hot? The answer is no! There is a temperature below which neither man nor nature can ever attain. As heat energy is taken out of a group of molecules, the average speed and, therefore, the average kinetic energy of the molecules decrease. If the withdrawal of heat energy is continued, the kinetic energy and speed of the molecules are further decreased until eventually the molecules are at rest. When the speed of the molecules reaches zero, the kinetic energy of the molecules becomes zero. If the kinetic energy of the molecules is zero, the molecules do not contain any more heat that can be withdrawn. Since the molecules do not contain any more heat, they must be as cold as they can ever become. The temperature at which molecules have zero kinetic energy is called *absolute zero*. Absolute zero is −459.7°F or −273.2°C. It is reasonable that a temperature scale be devised that uses this point of zero kinetic energy as zero temperature. Such a scale is called the *Absolute or Kelvin temperature scale*. Shown in Figure 4.4 is a comparison of the Absolute, the Celsius, and the Fahrenheit temperature scales. As can be seen from Figure 4.4, the Celsius and

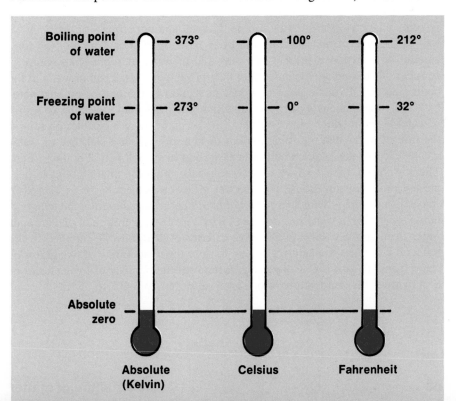

Boiling point of water — 373° — 100° — 212°

Freezing point of water — 273° — 0° — 32°

Absolute zero

Absolute (Kelvin)   **Celsius**   **Fahrenheit**

Fig. 4.4 **A comparison of the Absolute (Kelvin), Celsius, and Fahrenheit temperature scales.**

Absolute scales are readily interconverted. For instance, Celsius temperature is converted into an Absolute Temperature by simply adding 273° to the Celsius temperature. On the other hand, an Absolute temperature is converted to a Celsius temperature by subtracting 273° from the Absolute temperature. A Celsius temperature can be converted to a Fahrenheit temperature by multiplying the Celsius temperature by 9/5 and adding 32. For example, a temperature of 50°C is changed into a Fahrenheit temperature as shown below:

$$F = 9/5 \; C + 32$$
$$F = 9/5 \; (50) + 32$$
$$F = 90 + 32$$
$$F = 122°$$

A Fahrenheit temperature is changed into a Celsius temperature by subtracting 32 from the Fahrenheit temperature and multiplying by 5/9. A temperature of 50° Fahrenheit is changed into a Celsius temperature as shown below:

$$C = 5/9 \; (F - 32)$$
$$C = 5/9 \; (50 - 32)$$
$$C = 5/9 \; (18)$$
$$C = 10°$$

THE CALORIE AND CHANGES IN STATE

Heat, a form of energy, is referred to as *thermal energy*. It is necessary to distinguish between the temperature or degree of heat of a body and the amount of heat that a body possesses. The temperature of a body is a measure of how hot or cold the body is, but the amount of heat that a body possesses is dependent not only on its temperature but also on the total amount of mass of the body, as well as the type of material that the body is made of. For example, the total amount of heat that is present in a sink full of water at room temperature is considerably more than the amount of heat present in a small pan of water at the boiling point. Of course, the individual molecules in the pan of boiling water have more thermal energy per molecule than the individual molecules in a sink of water at room temperature, but there are so many more molecules present in the sink of water that the total amount of heat present in a sink full of water greatly exceeds the total amount of heat present in a small pan of boiling water. There is a need, then, to have some way to talk about the total amount of heat present in various materials. The amount of heat an object has is measured in *calories*. A calorie is defined as *the amount of heat required to raise the temperature of one gram of water one degree C*. Thus, if the temperature of one gram of water is raised from 40° to 45°C, five calories of heat must be supplied to the water. It also takes five calories of heat to raise the temperature of five grams of water from 40° to 41°C. Conversely, it requires the removal of five calories of heat to lower the temperature of one gram of water from 40° to 35°C.

Different substances require different numbers of calories to change the temperature of one gram of the substance. Thus, whereas one calorie is necessary to raise the temperature of water one degree C, one gram of iron requires only 0.108 calories to increase its temperature one degree, and one gram of aluminum requires 0.216 calories to increase its temperature one degree.

It turns out that the "calorie" that nutritionists talk about is actually a kilocalorie (1000 calories). Since we have become so conditioned in our society to "counting calories," it may be wise that the nutritionists state that a peanut butter sandwich contains 365 calories, since a person who is watching his diet might well go into cardiac arrest on learning that the peanut butter sandwich he has just consumed contained not 365 calories but 365,000!

It has been stated that the application of heat to a substance increases the temperature of the substance. However, if one observes a thermometer submerged in boiling water, he notices that the continued application of heat to the boiling water does not increase the temperature of the boiling water. Since the boiling water slowly decreases in volume, one must assume that the added heat energy is being used simply to convert the liquid water into gaseous water. In other words, the application of heat is necessary to bring about a change in the state of water from a liquid to a gas without a change in temperature. The amount of heat necessary to bring about a change in state of a substance from a liquid to a gas is called the *heat of vaporization* of the substance.

One gram of water at 100°C requires the application of approximately 540 calories of heat to vaporize one gram of liquid water at 100°C into one gram of gaseous water still at 100°C. Thus, the heat of vaporization of water is 540 calories per gram. Different substances require different quantities of heat energy to convert one gram of a liquid substance into one gram of gas.

Since the temperature of boiling water does not change with the application of additional heat, where does the heat go? If the applied heat energy increased the kinetic energy of the water molecules, the temperature of the water would have to increase. Apparently, then, the heat energy is not being used to increase the kinetic energy of the water molecules. Instead the energy is being used to overcome the intermolecular forces of attraction that hold water molecules together in the liquid state. Since it requires 540 calories to overcome the attractive forces of one gram of liquid water molecules, then when one gram of gaseous water at 100°C condenses into liquid water at 100°C, 540 calories of heat must be liberated.

The considerations of heat application to a change in states of a liquid to a gas are also applicable to the change in state of a substance from a solid to a liquid. Thus, in the case of water, one must supply eighty calories of heat to one gram of ice at 0°C in order to change its state from a solid at 0°C to a liquid at 0°C. Conversely, when one gram of water at 0°C is converted to one gram of ice at 0°C, eighty calories of heat are liberated. The amount of heat in calories necessary to bring about this change in state of a substance from a solid to a liquid is called the *heat of fusion* of a substance.

CHAPTER SUMMARY

Atoms consist of positively charged particles called protons, negatively charged particles called electrons, and neutral particles called neutrons. The nucleus of an atom contains protons and neutrons. The electrons of an atom surround the nucleus in various shells, and each shell can hold only a definite number of electrons. Atoms of a particular element always contain the same number of protons but may possess different numbers of neutrons. Such atoms are called isotopes.

In a solid, the intermolecular forces that hold the molecules together are strong enough to bind the molecules into a three-dimensional rigid lattice. In the liquid state, the intermolecular forces are reduced to the point that the molecules of the liquid are free to move in relation to each other. In the gaseous state, the molecules have sufficient kinetic energy and are far enough apart that the intermolecular forces are extremely weak; consequently, the intermolecular forces do little to restrict movements of gaseous molecules.

Matter is composed of discrete particles called atoms, which are in constant motion at any temperature above absolute zero. At the temperature of absolute zero, all molecular motion has ceased, and, therefore, the molecules of matter have no kinetic energy. The heat that a substance contains is a measure of the random motion of the molecules of the substance.

The heat of fusion of a solid represents the amount of energy necessary to overcome most of the intermolecular forces present in the solid with the result that disruption of the crystal lattice occurs and the solid is converted into a liquid. The heat of vaporization of a liquid represents the amount of heat necessary to overcome much of the remaining intermolecular forces of the molecules of the liquid so that the molecules are able to move with sufficient freedom to pass into the gaseous state.

QUESTIONS AND PROBLEMS

1. List various experiments that you could devise that would support the kinetic molecular theory of matter. (p. 63)
2. How much heavier is a proton than an electron? (p. 60)
3. If atoms of an element have a Z number of 79 and an A number of 197, how many neutrons do the atoms have? How many electrons? (p. 61)
4. Would increasing the temperature of a stove burner make vegetables cook faster in boiling water? (p. 69)
5. Explain why perspiration results in a cooling of the body. (p. 69)
6. On a hot summer day a Fahrenheit thermometer might read $95°$. What would a Celsius thermometer read? What would a Kelvin thermometer read? (p. 67)
7. How much heat would be given off when 5000 grams of water at $0°C$ is allowed to freeze at $0°C$? (p. 69)
8. In view of the kinetic theory of gases, explain why a liquid should expand when heated. (p. 66)
9. State the kinetic molecular theory of gases. (p. 65)

the structure of matter

some familiar areas of applied physics

5

MACHINES AND WORK
POWER AND ENERGY
PRESSURE AND ITS APPLICATIONS
HEAT, FRICTION, AND TYPES OF MOTION
EXPANSION OF LIQUIDS AND SOLIDS
AIR CONDITIONING
THE CHILL FACTOR

vocabulary

Chill Factor	Conduction
Convection	Energy
Heat	Kinetic Energy
Potential Energy	Radiation
Temperature	Work

The Educational Policies Commission of the National Education Associa-
ɔn proposes that all education be imbued with the spirit of science. A report
ɔblished several years ago by the NEA defines this spirit as longing to know
....d to understand; the questioning of all things; the searching for data and their
meaning; the demanding for verification; the respecting of logic; the considering
of premises; and the deliberating of consequences. The field of physics encompasses each phase of this spirit as do other sciences. Physics accomplishes this
claim through the investigation of gravity, motion, mass, energy, and the
world's manner of construction of the physical surroundings. It is the most
developed of the sciences and underlies the fields of biology, chemistry, engi-
neering, and medicine by serving as the basic foundation for each.

Other chapters within this textbook are concerned with the general role
of physics; however, this chapter is designed to permit the student to develop
an awareness of the degree to which physics is present in our everyday physical
surroundings.

MACHINES AND WORK

The numerous mechanical devices which aid man as he performs his
tasks are called *machines*. Machines of all sorts play an important and integral
part in the modern home. Appliances such as dishwashers, clothes dryers,
garbage disposals, and sewing machines are common to our society. Each device,
although in total a complicated bit of ingenuity, is simple in the design of its
individual component parts. The machine is intended to perform one of three
functions: (1) to lift a heavy load through the application of a comparatively
small force; (2) to move objects very rapidly although the applied force changes
position slowly; or (3) to cause work to be done in a manner more convenient
than would be the case if the applied force were applied directly to the object to
be moved.

When a weight is lifted away from the surface of Earth, *work* is performed,
whether the force is supplied by man or by his machine. The farther from
Earth's surface the object is lifted, the greater the amount of work done. In all
cases, two factors determine the amount of work done: (1) the amount of force
exerted and (2) the distance the force moves. In fact, a quantitative measure of
work is determined by the product of the force acting and the distance moved in
the direction the force acts. As an algebraic expression: work = force × distance

$$w = F \times d$$
$$w = Fd.$$

If the distance a person moves as he pushes on a wall is zero, he may become tired but no work has been done. If an object moves without a force acting on it, no work is done. The units of work contain a unit of distance and a unit of force. Work, therefore, is expressed as foot-pounds in the "*fps*" system of measurement. There are two common units by which work is measured in the metric system. In the "*cgs*" system, force is measured in dynes and the distance through which the force moves in centimeters. The unit of work becomes the dyne-centimeter, commonly referred to as the *erg*. In the "*mks*" system, force is measured in newtons and the distance the force moves in meters. The unit of work in this system of measurement is the newton-meter, known as the *joule*. The joule is 10,000,000 (10^7) times larger than the erg. In order to establish an appreciation for the size of the erg, imagine the amount of work required to lift two aspirin tablets approximately 0.5 inches (or 1 cm), a quantity of work equal to the erg.

In no case is the machine capable of doing a greater amount of work than the applied force is capable of doing on the machine. If more work could be done by the machine than was done on it, the machine would have the unprecedented ability to do work within itself. Any machine loses some of the energy applied to it because of friction among its moving parts. For this reason, no machine can be totally efficient; that is, it cannot yield the amount of work as output that is applied to the machine as input. The efficiency then is calculated from the following relation:

$$\text{Efficiency } (\%) = \frac{\text{Output work}}{\text{Input work}} \times 100.$$

Because of its increased number of moving parts, a complex machine has lower efficiency and a lower return from applied energy.

All machines are composed of a varying number and arrangement of fundamental components called *simple machines*. The origin of each—the lever, the wheel and axle, the pulley, the wedge, the screw, and the inclined plane—is long-lost history. The most commonly used types of simple machines are the lever, the pulley and its systems, and the inclined plane. The others mentioned above are in reality special applications of these three simple machines.

The lever in a basic form consists of a rod or pole that is rested on an external object at a given point called a fulcrum (F). The object to be lifted, commonly referred to as the resistance (R), is placed in various positions with respect to the fulcrum and the applied effort (E), a force which is designed to counteract and move the resistance. The three possible arrangements are illustrated in Figure 5.1 in addition to typical applications of each arrangement.

The calculation of the effects of varying the positions of the resistance and the effort in regards to the fulcrum is presented in Example 5.1. Similar analysis, regardless of the various arrangements of E, F, and R, follows the same pattern.

Example 5.1

The use of a hammer to remove a nail from a board is shown in Figure 5.2. The nail held in place by the friction between it and the board, offers re-

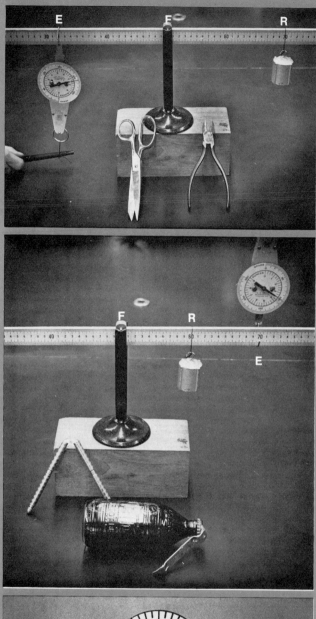

Fig. 5.1 The three types of levers and typical applications of each.

sistance at a point 3 inches (d_1) from the head of the hammer which serves as a fulcrum. The effort (E) is applied at the end of the handle, some 12 inches (d_2) from the fulcrum. If the effort required to remove the nail is 25 pounds, the resistance offered by the nail is determined by:

$$R \times d_1 = E \times d_2$$

$$R \times 3 \text{ in} = 25 \text{ lbs} \times 12 \text{ in}$$

$$R = \frac{25 \times 12}{3} = 100 \text{ lbs.}$$

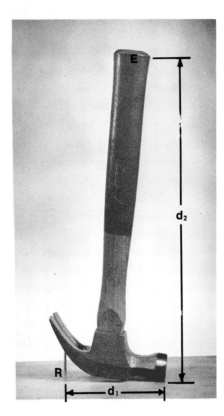

Fig. 5.2 The hammer is often used as a lever.

A system of pulleys is closely related to the lever. The resistance can be moved with less effort at a slow pace, or it can be moved a greater distance than the effort moves by the exertion of effort in excess of that actually required to move the resistance. A single fixed pulley, illustrated in Figure 5.3 (a), permits a change of direction. That is, the resistance (R) is lifted as the effort (E) is applied downward. The individual who applies the effort can use his weight to lift the resistance. If the resistance is greater than his weight, then the single fixed pulley is useless. A single movable pulley, illustrated in Figure 5.3 (b), can be used to lift an object with one-half the effort required with the single fixed pulley. The single movable pulley then produces a mechanical advantage (m. a.) of two. Combinations of pulleys permit decidedly greater mechanical advantages, displayed in Figure 5.3 (c). Pulley systems are quite common around the modern home. For example, the systems are component parts of radios, drapery rods, garage doors, and numerous other home appliances.

some familiar areas of applied science

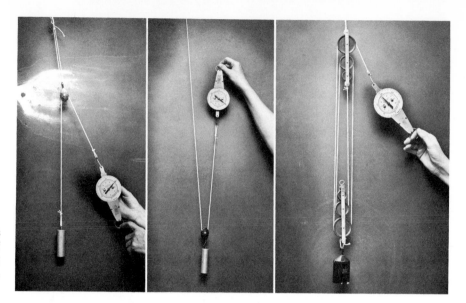

Fig. 5.3 **Various arrangements of pulleys and pulley systems.**

A third type of simple machine commonly used in the home is the inclined plane. In various forms, it is applicable to many situations. Generally, it is a ramp that connects two levels, whether stairs between floors or a sloping driveway. In a circular form it makes its appearance as a screw or a bolt. The efficiency is high since frictional forces are minimal and the mechanical advantage is simple to calculate:

$$\text{m. a.} = \frac{\text{length of plane}}{\text{height}} = \frac{l}{h}.$$

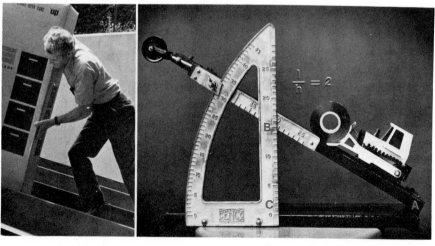

Fig. 5.4 **Two applications of the inclined plane.**

(Courtesy Central Scientific Company)

Figure 5.4 (a) and (b) illustrate the inclined plane. Typical applications around the home take the form of the ax, the door wedge, the letter opener, the stairs, and the knife. The screw is a special application of the inclined plane, as mentioned earlier. In order to see the manner in which a screw is in reality an inclined plane, imagine a piece of paper cut similar to the inclined plane illustrated in Figure 5.4 (a). If the piece of paper were rolled into a cylindrical shape, starting with side BC and progressing toward angle A, the line AB, though spiral, would still be an incline. Other examples of circular inclined planes are the spiral staircase and the typical highway through mountainous terrain.

Another simple machine known as the wheel and axle operates in principle as a continuous lever, displayed in Figure 5.5. In the ice cream freezer, the rotating can opener, and the door knob, the effort is applied in a circular path of large circumference as the resistance sweeps out a smaller circular path. The watch stem, the automobile steering wheel, and the screwdriver are other examples of the wheel and axle.

Fig. 5.5 The wheel and axle. The mechanical advantage is determined in the same manner as is the mechanical advantage for other simple machines.

In each type of simple machine, an invisible resistance to moving the object is always present. The resistant force is in reality the friction which exists between various surfaces in addition to the amount of inertia that the object possesses. Friction is found to vary between types and conditions of surfaces. For example, the effect of friction on the amount of force required to pull a block across a rough surface and across a smooth surface is displayed in Figure 5.6.

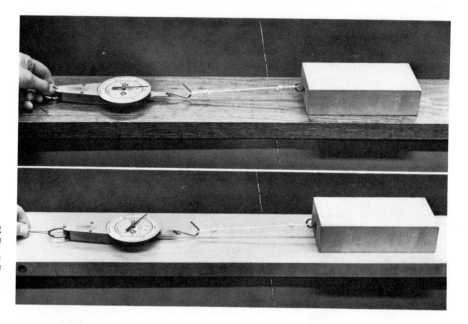

Fig. 5.6   The effect of smoothness of contact surface on the degree of sliding friction.

Frictional force is decidedly greater on the rougher surface; however, even extremely smooth surfaces have areas of irregularity revealed when they are examined under a microscope. Friction is present in fluids also since an object which moves through the fluid must push some of the substance aside.

Friction occurs whenever two bodies move relative to each other. If a lubricant is placed between the two surfaces, the frictional relationship that results is quite complicated. The purpose of the lubricant, however, is to lower frictional forces so that less force is required to move one surface across the surface of another. The lubricant tends to smooth out the irregularities in both surfaces by filling in the depressions which are innate in both bodies. When one body slides over the other, the type of friction that results is known as *sliding friction. Rolling friction,* generally a much smaller resistive force, results when one body rolls along the surface of the other.

The force required to slide one body over the surface of another is dependent on two factors: (1) the weight of the moving body, and (2) the type of surfaces involved. The magnitude of the force required to slide an object across a given surface compared to the force necessary to lift the object is known as the *coefficient of sliding friction.* Since friction varies between surfaces, the coefficient of friction assumes values commensurate with the surfaces involved. For instance, a dry wooden block that weighs 10 pounds requires an effort of 2.5 pounds to cause it to slide across a horizontal wooden surface. Therefore, the coefficient of sliding friction for wood on wood is 2.5 pounds/10 pounds or 0.25. If both surfaces are steel, only 1.5 pounds of effort are required to slide the 10 pound steel object. Various coefficients of sliding friction between surfaces appear in Table 5.1. Rolling friction as between a steel ball or cylinder and a steel plate is about 0.002. In other words, a 10 pound steel ball requires a force of only 0.02 pounds to keep the ball rolling once inertia has been overcome.

TABLE 5.1

Coefficients of sliding friction for smooth contact surfaces

Materials	Sliding Coefficients
Wood on wood, dry	0.25 to 0.50
Wood on wood, soapy	0.20
Metal on metal, dry	0.15 to 0.20
Metal on metal, greased	0.03 to 0.05
Rubber on concrete, dry	0.60 to 0.70
Leather on metals	0.56
Glass on glass	0.40

Man has learned to use friction to aid him in many ways. The friction between his shoes and the ground or between the tires on his auto and the highway has permitted him to move. He uses friction to bring his auto to a stop through the application of the brakes. Friction is also used by man as he builds his home since it provides the resistive force that actually holds the nails in the boards.

Friction is unwanted in many cases. If man could successfully eliminate it entirely, phenomenal efficiencies with various machines would result. There would be little challenge to devise an engine that would propel an automobile some 300 miles on a single gallon of gasoline. Parts in machinery would seldom wear out. Objects such as roller-coasters would be perpetual in that they could run endlessly once an initial force helped them to overcome inertia. Many lifetimes have been dedicated to the task of developing a frictionless machine, and all resulted in some degree of failure, although unwanted friction has been minimized through the application of ingenious suggestions inventors have proposed. The United States Patent Office still receives requests for patents on perpetual motion devices and has many seemingly working models on hand, but it discourages additional inventors of continuous motion devices as much as possible.

POWER AND ENERGY

Work, as stated previously, is the product of the force exerted on an object and the distance through which the force moves. The same amount of work is done in moving a heavy weight up a flight of stairs whether the person applying the force runs or walks. In order to differentiate between the time in which the work is done, a time rate unit is introduced called *power*. The term power is defined as the work done per unit time. As a formula, this relationship is stated:

$$Power = \frac{work}{length\ of\ time\ to\ do\ the\ work}$$

$$P = \frac{w}{t}$$

The common units by which power is measured are $P = \dfrac{joules}{sec} = watts$ and $P = \dfrac{ft\text{-}lb}{sec}$. Practical units of power are the kilowatt (1000 watts) and the horsepower (550 ft-lbs/sec). Since 1 ft-lb/sec is equivalent to 1.36 watts, the horsepower is equivalent to 746 watts of electrical power.

Example 5.2

A motor can lift a 300 pound object at a constant rate of 3 ft/sec. Determine the horsepower developed by the motor.

Solution:

$$Work\ per\ second = force \times distance\ lifted/second$$
$$= 300\ lb \times 3\ ft$$
$$= 900\ ft\text{-}lb$$

$$P = \frac{w}{t} = \frac{900\ ft\text{-}lb}{1\ sec}$$
$$= 900\ ft\text{-}lb/sec$$

$$Horsepower\ (H.P.) = \frac{Power\ (ft\text{-}lb/sec)}{550\ ft\text{-}lb/sec/H.P.} = \frac{900\ ft\text{-}lb/sec}{550\ ft\text{-}lb/sec/H.P.}$$
$$= 1.63\ horsepower\ (H.P.).$$

The greater the power of an engine, the faster it can do work. An engine which will propel a car to the top of a steep hill in thirty seconds develops greater power than an engine which takes a car of comparable weight to the top of the same hill in a longer period of time. One of the common units by which the power of an auto's engine is quoted by the manufacturer is the horsepower. (Some manufacturers also include the volume of the engine, measured in cubic inches.) The horsepower and its equivalency as a unit can be traced back to the days of James Watt and his steam engine. According to recorders of eighteenth century history, James Watt also attempted to sell his own invention. At the time, the horse was used in the coal mines to perform tasks that his steam engine could perform. Watt was asked by numerous mine operators the number of horses he thought his engine could replace. Eventually he realized that he needed to determine the equivalent of his engine to the horse. To accomplish this task, he used some of the horses which were on hand at one of the mines to find how long it would take each to raise a heavy weight out of a deep well. He soon determined that an average workhorse could perform about 375 foot-pounds of work per second over a typical work day. In order to facilitate the sale of his engine, he allowed himself room for experimental error and rated each of his steam engines in term of the power of a horse, but at 550 foot-pounds per second. The miners were elated with the efficiency and operation of the engine, and sales soared, particularly in the 20-30 horsepower range. The horsepower unit is also commonly expressed in unit time of minutes, which based on this length of time would be 60 sec $\times$ 550 foot-pounds/second = 33,000 foot-pounds/minute.

In order for a machine of any sort to do work, energy is required. Energy is the ability to do work, whether the work is done by machine, horse, or man. If something has energy, it can exert a force upon other objects and perform work in doing so. If, however, a force is applied to an object and work is done, energy is absorbed from the applied force and the object has been given the ability to do work. For example, a cuckoo clock is wound by a force that lifts the suspended weight. As gravity pulls on the weight, it moves downward. The attached gear system turns and thus the clock keeps time.

There are two forms of mechanical energy; that is, the energy concerned with motion and the action of force on bodies. Energy can be stored by an object by virtue of its position. Energy stored in such a manner is referred to as *potential energy*. A book which is lifted from its position on the floor and placed on a table has been given potential energy, since if the book were to fall to the floor it could do work. Work is performed in order to lift the book to the top of the table from its original position. Whenever work is done on a body, the body gains energy. The potential energy the book gains is equal to the work done in lifting it from the floor to the tabletop, a distance h. From the equation, work = $Fd = mgh$, the potential energy that is given the book is: $P.E. = mgh$ (or weight $\times$ height, if the weight is known rather than the mass). If the book has five joules of work done on it, the book gains five joules of energy.

The water at the top of Niagara Falls has potential energy because of its position, as does a skier poised on a mountain slope. Other objects which have potential energy because of energy stored include such things as a cocked pistol, a stretched spring or rubber band, or a poised hammer ready to strike a nail. Energy stored in an auto battery, though chemical in nature, is also potential energy. Other sources of chemical energy are gasoline, dynamite, coal, and food. All are capable of doing work as potential energy is released.

Energy, then, can assume many roles. Most common is the energy an object has because of its motion. To give motion, hence velocity, to an object, it must be accelerated. By Newton's Second Law, a force must be applied to the object, and by Newton's Third Law an equal and opposite force acts upon whatever applies the forward force. To give a body a velocity, work must be done to move the object against the reaction force. Any object which is moving has the capability of doing work as it strikes another object and causes the second object to move. Though the second object may roll or slide to a stop without doing work, the ability remains as long as the object is in motion. The ability a moving object has to do work is called *kinetic energy*. Kinetic energy is equal to one-half the product of the object's mass and the square of the velocity by which the object moves; that is: $K.E. = \frac{1}{2}mv^2$.

A sledge hammer poised over a nail has potential energy only. Once the hammer starts downward, kinetic energy is present that reaches a maximum value the instant before the hammer strikes the nail. The nail in turn is driven into the board as far as friction will permit. But the potential energy that the sledge hammer has is not instantaneously lost, nor is the maximum kinetic energy that the hammer develops immediately gained. Rather, as kinetic energy is gained, potential energy is proportionally lost.

Consider a child who is swinging at a playground. As the child swings back and forth, she alternately gains and loses both potential and kinetic energy. At the peak of her swing, potential energy is maximum, and for an instant, kinetic energy is zero. As she starts her swing downward, potential energy decreases and kinetic energy increases. At the lowest point where she is nearest the ground, kinetic energy is maximum and potential energy is zero. At points between the peak and the lowest part of her swing, the total amount of energy is the same as at any point in the arc the swing makes. Thus:

$$\text{Total Energy} = P.E. + K.E.$$

Energy can be transformed from one type to another, and all types can be transformed into mechanical energy, as is illustrated in Figure 5.7.

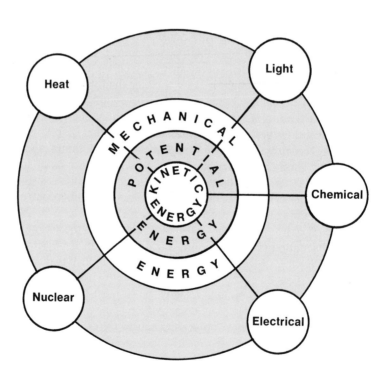

Fig. 5.7 All types of energy are interrelated. Each type may be transformed into any other type and all types can be transformed into mechanical energy.

some familiar areas of applied science

No machine is perfectly efficient; thus energy is seemingly lost. However, close observation can account for the entire amount of initial energy, since energy is always transformed into at least two kinds and generally more, one of which is typically thermal energy (heat). Energy is always totally conserved; in fact, an important principle known as the *law of conservation of energy* is involved. "Energy cannot be created or destroyed but may be transformed from one type into another; the total amount of energy available remains constant."

Let's look more closely at the energy Earth receives from the sun. Part of the energy directly heats Earth and gives us light. Other portions of the total energy are used to evaporate the water from our rivers, lakes, and oceans. As the water falls back to Earth in the form of rain or snow, much of it is redeposited in rivers and man-made lakes. The water then flows downstream or over spillways and, therefore, has the ability to do work by virtue of its original position. Man uses the water's kinetic energy to turn turbines which produce our electricity. Then electricity is used to heat and light our homes. Nuclear energy is also interrelated and is capable of being transformed into all other types of energy. As is discussed in other chapters, nuclear energy is the process by which our sun obtains its energy. The light or rather radiant energy which results from nuclear reaction then furnishes the energy necessary for photosynthesis, the synthesis of compounds which comprise our plant life. As the plants die and decompose, coal and petroleum are formed. The process is continued as other kinds of energy are provided from the coal and gasoline. The availability of energy is governed by the sun, whose life, we hope, will be a long one.

PRESSURE AND ITS APPLICATION

The modern home is generally equipped with devices that make the drudgery of cooking more pleasant to the housewife. Among these many devices is the pressure cooker. This appliance permits foods to be cooked under increased pressure and permits them to reach a state of doneness long before similar foods which are being cooked in conventional manners. Keep in mind that "cooking" is in effect a process by which kinetic energy is transferred from one body to another because of temperature differences. Also, temperature is a measure of the average kinetic energy among molecules in a body. Water boils at a much higher temperature when heated under the pressure of the pressure cooker than when it is heated in an open and conventional cooking utensil. Water has been observed to reach its boiling point at different temperatures because of various altitudes, thus differences in air pressure. On Pike's Peak, water boils at 196°F (91.1°C) and at Denver, Colorado, at 202°F (94.4°C). At sea level, water has been found to boil at 212°F (100°C). The effect of air pressure on water's boiling point is displayed in Table 5.2.

The length of time required to cook vegetables and meat in boiling water is much longer at high elevations than at low elevations, as many families who seek outdoor recreation have discovered. The wise camper who anticipates spending some of his vacation above normal elevations certainly includes a

some familiar areas of applied science 85

TABLE 5.2

The effect of pressure on the boiling point of water

Pounds per square inch	Barometer Reading		Boiling Point of water	
	Inches	Centimeters	Centigrade	Fahrenheit
Pounds			Degrees	Degrees
0.09	0.18	0.46	0	32
0.34	0.69	1.75	20	68
1.00	2.04	5.17	39	102
7.35	14.96	38.00	81	178
10.00	20.35	51.70	90	193
14.13	28.74	73.00	98.88	210
14.32	29.13	74.00	99.26	210.7
14.51	29.53	75.00	99.63	211.3
(sea level) 14.70	29.92	76.00	100	212
14.89	30.32	77.00	100.37	212.7
15.08	30.71	78.00	100.73	213.3
20.00	40.71	103.40	109	228
30.00	61.06	155.10	121	250
40.00	81.42	206.80	131	267
50.00	101.77	258.50	138	281
250.00	508.86	1292.50	208	406

pressure cooker among his necessities. A typical pressure cooker, displayed in Figure 5.8 (a) and (b), closes the escape route of the steam produced by heat and increases the internal pressure very rapidly, a feature which raises the temperature at which water boils. Meats that are typically tough when boiled under ordinary atmospheric pressure become tender and tasty when cooked in water which boils under increased pressure at higher temperatures. The effect of increased pressure on cooking times of some foods is displayed in Table 5.3.

The scientist realizes that we are exposed to tremendous air pressures constantly. At sea level, each square inch of our body is exposed to about 14.7 pounds of air pressure, the weight of the air at this altitude. For instance, an object whose surface area is 1000 in² is subject to a total pressure of 14,700

Fig. 5.8 **(a) The pressure cooker is a common cooking utensil in today's modern home. (b) The components of the pressure cooker are standard in practically all models.**

(a)

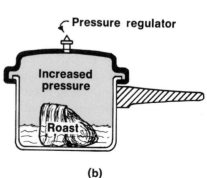

(b)

some familiar areas of applied science

TABLE 5.3

*The comparison of cooking times of common foods
as affected by increased pressure*

	Cooking Time	
Food	Regular	Pressurized
Beans	50 min	15 min
Chicken	90 min	30 min
Ham	240 min	45 min
Potatoes	30 min	10 min
Pot Roast	120 min	35 min

pounds on its outer surface. If this great pressure prevails on our bodies, what mysterious force keeps us from being crushed under the sheer weight of the air? The countering force of air pressure inside our bodies keeps us from being crushed to death and will continue to do so as long as we can maintain the two forces approximately equal. When we gradually change altitudes as we ride in an airplane or automobile, we become aware of this difference. We can compensate for the difference by equalizing the pressure by yawning or swallowing. What, then, is meant by an automobile tire gauge that indicates tire pressure at 30 pounds? The pressure inside the tire is 44.7 lbs/in² or 44.7 lbs/in² − 14.7 lbs/in² = 30 lbs/in² greater than the outside pressure on the tire.

Not all pressure to which we are exposed is due to air. Water, since it is obviously heavier than air, exerts much greater pressure than does air. A cubic foot of water weighs 62.4 lbs so each square inch of area on the bottom of a square container one foot in all dimensions, which has a volume of one cubic foot, is $\frac{62.4 \text{ lbs}}{144 \text{ in}^2}$ = 0.433 lbs/in². For every foot an object is submerged, water pressure increases 0.433 lbs/in². As a diver reaches a depth of 100 feet, he must contend with a water pressure of 43.3 lbs/in² in excess of normal air pressure on his body. An object submerged to depths greater than 100 feet must be strongly built and designed to withstand this crushing force. The shape which withstands the greater pressures has been found to be the sphere. Beebe's bathysphere, submerged in salt water whose weight per cubic foot or *density* is 64 pounds, withstood a total pressure of 192,000 lbs/ft² or 1340 lbs/in² as it ventured to an ocean depth of 3000 feet. Density is commonly expressed as weight per unit volume in the English system of measurement rather than mass per unit volume because the unit of mass in the English system is uncommon to most people.

The pressure exerted by liquids or air is constant in all directions at a given depth. A submerged object that cannot equalize internal pressure with that on its outside is subject to collapse if the pressure is great enough. The pressure of liquids is evident when an observer notes that the liquid in connected vessels, illustrated in Figure 5.9, reaches a common height in each vessel. Liquid placed in container A will flow through the connecting trough into B, C, and D until the

Fig. 5.9 The effect of the shape of the container on height of the liquid in connected vessels.

height in each is identical. The pressures at the bottom of each container are equal and indicate that the shape of the container has no effect on pressure. The factors which affect pressure are column height and liquid density. The application of the observation that liquids seek their own level in connected vessels has led man to the method by which he transports water into his home from distant storage tanks or reservoirs. The force by which the water is available through the conducting pipes is dependent on the variation in reservoir height compared to the height of the house. If the reservoir is not high enough to yield sufficient water pressure, a water tower must be included in the system into which the water is pumped before it is made available to the consumer (note Fig. 5.10).

From the discussion of pressure as it is related to liquids, the reader should be aware that pressure is equal to the product of density and depth, as obtained from the definitions of density and pressure. Pressure, force per unit area, results from the weight of the column of liquid directly above a given area, such as the area at the bottom of column A in Figure 5.9. Since the column is cylindrical, the column's cross-sectional view would be a circle. The area of a circle is calculated from the expression $A = \pi r^2$ where π is a constant equal to about 3.14 and r represents the radius, the distance from the center of the circle to its periphery. Density is the weight (or mass) of a liquid or solid per unit volume, and, therefore, weight is the product of density times volume. The volume equals the area multiplied by the height of the column of liquid. Summarizing,

$$\text{Pressure} = \frac{\text{force}}{\text{area}} = \frac{\text{weight}}{\text{area}} = \frac{\text{density} \times \text{volume}}{\text{area}}$$

$$= \frac{\text{density} \times \text{area} \times \text{column height}}{\text{area}}$$

$$= \text{density} \times \text{column height (depth).}$$

some familiar areas of applied science

Fig. 5.10 **The water tower supplies sufficient water and water pressure for rural America and other remote regions.**

Not all exerted pressures result from gases or liquids. One common source of pressure found in the home is a result of fashion. On occasion, many ladies enjoy wearing shoes with high heels that have a small contact area with the floor. The pressure produced by the heel of the shoe and calculated in pounds/in² is destructively great. For instance, if the wearer of the shoe weighs 130 pounds, about eighty percent or 104 pounds of her weight is centered on one heel as she walks. The 104 pounds is likely to be supported on a square heel as small as one-fourth inch on a side. The calculation continues and points out that the heel whose contact area is $\frac{1}{16}$ in² supports 104 pounds. The pressure exerted on the floor in conventional units would be 1664 lbs/in², a pressure as great as that exerted by the mightiest army tank. There is little wonder that many home floors are pock-marked by small square or round indentations.

The same principle, that is, the smaller the area of contact the greater the pressure which results, accounts for the reason that nails and stakes designed to be driven into the ground are sharpened to a point. Of course, the force which drives the nail or stake is multiplied in the same degree relevant to the amount the object is sharpened. The snowshoe applies the principle in reverse as it increases the contact area of the wearer's foot and thus lowers the pressure exerted on the snow proportionally. The same application explains why metal boats and their cargoes float in water.

Investigators have found that they can control pressure in the opposite extreme. Objects can be subjected to extremely low air pressure, a condition which reveals characteristically different properties in various substances and systems. For example, the reduction of air pressure in a sealed container naturally leads to less air friction. Prepared food that is maintained under reduced air pressure and the accompanying sterilized conditions keeps its taste and does not spoil, even though the food is not refrigerated.

The absence of air pressure is generally known as a vacuum and has been assumed to exist in outer space. The feasibility that a complete vacuum exists is virtually impossible since the most void areas of outer space are estimated to contain ten to twenty atoms per cubic inch. Perhaps the most void space in nature exists among the atom itself.

HEAT, FRICTION, AND TYPES OF MOTION

The concept of heat has been extremely difficult for man to comprehend. There has been a constant awareness of the presence of heat and its complexities since the days of ancient man. Long ago he observed that heat could be produced by friction. Two objects rubbed briskly together increase in the amount of heat they contain. If a wire is bent back and forth, the wire gets warm in the immediate region where it is being bent. Similarly, if a nail is driven into a board, the nail increases in the amount of heat it contains. And, if an object is exposed to the light energy from the sun, the object increases in temperature as the sun's rays are absorbed.

In Figure 5.11, evidence of the heat produced by friction between the grinding wheel and the axe is provided by the sparks produced, fragments heated to the point of incandescence. The same phenomenon is visible when a

Fig. 5.11 The grinding wheel uses friction to sharpen and shape the tools of modern man.

train is brought to an abrupt halt as the wheels are locked and thus slide along the tracks. Similarly, many observers have seen a sleek jet plane as it approaches the landing site with the front of its wings glowing with a red hue, a condition produced by the friction between air molecules and those of the wing. In order to prevent a space capsule from heating to the point of incandescence as it enters Earth's atmosphere, the vehicle is equipped with a heat shield.

Objects, regardless of their respective temperatures, have their molecules in a state of constant motion. This motion can be placed into three separate categories: *vibratory, rotary,* and *translatory.* Vibratory motion can be readily explained by considering two massive objects attached to each other by means of a common spring. If the two masses are each supported by a string and are then pulled apart, as illustrated in Figure 5.12 (a), the two masses move toward each other, then apart in the opposite direction. This motion is generally known as vibration.

Rotary motion is displayed in Figure 5.12 (b). This type of motion is evident everywhere in the universe, from the largest of stars to the smallest of atoms. Many objects and systems rotate; that is, they turn on their own axis.

The third type of motion is shown in Figure 5.12 (c). Translatory motion is the type which permits travel between streets, cities, and even bodies in or beyond our solar system. Generally, translatory motion involves movement along a straight line, but close observation reveals that we turn many corners and manipulate many curves in driving from city to city. Translatory motion is then qualified to mean that as a result of this motion, the ultimate direction of travel is a straight line. The same concept is prevalent in nature, even at the microscopic level and is described as *Brownian Motion.* Figure 5.13 reveals the

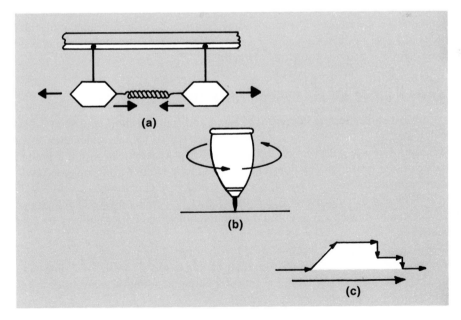

(a)

(b)

(c)

Fig. 5.12 **The three types of motion which an object can possess.**

some familiar areas of applied science   **91**

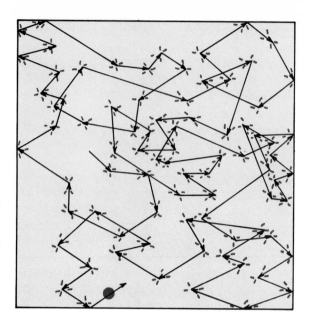

Fig. 5.13 Molecules move in a random direction because of multiple collisions with other molecules; a phenomenon revealed by close observation.

typical random path a molecule may assume as a result of numerous collisions with other molecules in a confined area. The molecules present in the confined area gain kinetic energy as heat is transferred from another body to it. Although much of the energy absorbed causes the molecules to vibrate more rapidly as well as perhaps to increase their rate of rotation, part of the absorbed energy causes the molecules to increase in translatory motion. The ultimate result is an increase in the number of collisions between molecules per unit time. *Heat* is the thermal energy which is transferred from one body to another because of a temperature difference. Thermal energy is a measure of the total molecular (kinetic) energies in a body. The measure of the average kinetic energy of each molecule in a body is known as *temperature.*

EXPANSION OF LIQUIDS AND SOLIDS

Most substances, because of the rapid and frenzied motion of their molecules, expand as they absorb heat or decrease in size as the molecules slow down because of loss of heat energy. Water is one of the exceptions within a limited range of temperatures. This substance is found to react in the same manner as do other substances when its temperature is above 4°C. However, below 4°C (39°F) water actually expands, a feature which is readily observed by the shape of a carton of milk if it is not brought into the house soon after delivery on cold mornings.

The expansion of water is a powerful force that has served as an agent which removes layers of rocks from their resting place. Gigantic land slides and cracked sidewalks result from the freezing and expanding of water after it has penetrated beneath the surface of Earth.

some familiar areas of applied science

The peculiar manner in which water reacts to changes in temperature has permitted aquatic life to thrive in areas where freezing temperatures are reached. Without the property of expanding on cooling, water would freeze solidly at all depths and most life in lakes and rivers would end. Near the temperature at which water freezes, 0°C, the water which is coldest has a lower density than the rest of the water in the stream, so it rises to the surface and causes water to freeze first at this point. The surface ice forms a protective and insulating cover which retards the rate by which the lower levels of water freeze. Generally speaking, the lower levels of water stabilize at 4°C (39°F), illustrated in Figure 5.14. The strange effect of temperature on water produces an unusual situation. Like volumes of water whose temperatures are 2°C and 6°C weigh exactly the same. Few substances, none common to man, react in the same manner.

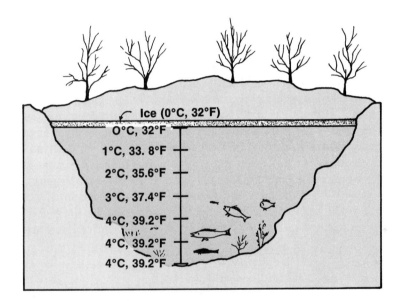

Fig. 5.14 **The aquatic life survives cold winters because ice is less dense than water.**

The expansion of all materials, regardless of the state of matter in which the substance exists, does not affect their masses. The volume, however, is increased by virtue of the added heat absorbed, so the density of the substance is decreased. For this reason, warmer air in a room is located near the ceiling. The hot water tank in our homes contains the hottest water at the top since hot water responds to heat just as do other liquids. An automobile tire filled with air on a cool morning to indicate an internal pressure of 30 lbs/in² may increase in air pressure to 34 lbs/in² without additional air as the tire temperature increases to 25°C (room temperature). The increase in pressure and temperature is caused by the exposure of the tire to either the sun or to the heat produced by friction as the car is driven. A similar increase in pressure causes a cake to expand in volume as the gas bubbles are formed during baking. The cake is caused to "rise" and thus become light and porous as heat is applied.

Solids which are exposed to heat expand in all dimensions: length, width, and height. The expansion in all three directions thus causes a considerable increase in volume, since volume is equal to the product of the three dimensions. For this reason sidewalks, streets, and bridges are constructed in small sections and are so placed in conjunction with the surrounding sections as to allow room for expansion. If the material experiences a drastic temperature increase due possibly to air temperature but usually to a catastrophe such as fire or explosion, the rate of expansion becomes an agent of destruction. Many streets and bridges have been extensively damaged by unusual amounts of expansion. Of course, cold temperatures cause materials to contract, a phenomenon which could produce vast destruction as the materials in bridges or in railroad tracks pull apart and leave behind gaping spaces not compensated for by the design engineers.

Solids expand at different rates, a property which makes the mixing of some materials somewhat limited in value, though the mixture may have many decided advantages. No force as yet discovered by man can influence the rate by which materials expand. There are various rates of expansion among the metals, the plastics, and other substances. Measures of this rate of expansion have been undertaken for most substances and are known as the *coefficient of linear expansion* and the *coefficient of volumetric expansion*. These unique measures reveal the amount of expansion that occurs in a given length or volume of a substance as the temperature increases per unit degree. To clarify the manner in which the coefficients of linear expansion and volumetric expansion are determined, let's examine the coefficient of linear expansion more precisely. Almost all solids expand in a uniform manner as they are heated. If we assign the length of a metal rod the value "l" and the temperature the value "t", then the small increase in length, Δl, with a small increase in temperature, Δt, indicates the rate of expansion. To express the relation algebraically, the coefficient of linear expansion

$$\alpha = \frac{l_2 - l_1}{t_2 - t_1} = \frac{\Delta l}{\Delta t}.$$

Generally, the temperature is expressed in centigrade degrees, C° and the unit of length and its corresponding increase (or decrease if the temperature is decreased) are expressed in inches, meters, miles, or any other convenient unit of linear measure. Some common substances and their coefficients of linear expansion (length, width, or height) are listed in Table 5.4. The unit increase is in proportion to the unit by which the original length is expressed. For example, if the coefficient of linear expansion of a substance is 0.000017 (read seventeen millionths), a change in temperature of one centigrade degree would produce a change in original length from 1.000000 to 1.000017, regardless of the unit chosen to make the original linear measure.

Many applications of the variation in expansion rates among materials are vivid in our everyday lives. For example, the housewife finds that on glass jars the metal lids that are too tight to be removed readily can be easily loosened by pouring hot water over the lid. In addition, household repairs and the problems which involve the removal of a rusty nut and bolt are greatly resolved by heating

TABLE 5.4

*The coefficient of linear expansion
per centigrade degree of some
common substances*

Aluminum	0.000026
Brass	0.000019
Copper	0.000017
Glass	0.000009
Iron	0.000011
Lead	0.000029
Pyrex	0.000003
Steel	0.000012

only the nut and then removing it by a conventional manner. On the other hand, electric stoves and other devices which make use of a heating element can be ruined through being bent out of shape if the device is heated above the recommended temperature or if the element is jarred or otherwise disturbed while it is excessively hot.

For what purpose then do construction workers heat rivets to glowing before they install them in steel beams? The reason, of course, is to pull the steel beams more closely together as the rivets contract on cooling. Why does a cold glass crack when suddenly submerged in boiling water, or a hot light bulb shatter when touched by a drop or two of cold water? Obviously, expansion or contraction is involved. Similarly, the canning of foods meets with success only if the containers are sealed tightly while hot so that they contract on cooling. The pressure inside the containers is lowered below normal atmospheric pressure as the lid becomes airtight. Many mothers have discovered that if the baby's prepared milk formula is placed in bottles when it is cold, the milk will overflow as the bottles are heated in preparation for the baby's consumption since the milk expands at a greater rate than does the plastic or glass.

AIR-CONDITIONING

The conditioning of the air, in a broad sense, refers to man's attempt to control his comfort, health, and efficiency. Air-conditioning is installed in stores to increase sales through customer comfort. In offices or schools, air-conditioning is used to increase the efficiency of the individuals concerned. Air-conditioning is a unique application of two areas of physics known as *thermodynamics* and *hydrodynamics*. From the former, basic data are accumulated concerning the thermal properties of gas mixtures and vapors such as would be found in an average environment of dry air and related water vapor. Hydrodynamics concerns itself with fluid motion, both liquid and gaseous, which occurs in fans and in the duct work. This area of science also considers conditions which influence turbulent flow.

In air-conditioning, the variables which must be controlled include the following: temperature; moisture content or humidity; gas composition, which includes carbon dioxide; and freshness, an undefinable concept. The comfort air-conditioning devices are primarily concerned with temperature and humidity and display little concern for the other variables. The range that is considered optimum is presented in Figure 5.15. Commercially, the installation may neglect temperature and humidity, but may concern itself with the removal of toxic vapors and particulate matter. This type of air-conditioning is prevalent in plants which make constant use of ovens, welding devices, or furnaces in their operation.

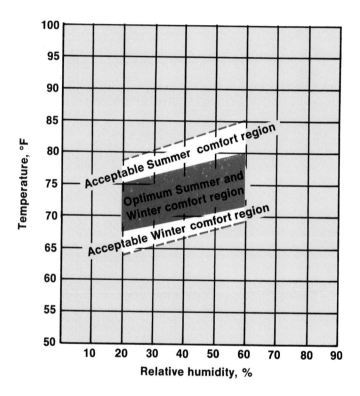

Fig. 5.15 The range of optimum conditions for summer and winter comfort overlap, an observation which provides a range for control.

The basic requirement of comfort conditioning is the maintenance of an indoor environment which helps the occupant develop a thermal balance between body energy production and body heat loss. This balance must be provided without subjection of the individual to unnecessary strain or without the enforcement of excessive operation of body mechanisms such as sweat glands or the muscles and nerves which induce shivering. The body is similar to the internal-combustion engine, since it receives energy from fuel (food) and dissipates it as work and heat energy. Heat energy is measured in kilocalories, a unit which is the amount of heat required to change the temperature of one kilogram of water 1 °C. The *calorie,* used in rating the thermal energy of foods, is

some familiar areas of applied science

actually a kilocalorie (kcal). The average adult requires about 100 kcal/hr because of the internal work necessary to maintain life, such as breathing, beating of the heart, and releasing of wasted heat energy. This loss is given to the surrounding air if the body is to maintain equilibrium. If the rate of total energy loss is less than the production rate, the body temperature increases. This rise in temperature produces a maladjustment of the physiological mechanisms and creates a possible hazard to the individual's health. A man who weighs approximately 175 lbs develops a fever of some 3°F in about one hour if he is prevented from losing heat energy by wearing excessive clothing or being confined to a warm area with insufficient ventilation.

Generally speaking, the heat energy of the body is dissipated at fixed rates through (1) radiation from exposed skin or clothing to the surrounding air which involves a rate of about forty-five kilocalories per hour at a normal room temperature of 70°F (21°C); (2) convection—the transfer of heat to ambient air (the most exacting to control) at the rate of thirty kilocalories per hour at normal room temperature; (3) latent heat of vaporization, which accounts for twenty-five kilocalories per hour through the evaporation of about forty-five grams of moisture hourly in the form of perspiration; and (4) some small amount as direct evaporation from living surfaces to respiratory air. The latter two methods of body heat loss are essentially constant and, therefore, are most subject to adjustment through air conditioning. Another method by which heat is transferred from one object to another is through conduction that involves the actual contact of the two surfaces. Conduction is somewhat involved in both (3) and (4) above. The three methods of heat transfer then are conduction, radiation, and convection, as illustrated in Figure 5.16.

The objectives generally taken into consideration for comfort air-conditioning follow. First, humidity is controlled to prevent the sensation of dryness

Fig. 5.16 **Heat is transferred from one body to another by the three methods illustrated.**

or clamminess of the skin. However, within the comfort region, variation will not compensate for major changes in temperature, as is displayed in Figure 5.15. Second, temperature is controlled to provide an average of 70°F. The comfort equation, one in which the sum of the air temperature and the immediate wall temperature is 140°F (60°C), explains why, for proper control, a temperature of 70°F cannot be fixed. As one would expect, the surface temperature of the walls varies directly, to a large extent, with the outside air temperature.

Our homes require a manner of controlling the temperature in them for comfort's sake. In some homes which use a floor furnace located near the center of the living area, the air over the furnace is heated so that it increases in kinetic energy and thus expands. This less-dense air is forced upward by the colder, more dense air moving along the floor. As the warmer air reaches the ceiling, it spreads out toward the walls and loses kinetic energy by virtue of contact with colder air, so it returns to lower room levels. From this transfer convection currents are created by unequal air temperature, and heat is transmitted throughout the home as a result.

The process of convection permits the ventilation of the home. The air we breathe becomes increasingly warmer so it becomes less dense than the air from outdoors. Therefore, with proper arrangements of the windows, air in the rooms is easily changed. A window opened near the top of the room will permit the less dense air to be forced outward if a window is also opened near the floor. The fan installed over many kitchen cooking units exhausts the warm and less dense air which contains cooking odors without relying on warm air and cold air initiated convection currents.

Convection plays a major role in the heating of homes which have hot and cold air returns distributed throughout them. The hot air vent is placed under a window when at all possible. This arrangement causes convection currents because of the cold window glass coming in contact with the warm air of the room. Another popular use of convection is displayed in the use of fireplaces in the home. The heated air in the fireplace's chimney expands and, in conjunction with the smoke and gases, is forced upward by the cooler, denser air present in the room along with that air coming down the chimney. A tall chimney is more efficient than a shorter one due to the greater difference in weight between the hot air and gases in the chimney and an equal volume of cold air outside it. Still another obvious application of the principle of convection is the number of land breezes, sea breezes, windstorms, tornadoes, and hurricanes produced by nature.

Scientists have determined that control of the air is necessary for better efficiency, comfort, health, and possibly survival. They are also aware that man is pouring some 390,000 tons of waste into the atmosphere over the United States annually. Unless the amount of waste products, gases, and particulate matter is reduced drastically, man may in the future have to carry his own supply of air as he would upon visiting the surface of the moon. The task would be virtually impossible, since an average adult breathes some 35 pounds of air daily or over 6.5 tons of air per year!

THE CHILL FACTOR

As the scientist has learned to be more precise in his description of various scientific endeavors, he has called our attention to a rather new phenomenon involved in the recording of weather observations. The term *chill factor* has come to represent man's attempt to identify an atmospheric condition which results when observations of temperature and wind velocity are combined. The effect is one which points out a lower equivalent temperature than that observed on a typical thermometer.

There are many cases of people, particularly sportsmen such as hunters and skiers, who have been seriously frostbitten when suddenly exposed to increased wind velocity. This condition brings about extreme discomfort even though the individual is dressed properly for exposure to the temperature indicated on the thermometer. Since winter outdoor recreation is growing in popularity at a rapid pace, an awareness of the chill factor is increasingly important.

In Figure 5.17, the effect of wind velocity on temperature is displayed in terms of equivalent temperature. In addition, the viewer can ascertain the various degrees of danger to face, hands, and other exposed areas of his body that result

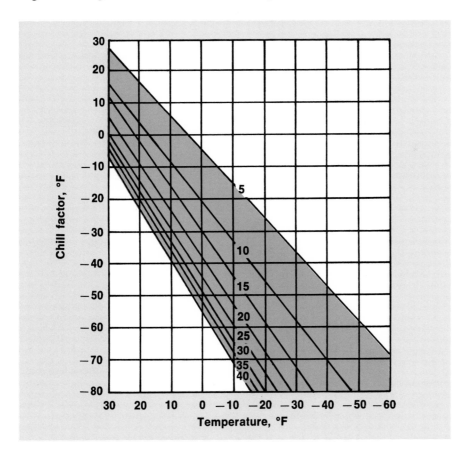

Fig. 5.17 The effect of wind velocity is graphically represented in a manner which has come to be known as the chill factor.

some familiar areas of applied science   **99**

from various conditions of the atmosphere. As an example, if a person were to attend a football game in which the gametime temperature were to drop to 15°F with gusts of wind that reach twenty-five miles per hour, how warmly should he dress? As is obtainable from the table, the spectator, as well as the players, can expect to be exposed to an equivalent temperature of 22° below zero! The danger from frostbite is considerable to all exposed areas of the body as well as to those areas which have improper or insufficient protection. If an individual were exposed to such conditions as above while riding on a snowmobile or while skiing, the equivalent temperature would need to be adjusted to compensate for the additional wind velocity to which he would be exposed. If the snowmobile were moving into the wind at a velocity of ten miles per hour, the equivalent temperature would drop from −22°F to −28°F with the probability of receiving frostbitten tissue increasing accordingly. Conversely, if the snowmobile were moving in the same direction as was the wind, the chill factor would be lowered because of decreased resultant velocity. A knowledge of the inherent dangers that result from this concept of chill factor can aid in preventing a pleasurable outing from being turned into one of disaster.

CHAPTER SUMMARY

Machines are designed to make man's tasks easier to perform. Machines *transfer energy, transform energy, change direction* of an applied force, and multiply force or speed. Regardless of complexity, all devices are composed of one or more levers, pulleys, inclined planes, or wheel and axles. Both man and his machines can do work. The amount of work done is dependent on the amount of force exerted and the distance the force moves. Quantitatively, *work* is equal to the product of the force exerted and the distance the force moves. Work is measured in ergs, joules, and foot-pounds. The *efficiency* of a machine is a comparison of the output work to the input work. Efficiency is inversely related to the friction present between the machine's moving parts.

The work done per unit time is called *power*. The common units of power are *watts* and *horsepower*. In order for work to be done, energy must be consumed. Stored energy is referred to as *potential* energy. Potential energy is present in a battery, in gasoline, or in a hammer poised over a nail. *Kinetic* energy is the energy of motion. As an object loses potential energy, the object gains kinetic energy proportionally. Since energy is always totally conserved, the total energy a body contains is a sum of its potential and kinetic energies.

Increased pressure raises the boiling point of liquids. Conversely, the boiling point of a given liquid is less if the liquid is exposed to a decrease in pressure. In order to cook food more rapidly, pressure cookers, devices which increase the pressure to which the food is exposed, are used. Liquids and gases both exert pressure as a result of their respective weights. For example, the pressure on a body submerged in water is increased 0.433 lbs/in² per foot of depth. The pressure a fluid (gas or liquid) exerts is equal to the product of the density of the fluid and the depth to which an object is submerged in it.

Heat is the total kinetic energy the molecules of a given body contain. Heat may be increased in a body by friction. The motion of molecules in a body consists of three possible types: *vibratory, rotary,* and *translatory.* Molecular motion is often random because of collisions between molecules.

Generally speaking, all solids and liquids expand as the kinetic energy of the molecules of the substance increases. Water, for some undetermined reason, expands rather than contracts as its temperature is decreased below 4°C. This strange property of water permits aquatic life to survive in cold areas. All substances vary in their rates of expansion, a measure of which is known as the coefficient of linear or volumetric expansion.

Air-conditioning is an application of thermodynamics and hydrodynamics. The variables in air-conditioning include temperature, humidity, gas composition, and freshness. Air-conditioning is designed to aid an individual maintain a thermal balance between body energy production and body heat loss.

An atmospheric condition that results from a comparison of air temperature and wind velocity is called *chill factor.* A temperature equivalent for a given air temperature and wind velocity indicates possible dangers to exposed areas of the body.

QUESTIONS AND PROBLEMS

1. What vertical force must be exerted on the handles of a wheelbarrow 3.5 feet from the axle of the wheel in order to lift a load of dirt that weighs 200 pounds and whose center of gravity is concentrated at a point 1.2 feet from the axle? (p. 77)

2. In the construction of the pyramids of Egypt, blocks of stones are said to have been pushed up an inclined road. How many laborers, each who could exert a force of 50 pounds parallel to the road, would have been required to push a 10-ton block of stone up a road which had a 3% grade (a 3% grade would indicate that the incline increased vertically 3 feet per 100 feet horizontally)? Would the magnitude of the net force be affected if the blocks were rolled on tree trunks up the incline? Support your answer. (p. 78)

3. Are the so-called simple machines of very ancient or of comparatively recent origin? Do these machines really save us work? Discuss your conclusions briefly. (p. 75)

4. What is a perpetual motion machine? Can you suggest some models which might be investigated? Is an artificial satellite or the moon a perpetually moving object as either orbits Earth? (p. 81)

5. How are the members of each pair of concepts similar and how are each of the members in the pair different? (p. 81)
   a. work and power
   b. potential energy and kinetic energy
   c. friction and heat
   d. power and energy
   e. pressure and force

6. What effect does lowering the pressure above a liquid's surface have on the liquid's boiling point? How might this effect be demonstrated? (p. 85)

7. Should the sides of a large storage tank filled with water be stronger near the bottom or near the top? Support your answer. (p. 88)

8. How is high pressure developed in a pressure cooker? How does the increased pressure lower necessary cooking time for foods? (p. 86)

9. The stewardess on a passenger airplane discovered that she could boil eggs for the passengers as easily at an altitude of 50,000 feet as she could at her home in New York. She also found that the boiling of eggs to the same degree of doneness was an impossible task in her hotel room in Switzerland. What principles which are concerned with boiling points are involved? (p. 86)

10. An inflated balloon expands as the air inside it is heated. Does the conclusion follow that if a volume of air is expanded it warms? Defend your answer. (p. 92)

11. Does a 10-lb object 20 feet above the ground have more potential energy than a 20-lb object 10 feet above the ground? Support your answer. (p. 83)

12. A man climbs a flight of stairs to the second floor in ten seconds and runs up the next flight of stairs to the third floor in four seconds. Compare (a) the force, (b) the work, and (c) the power involved in both cases. (p. 74)

13. Compare the amount of heat contained in a cup of hot coffee with that amount of heat contained in Lake Erie. (p. 92)

14. Does increasing the size of a gas flame under a pan of boiling water cause a submerged egg to cook faster? Defend your answer. (p. 85)

15. Does a radiator in a room heat the room primarily by radiation? Support your answer. (p. 98)

some familiar areas of applied science

the science of
of
sound and music
6

THE VELOCITY OF SOUND
KINDS OF WAVES
CHARACTERISTICS OF SOUND
SUBJECTIVE MEASURES OF SOUND
THE LAWS OF VIBRATING STRINGS
RESONANCE
THE MUSICAL SCALES
OTHER CHARACTERISTICS OF SOUND

vocabulary

Amplitude	Compression
Doppler Effect	Longitudinal Waves
Overtone	Period
Rarefaction	Sound
Transverse Waves	Wave length

A traditional question is often asked that concerns the nature of sound. "If a tree falls in a forest where there are no ears to hear, is there a sound?" Those who consider sound as a subjective measure typically respond in a negative manner, since they conclude that there is no sound without a listener. Sound to them is a sensation perceived by our sense of hearing, therefore, an *effect*. Another group responds affirmatively because they consider sound as an objective measure; therefore there is sound produced without a listener since there is a *cause*. The difference in opinions of the two factions results from their accepted definition of sound.

The scientist defines *sound* as a form of energy. The various characteristics of the sound determine if it is heard. Sound may be produced by nature, by fellow beings, or by ourselves. We have learned to know sounds as pleasing or displeasing to us. Music is an art in which intelligible combinations of sounds are structured to produce pleasant listening. But for one to understand music, he must first understand sound. In the pages that follow, the background presented on the subject of sound will prove helpful in understanding the science concepts that provide the basis for music.

The relationship between sound and music has fascinated man since the contributions made to the field of music by Herman Ludwig Ferdinand Von Helmholtz, German physiologist and physicist of the early nineteenth century. Dr. Helmholtz's book, *Sensation of Tone,* described such musical concepts as pitch, intensity, quality of musical notes, and general accoustics. His contributions to the understanding of sound have led to the relative ease with which scholars of music arrange and compose.

Man has realized that sound has its origin in the vibration of matter. The vibrating object disturbs the air in all directions, a disturbance that creates concentric spherical shells of waves in the air that center about the vibrating object (see Fig. 6.1). As these disturbances reach the ear, they cause the eardrum to vibrate, and the brain interprets these vibrations as sound. The musical

Fig. 6.1 (a) A representative view of air when no disturbance is present. (b) The spherical shells of disturbance formed in air about a vibrating object.

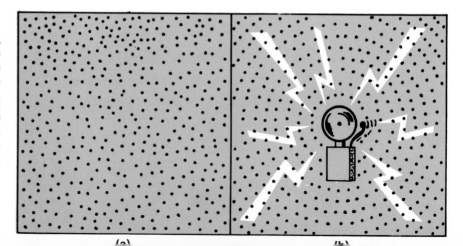

(a)   (b)

sound created by a violin's being played is caused by the strings being vibrated by the musician. The sound of a saxophone results from the air's being caused to vibrate in the hollow curved tube, a vibration brought about by the flexible reed in the mouthpiece of the instrument. If the speaker of a record player is touched while a record is playing, the components of the speaker can be felt to vibrate as the speaker is transforming electrical impulses into sound. Also, sound is produced by one object's striking another, such as the hammer's striking the gong in a telephone or doorbell.

THE VELOCITY OF SOUND

Sound waves must have some material medium through which they can be transmitted. Generally, this wave motion is produced in and transmitted through air. The disturbances produced in the air are illustrated in Figure 6.2. In Figure 6.2 the air is disturbed by the vibrations produced through electrical impulses in the speaker and is suddenly pushed and thus condensed in a conical pattern issuing from the front of the speaker. Since the air is condensed, other air in the immediate vicinity must lose its density and become rarefied. The areas of concentric circles of such disturbances move from the source at a pace commensurate with the velocity of sound in the medium, generally air. If the disturbance of the air is created by an explosion, the sudden and forceful condensation and resultant rarefaction of the air is transferred to an object, such as window glass in distant homes. The glass is caused to vibrate to such a violent degree that it is often shattered.

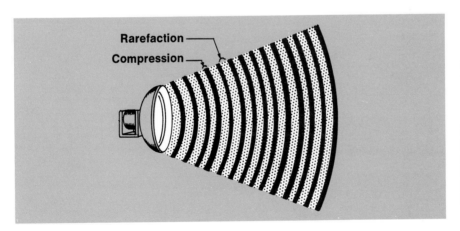

Fig. 6.2 **The disturbances created in the air by a speaker as is used in a radio or television set.**

The velocity of sound in air varies directly with the temperature. At 32°F (0°C) the velocity of sound is 1087 ft/sec (331 m/sec or 740 mph). As the temperature rises, the velocity of sound increases, or if the temperature drops, the velocity of sound decreases. On the Centigrade scale, the velocity of sound varies 2 ft/sec (0.61 m/sec) per degree. If the temperature is measured on the Fahrenheit scale, sound varies in velocity 1.1 ft/sec per degree.

Since light travels at 186,000 miles per second (300,000,000 m/sec) in air, light which originates on or near Earth reaches us almost instantly. Sound, however, travels about one fifth mile per second, so if we count the number of seconds from the time we see lightning until we hear the thunder that accompanies it and divide the result by five, we will approximate the distance in miles the storm is from us. For example, if ten seconds lapse from the time we see the lightning until we hear the thunder, the storm is approximately two miles from us.

Man has often applied his ability to measure the velocity of sound accurately. During World War I, for example, a technique was developed to determine the position of enemy gun emplacements by using the sound of the cannon in action. Sound ranging, as the technique was called, made use of microphones placed at strategic intervals over an area with the position of each microphone oriented very carefully on a map. The exact time-second was recorded at each spot when the sound of the cannons reached it. Through the use of triangulation, a common trigonometric technique for studying triangles, the gun's position was accurately located and immediately placed under bombardment by the guns of the allies.

The velocity of sound is generally greater in solids and in liquids than in gases, because of the closeness of the molecules and thus the greater density of substances which exist in the solid or liquid state. For example, sound travels about 4.5 times as fast in water as it does in air, almost one mile per second. Some examples of the approximate velocities of sound in various mediums appear in Table 6.1. If one observes the sound produced by the whistle of an approaching train, he may be able to hear the same sound twice—as the sound

TABLE 6.1

The velocity of sound in various substances

Substance	Velocity (ft/sec)	(m/sec)
GASES:		
Air (at 0°C)	1087	331
Carbon dioxide	846	258
Carbon monoxide	1106	337
Hydrogen	4160	1268
LIQUIDS:		
Alcohol	3890	1186
Benzine	3826	1166
Turpentine	4351	1327
Water	4708	1435
SOLIDS:		
Aluminum	16704	5093
Glass	18050	5503
Iron	16820	5128
Wood (Oak)	12620	3848

the science of sound and music

reaches him through the train track and again as it reaches him through the medium of air. The velocity of sound and its relationship to various mediums bring about other interesting features which will be considered later in the chapter.

KINDS OF WAVES

If an investigator fastens both ends of a stretched spring and plucks the spring at right angles to its length, he notes a disturbance in the form of a wave that travels along the length of the spring and back to the origin of the disturbance. Such a wave in which the particles vibrate at right angles to the path the wave travels is known as a *transverse wave,* illustrated in Figure 6.3 (a). Some classic examples of such a wave disturbance in which there is little or no actual forward movement of the medium are the ocean waves and those waves produced by the wind blowing across fields of corn or grain. The only movement created in each case is primarily back and forth or up and down, but not forward.

If several turns of the stretched spring are compressed and then suddenly released, the compressions are observed to move along the length of the spring in a longitudinal wave, illustrated in Figure 6.3 (b). Close observation reveals rarefied areas (R) immediately following the compressed region (C) and preceding it. As in the transverse wave, little movement occurs in the medium along the length of the spring. Sound, of course, is an example of this longitudinal wave as it moves through the interior of a given medium. A visual comparison of a transverse wave with a longitudinal wave, created by a tuning fork, is displayed in Figure 6.3 (c). Sound, then, is created by both transverse and longitudinal wave patterns. For instance, if an investigator strikes the surface of a

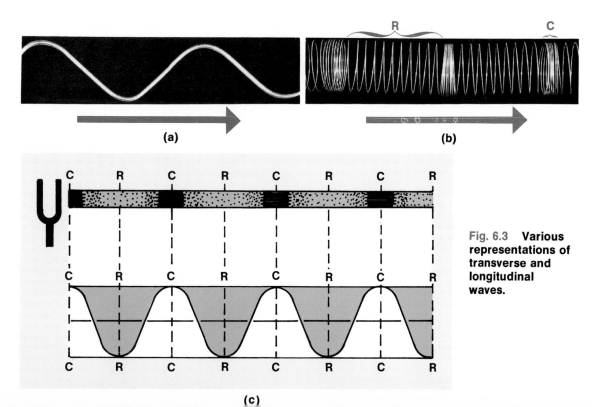

Fig. 6.3 **Various representations of transverse and longitudinal waves.**

taut trampoline and then observes the disturbances so created, he notes a longitudinal wave set up by the transfer of disturbance from particle to particle. Also, the entire length of the trampoline vibrates up and down in the form of a transverse wave. An observer becomes aware of this motion if he views the surface from a position level with the surface of the trampoline.

CHARACTERISTICS OF SOUND

Sound waves, like other waves, have three outstanding and fundamental physical characteristics. Each wave produced by a vibrating object can be measured in terms of (1) its *frequency,* (2) its *wave length,* and (3) its *amplitude.* Each characteristic is illustrated in Figure 6.4. Frequency is a concept measured in terms of the number of times or fraction thereof which an object vibrates in a given length of time, usually the second.

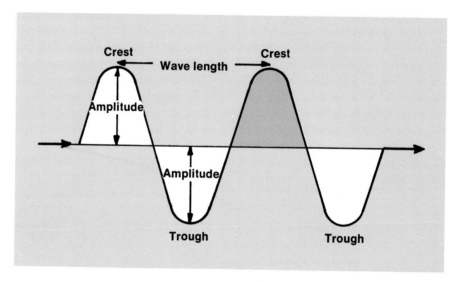

Fig 6.4 The quantitative characteristics of transverse waves.

Typical measurements of frequency are recorded in vibrations per second (vps or vib/sec); in cycles per second (cps or c/s); or in Hertz (hz), a unit of frequency named for Heinrick R. Hertz, a nineteenth century physicist, which is equal to vibrations per second. The latter is noted as growing rapidly in popularity. A measure of the number of times a pendulum bob swings back and forth in one second denotes its frequency. No sound is produced by a swinging pendulum bob because its typical frequency is well below the minimum frequency of vibrating matter our senses interpret as sound. The pendulum is similar in principle to vibrating matter which does produce sound and is therefore useful in permitting a student to relate characteristics of the pendulum to vibrating objects. The *period,* for instance, is a reciprocal measurement of frequency which permits an observer to determine the length of time a pendulum requires to swing back and forth or to complete one vibration. The frequency of five vibrations per second indicates a period of ⅕ or 0.2 second for a given pendulum. If the frequency of a vibrating object is 60 vps, the resultant period is calculated as ¹⁄₆₀ second.

In Figure 6.5, if one measures the straight-line vertical distance that a point at the center of the pendulum bob moves from its position at rest (0) to either maximum distance (D, D¹) that the point moves upward during its vibration, he has determined the amplitude (h) of the pendulum's swing.

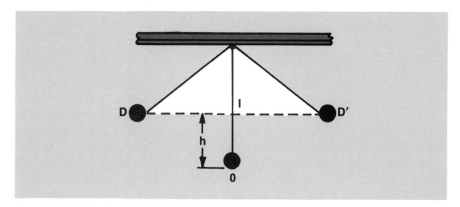

Fig. 6.5 **The length of the pendulum is represented by *l*; *h* represents amplitude, the maximum distance the center of the bob moves vertically, and *o* shows the position of the bob at rest.**

The length of the string or other device which supports the pendulum bob is related to the last physical characteristic under discussion. Wave length is related to the frequency of the pendulum, but in an inverse manner. If one shortens the string which supports the pendulum bob, the frequency of the pendulum increases and its period decreases. If one lengthens the string, the period increases and the frequency decreases. Thus the period of the pendulum is directly related to the length of the string, and frequency is inversely related to the length of the string.

As in Figure 6.4, the number of vibrations created in the string is referred to as the frequency (f). The amplitude, as illustrated in Figure 6.5, is the distance the string deviates from its normal position at rest, and is measurable above or below the imaginary position of rest, called the base line. Wave length of transverse waves is generally measured as the distance from the crest of one wave at maximum amplitude to the exact point on the preceding wave, measured parallel to the base line. In the case of longitudinal waves, illustrated in Figure 6.6, wave length is measured as the distance from the center of the compression area to the center of the one that immediately follows. Wave length, in either type of wave, is represented by the symbol "l" or "λ" in formulas and equations.

The relationship that exists between the frequency of a wave and its wave length permits us to determine the velocity of a wave even though most waves travel too fast through the medium to permit their velocity to be directly measured. The dependency of one concept on the other permits the measure of wave velocity when waves other than sound or light are considered, such as

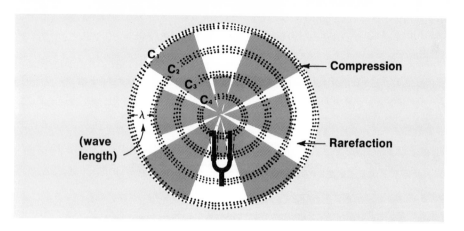

Fig. 6.6 Representation of a series of longitudinal waves created by a tuning fork. The wave length (λ) is a measure of the distance between C₁ and C₂, C₂ and C₃, and so forth.

waves produced in water or other mediums. The velocity of such a wave can be calculated by determining the number of vibrations produced in a given period of time that seem to move through the fixed medium. Thus, the velocity of such a wave equals the product of the number of vibrations produced in a given period of time and the wave length of the vibration. That is, $v = f \times l\,(\lambda)$. This relation reveals that as the frequency of a wave increases or decreases the wave length changes in proportion, since investigation of various frequencies in a given medium permits one to establish that velocity is constant for a specified medium. In the study of sound, waves of extremely short wave lengths are indicative of high frequencies and are classed as *ultrasonic*. Sounds with low frequencies, hence long wave lengths, are classed as *infrasonic*. A third class, *audiosonics,* deals with waves which have wave lengths between about 0.055 feet and 68.75 feet as produced in an air medium. The corresponding range of frequencies lies between 16 vps and 20,000 vps, the subjective "range of hearing" for man. The human ear seems to be more sensitive to frequencies produced between 2000 and 4000 vps. This sensitivity diminishes rapidly for most individuals as frequencies increase or decrease (see Chapter 24).

SUBJECTIVE MEASURES OF SOUND

When a sound is produced, is it one of music or is it one of noise? The distinction between the two categories has always been one of culture and is fast becoming one of generation as well. If the sound is considered pleasant to the listener, it is thus one of music, but if it is considered displeasing to the observer, the sound is classed as noise. Therefore, one can assign either category to a given sound, depending on the culture and mood of the listener. In a broad sense, the general attributes of a musical sound help discriminate between it and one of noise. These attributes are *pitch, loudness,* and *quality.* All are psychological terms which indicate the sensation produced on our senses. In addition, one could include the length of time that the sound lasts—its *duration.* Each of the attributes has a counterpart that is physical in nature. These counterparts

the science of sound and music

are frequency, amplitude or intensity, and structure of overtone (a topic for later discussion). All are material in nature and thus are capable of being measured.

Pitch, as it applies to music, is a subjective measurement and thus varies among individuals. The variation in each listener's degree of sensitivity creates the sensation of pitch. This sensitivity to pitch is somewhat inherent but can be enforced by proper and intense training in others. Frequency, on the other hand, is a physical measurement of the number of times the object that creates the sound vibrates in one second.

The sensation of pitch, then, is very similar to frequency in that high-frequency sounds produce high pitch, and low-frequency sounds produce low pitch. Some musicians contend, however, that loudness affects the sensation of pitch to some degree. This group feels that as an object is caused to vibrate more vigorously but at a low constant rate, the tone produced not only becomes louder but also changes to a lower pitch. Conversely, as the object is forced to vibrate at a relatively high rate, the trained listener can note an increase in the pitch as the sound's intensity is increased. Because of these observations and the result of various studies in which trained and untrained observers found significant differences when they were asked to compare sounds of varying degrees of loudness, pitch and frequency are not acceptable as identical terms.

In order to summarize the concept of pitch, let us review its relation to frequency. As has been pointed out, the greater the frequency of a vibrating object, the higher the pitch produced. In addition, the wave length has been found to be inversely proportional to the frequency; that is, the shorter wave length has higher frequency and the longer wave length lower frequency, each varying in an inverse manner because of the constant velocity of the wave in any given medium.

Amplitude, the second term, is the counterpart of the subjective characteristic of sound known as loudness. Amplitude, as is indicated in Figure 6.7 (a) and (b), is a measure of the deviation of the crest of the wave from an imaginary line that represents the surface of the medium when all disturbances cease, such

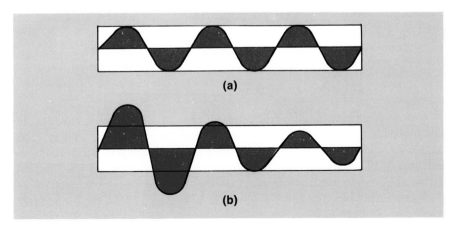

(a)

(b)

Fig. 6.7 (a) A representation of constant amplitude and (b) one of decreasing amplitude.

as the surface of a body of water. Amplitude is also, and more commonly, represented as a deviation of an object such as a violin string from its position of rest. The greater the deviation of the string or the greater the disturbance on the body of water, the greater the amplitude of the wave produced, hence the louder the sound it creates.

Frequently, intense sounds do not appear very loud because of the poor transmission of the sound through a given medium or because the sounds are produced far from the observer. Loudness is a property of sound that is relative to the effect it has on the observer's ears and, therefore, is subjective. The intensity of a sound is dependent on the energy of the sound waves and is amplified by increasing the amplitude and the area of the object in vibration.

One can conclude, then, that loudness, expressed in a subjective manner, is a measure of sound power based on the effect the sound produces on the listener's senses. Intensity can be measured objectively as the amount of sound power which is produced. A convenient method of measure is to compare the power produced by one sound with that produced by another. The unit that is conventionally used is the "bel," in honor of Alexander Graham Bell, inventor of the telephone. The ratio of intensity of a sound ten times that of another has an intensity difference of one bel; 100 (10×10 or 10^2) times as intense signifies a difference of two bels; 10,000 ($10 \times 10 \times 10 \times 10$ or 10^4) times as intense is equivalent to four bels, and so forth. For convenience's sake, the bel is divided into ten equal parts called "decibels" (db), the unit of loudness commonly used. Some comparative values follow in Table 6.2.

TABLE 6.2

Representative values of loudness of sounds.
Additional values are listed in Figure 24.3.

Source of Sound	Sound Level (b)	Sound Level (db)
Threshold of hearing	0	0
Rustle of leaves	1	10
Air-Conditioning, Window Unit	4	40
Vacuum sweeper on rug	6	60
Modern Automobile	7	70
Pneumatic drill	8	80
Airplane engine	11	110
Threshold of pain	12	120

As one can note in all measures of intensity, when energy is moving and thus dissipating through a medium, sound loses its power very rapidly. The manner in which it loses its intensity involves a relationship in which loudness and intensity of a sound are inversely proportional to the square of the distance measured from the source. Loudness also varies with the medium through which the sound is moving. The inverse square law, as it applies to the intensity of sound, can be stated as:

the science of sound and music

$$\frac{I_1}{I_2} = \frac{d_2{}^2}{d_2{}^1}$$

where I_1 and I_2 are intensity measures of the sound in decibels at distances d_1 and d_2 from the sources, respectively. An illustration of the inverse square law appears in Figure 6.8.

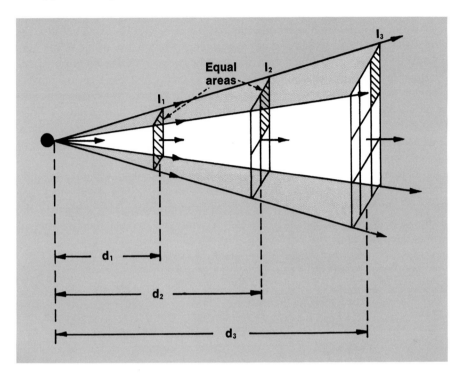

Fig. 6.8 **The inverse square law as it applies to sound intensity.**

As has been mentioned, sound waves spread in all directions from the vibrating source. If one uses a megaphone, however, the sound is projected toward the direction in which the larger end of the megaphone is pointed. Doubling the distance, in this case, does not decrease the loudness to one-fourth the initial measurement for those observers in front of the person using the megaphone. Cheerleaders make use of megaphones in an attempt to make themselves heard above the noise of the crowd at athletic contests. The principle of the megaphone is also involved in the design of most musical instruments which are intended to be somewhat directional. One can observe the effectiveness of design by noting the decrease of loudness as a band performing at the half-time of a football game pivots away from him to play for the fans in the opposite stands.

The last of the three fundamental characteristics, quality or structure of overtone, is one that requires the greatest amount of discussion, since it is the property that permits a listener to discriminate between sounds that otherwise seem identical.

the science of sound and music   **115**

Two objects vibrating at identical rates may produce sounds that differ in the degree to which they are pleasing to the ear. Two instruments or two voices that produce the same pitch are distinguishable, even though the pitches are the same and each is equally loud. The manner in which the two sounds differ is in quality. If a bow is applied to a string on a violin near the middle and thus causes the string to vibrate, the sound produced is known as the *fundamental* tone for that string. If the string is touched near the middle very lightly while it is vibrating, both the fundamental tone and the first *overtone* can be heard because the string is caused to vibrate in two equal segments. The quality of the new sound produced is richer and fuller than the former because of the addition of the overtone to the fundamental. Violins, guitars, violas and the like are bowed, plucked, or struck near the end of the string and thus produce overtones as well as the fundamental tone and hence tones of higher quality. A well-played violin is bowed so as to produce the fundamental and some five overtones. The conclusion can be reached that the quality of a sound is dependent on the number of overtones present as well as their prominence to each other.

The sonometer, an instrument found very useful in the laboratory to study vibrating strings, is composed of a hollow rectangular box over which are stretched several strings of varying size and composition. The tension in the strings is adjustable by means of adding or removing weights. The instrument is illustrated in Figure 6.9.

Fig. 6.9 The sonometer is an ideal device for studying the laws of strings.

The meaning of the term *overtone* is easily demonstrated. It is possible to show that a string can be made to vibrate as a whole as well as in segments simultaneously, as illustrated in Figure 6.10. If the string vibrates in two equal segments, the first overtone is produced which is twice the frequency or one *octave* above the fundamental, and if the string is made to vibrate in three equal segments, the sound produced is three vibrating rates above the fundamental. For example, the first overtone of a C string is C', a note twice the vibrating rate of C; $256 \times 2 = 512$ vps (hz). The second overtone of C would be $256 \times 3 = 768$ vps, a note corresponding to the note G', a *harmonic*. The third overtone of C has $256 \times 4 = 1024$ vps, and thus is the note C'', the second octave of C. The fourth overtone of C is E'', $256 \times 5 = 1280$ vps. A listener with a minimum of training can distinguish between an octave change and a harmonic.

The conclusion can be reached from this discussion that a note which has a vibrating rate that is a whole number (or integral) multiple of the fundamental

the science of sound and music

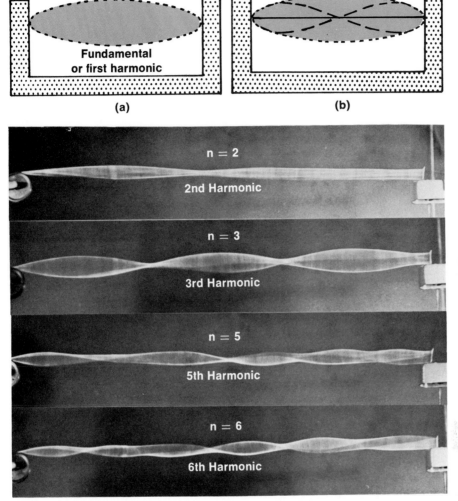

n = 1

Fundamental or first harmonic

(a)

(b)

n = 2

2nd Harmonic

n = 3

3rd Harmonic

n = 5

5th Harmonic

n = 6

6th Harmonic

(c)

Fig. 6.10 Standing waves in a vibrating string. (a) The string vibrating as a whole. (b) The string vibrating as a whole and in two equal segments. (c) The string vibrating in equal segments to produce various harmonics.

frequency is known as an overtone or a harmonic. In mathematical terms, the frequency (f) of each overtone or harmonic is expressed as $f = nf_1$, where f_1 is the fundamental frequency and n is the order of the whole number multiple. (Note Fig. 6.10 (c)).

The quality of a tone is subjective in that the fundamental frequency and its first overtone, illustrated in Figure 6.11 (a) and (b), combine in (c) and thus reach the listener and thereby cause his eardrum to vibrate accordingly. Some sound wave patterns appear in Figure 6.12 as they are produced by a wave analysis device called an oscilloscope. The instrument interprets variations in fluctuating electric current, converts them into wave forms, and projects them on a screen much like that of a television set.

the science of sound and music

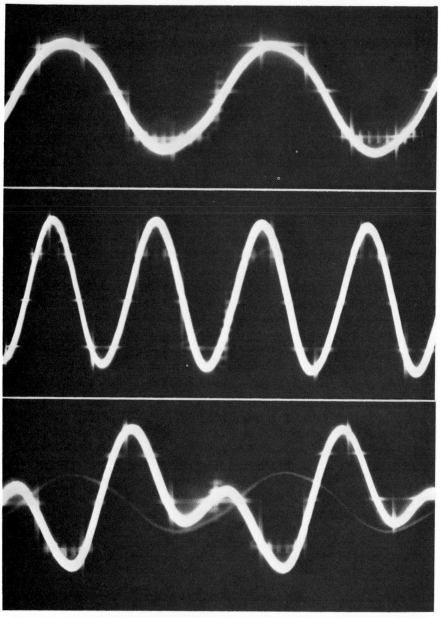

Fig. 6.11 Oscillo-
scope patterns of
fundamental tones
and overtones.

the science of sound and music

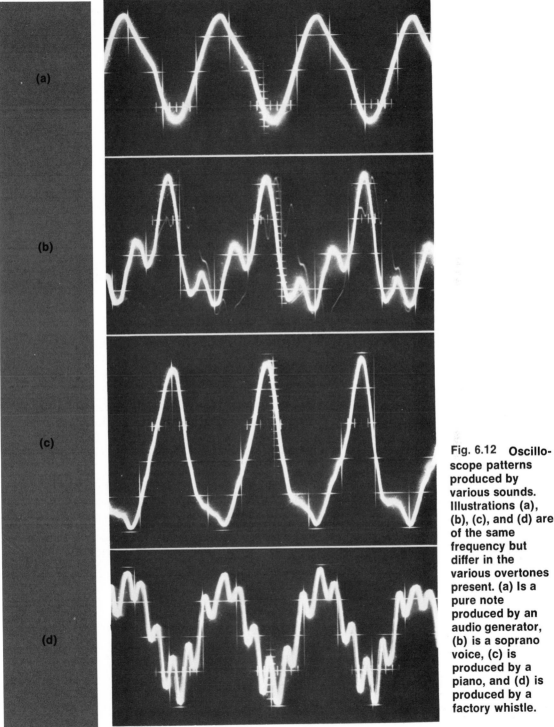

Fig. 6.12 Oscilloscope patterns produced by various sounds. Illustrations (a), (b), (c), and (d) are of the same frequency but differ in the various overtones present. (a) Is a pure note produced by an audio generator, (b) is a soprano voice, (c) is produced by a piano, and (d) is produced by a factory whistle.

the science of sound and music

THE LAWS OF VIBRATING STRINGS

There are several laws that govern the rate by which strings vibrate. If one investigates the various strings within a piano, he notes that they differ in several ways. The strings vary in length, in diameter, in tension, and in composition, hence density. Through the investigation of each variable, each factor can be shown to affect the sound produced by a given vibrating string. For instance, if all factors other than length remain constant, the vibrating rate of a string is found to be inversely proportional to its length.

Example 6.1

If the D string on a violin is 15.00 inches long, how much must it be shortened to produce the note C'?

Solution: The note D vibrates 288 times per second, C' at 512 vps. The proportion exists:

$$\frac{288 \text{ vps}}{512 \text{ vps}} = \frac{X \text{ in}}{15 \text{ in}}$$

$$512 \times X = 288 \times 15$$

$$512 \, X = 4320$$

$$X = 8.42 \text{ in.}$$

Thus for the musician to produce the note C' on the D string of his violin, he must touch the string to the fingerboard in such a way that the string is shortened by $15.00 - 8.42 = 6.58$ in.

On a guitar or viola, the strings vary in diameter. The larger strings produce lower frequencies than the smaller strings. A string twice the diameter of another vibrates about one-half the rate of the smaller. If all other variables remain constant, the vibrating rate of a string is inversely proportional to its diameter.

If a musician wishes to tune his stringed instrument, he raises the frequency of the vibrating string by increasing the tension on it or by tightening the string, and to lower the frequency, he loosens the string. The change in frequency resulting from a variation in tension is not a constant ratio. Assuming that all other factors remain constant, the vibrating rate of the string is found to be directly proportional to the square root of the tension on the string.

Example 6.2

A 25-lb force attached to one end of a string on a sonometer produces the note D, 288 vps when the string is struck. In comparison, what is the value of the total force required to produce the note D', 576 vps in the same string?

Solution:

$$\frac{288 \text{ vps}}{576 \text{ vps}} = \frac{\sqrt{25} \text{ lbs}}{\sqrt{X}}$$

$$288 \times \sqrt{X} = 576 \times \sqrt{25}$$

$$\sqrt{X} = \frac{576 \times 5}{288}$$

$$\sqrt{X} = 10$$

$$X = 100 \text{ lbs.}$$

Thus, a 100-lb weight (force) attached to the string causes it to vibrate 576 times per second.

Last, if one string is heavier than another, the denser or heavier string vibrates more slowly. Guitars, harps, and other instruments have some of their strings wrapped with copper wire for the purpose of increasing the string's overall density. Close observation permits one to conclude that the frequency of a vibrating string is inversely proportional to the square root of its density, if all other factors are constant. So a string four times as dense as another will vibrate at one-half the rate of the less dense, if all other factors hold constant.

RESONANCE

Have you ever observed a small child attempting to swing a playmate who is much larger? The small child can transfer energy in small quantities to the larger child by pushing at the right moment, thus building up a vibration of great amplitude for the child that is swinging. This energy transfer by small increments is commonly referred to as *resonance*: the building up of a large vibration by small impulses, the frequency of which equals one of the natural frequencies of the resonating body. As to the large child in the swing, a large amplitude results only if the small impulses synchronize with the natural period of vibration of the swing. A small variation in synchronization results in a decrease in amplitude of vibration of the swing.

All objects have a natural frequency at which they vibrate most willingly. If a sound wave of the same frequency as the natural frequency of the object comes in contact with the object, the object vibrates with a sympathetic vibration. If one depresses the loud pedal on a piano and hums, whistles, or otherwise creates a sound of the same frequency as the fundamental frequency of the piano strings, the string will be set in vibration by the sound and will remain so after the initial sound ceases.

Bridges have collapsed due to mechanical resonance built up by gusts of wind, by marching soldiers, by idling trucks, and by other objects vibrating at some constant rate. A certain rattle or vibration in the family auto occurs at 44 mph, but not at 34 mph or 54 mph, or most other speeds.

But how do two sounds perhaps produced by two identical tuning forks and therefore having exactly the same frequency affect each other? An experimenter will conclude that the two forks sounded simultaneously will reinforce the amplitude of each other in a form of constructive interference. Destructive interference can be created by causing the crest of one wave to coincide with the trough of the other, thus cancelling each other in terms of amplitude (see Fig. 6.13).

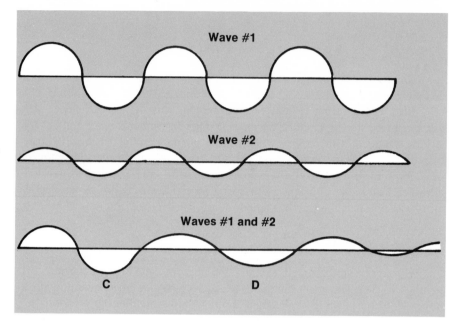

Fig. 6.13 Waves of varying length can interfere with each other both constructively and destructively. The interference is constructive at C, destructive at D.

If the two tuning forks vibrate at slightly different frequencies, a third phenomenon known as *beats* results when the two forks are simultaneously struck. The new wave that results is sometimes reinforced and sometimes diminished, displayed in Figure 6.14. The variation in loudness that results, varying between loud and soft, causes a throbbing sound instead of one reinforced when the two frequencies are identical. If a C tuning fork which vibrates 256 times per second is struck at the same time as a D tuning fork of 288 vps, 32 beats per second are produced, creating a sound displeasing to the ear. We obtain the number of beats produced per second by the two forks by determining the difference in the separate frequencies, that is, $f_1 - f_2 = \Delta f$ or 288 vps − 256 vps = 32 vps.

Almost everyone is fascinated by the magnificent and voluminous sounds which ring forth from the mighty pipe organ. The organ produces sound by virtue of a stream of air which is directed against the holes or openings in the pipes. The flute and the oboe are related in principle to the organ since sound is produced in all three instruments by a vibrating air column. Other instruments, such as a saxophone and a clarinet, produce sound by a vibrating reed on the mouthpiece.

The pitch of the sound produced by a musical instrument can be controlled by two factors: the length of the tube used and the number of open ends the tube has. A tube or pipe closed at both ends, as in Figure 6.15 (a), produces a note which has a wave length four times the length of the pipe itself. Thus, a closed

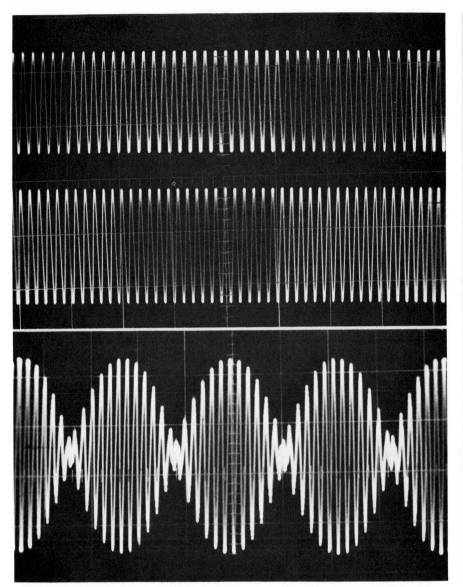

Fig. 6.14 **The waves of similar frequencies may interfere with each other and produce beats.**

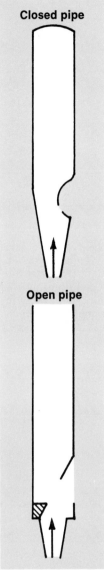

Closed pipe

Open pipe

Fig. 6.15 **The two different kinds of pipe arrangements. Both types are necessary to permit musicians to have a full range of harmonics at their disposal.**

pipe two feet long produces a note whose wave length is eight feet. An open pipe, displayed in Figure 6.15 (b), produces a note whose wave length is twice the length of the pipe; thus the two-foot pipe open at both ends produces a note twice the frequency of the closed pipe also two feet long. In order to construct the pipes to be able to sound a complete octave, one needs to be aware of the sounds that compose the various musical scales, a topic for discussion later in the chapter.

The lips of the player serve as the vibrating object which produces the sound for the trumpet, the trombone, the baritone horn, and the tuba. The frequency produced, however, is dependent on the length of the tube, regardless of the amount the tube is curved. In the trumpet, the holes or openings in the valves align with the curves of the tube in one route when all valves are up, and in different routes when the various valves are depressed. The length of the tube so constructed varies whether the first valve, the second, the third, or any combination is depressed, and thus alters the pitch produced (note Fig. 6.16).

Fig. 6.16 The popular cornet. Consider the length of tubing necessary to construct such an instrument.

In other instruments, the vibrating part is a membrane stretched over a hollow cylinder such as is found on a drum or a tambourine. The sound is produced by striking the membrane a blow with a drum stick or hand and causing the membrane to vibrate. The air column inside reinforces the loudness of the sound produced. Cymbals and bells, along with drums, do not produce overtones which are harmonics of the fundamental note. The notes produced are dependent on the shape of the cymbal, bell, or membrane as well as on the overall dimensions and type of material from which the instrument is constructed. The human voice also belongs in this category. Its sound is produced by means of vocal cords which vibrate when air is blown through them. This pair of membranes is stretched across the larynx (or Adam's apple) and is tightened, producing a high pitch when caused to vibrate. The sound produced when the vocal cords are relaxed due to muscular action is of lower pitch. The range between these two pitches varies with the individual, since the length, diameter, and tension of the vocal cords are different for everyone. On the average, the male speaking voice is 150 vps, with that of a woman being 230 vps. As children grow, the vocal cords increase in all dimensions; thus the range of their voice increases. But, where is the resonating air column that causes the sound produced to be much louder and of higher quality than that produced by the vocal cord membranes alone? Of course, one makes use of the passages in his chest, head, nose, throat, and mouth. The size and shape of each passage also determine the quality of the individual's voice. The various sounds produced when an

the science of sound and music

individual is engaged in ordinary conversation are generally all at the same frequency. The noticeable difference in the sounds results from the individual's changing the position of his tongue, lips, and cheeks. He thus creates passages of various size through which his voice can resonate.

THE MUSICAL SCALES

The scholars of music have developed several scales in which each note produced varies in some subjective manner from the standard. Most popular among the scales are the *diatonic,* the *chromatic,* and the *tempered* scales, all which have their proper place in the field of music.

As a matter of review, two sounds are said to differ by one octave when their frequencies are in a ratio of 2:1. Almost everyone has struck the keys of a piano and thus realizes that the piano has a range that is practically inclusive of all frequencies that fit into the realm of musical sounds, amounting to over seven octaves. The lowest C found on the piano has a frequency of 32.7 vps. Seven octaves above this C, one would find C^{7th} having a frequency of $2 \times 2 \times 2 \times 2 \times 2 \times 2 \times 2$ or $128 \times 32.7 = 4185.6$ vps. The range of various instruments is displayed in Figure 6.18. It is common knowledge that there are 88 keys on a piano. How, then, do the 88 notes differ? How is each related? The answer comes from a brief study of the common musical scales.

The diatonic scale is composed of the eight standard major tones taught to us as students in elementary school. The notes, syllable names, and comparisons of their vibrating rates are illustrated in Figure 6.17.

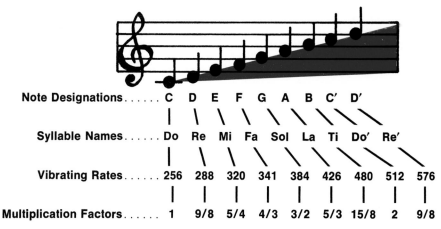

Note Designations	C	D	E	F	G	A	B	C'	D'
Syllable Names	Do	Re	Mi	Fa	Sol	La	Ti	Do'	Re'
Vibrating Rates	256	288	320	341	384	426	480	512	576
Multiplication Factors	1	9/8	5/4	4/3	3/2	5/3	15/8	2	9/8

Fig. 6.17 **The major diatonic scale presented in the key of C.**

From the figure, one can see that D is 288/256 or 9/8 times the frequency of C. The note E is 5/4 times the frequency of C. The note G, being 3/2 the frequency of C, produces a pleasing sound when sounded with C and thus is known as the "fifth" interval, the interval which encompasses five diatonic notes: C, D, E, F, and G. Another ratio, 5/4 is known as a "major third," and

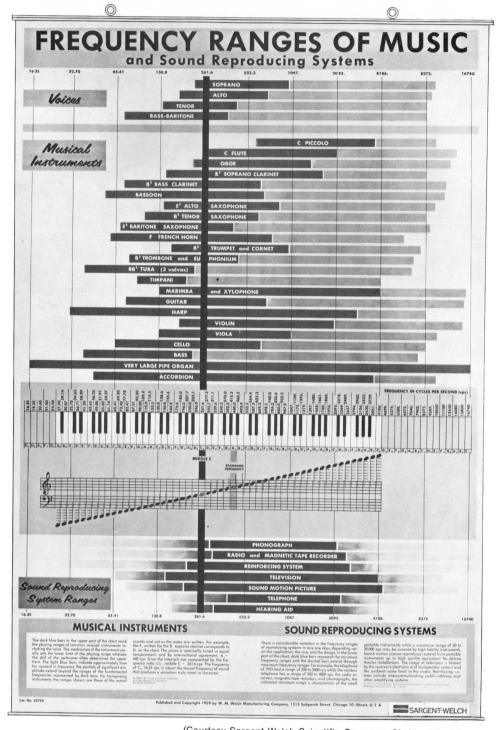

Fig. 6.18 A comparison of the frequency range of various musical instruments.

(Courtesy Sargent-Welch Scientific Company, Skokie, Illinois)

the science of sound and music

that of 6/5 as a "minor third," both so known because they are formed from two notes a third apart. The chord composed of "do-mi-sol," *C-E-G*, contains a fifth, *C-G*; a major third, *C-E*; and a minor third, *E-G*, and produces a tonic chord whose relative frequencies are 4, 5, and 6. Any three notes whose vibrating rates are of this ratio produce a major chord when sounded together. Another example is sounding in unison the notes *G*, *B*, and *D'* or *F*, *A*, and *C'*.

	C	D	E	F	G	A	B	C'	D'
Key of C	256	288	320	341	384	426	480	512	576
Multiplication Factor		1	9/8	5/4	4/3	3/2	5/3	15/8	2
Key of D		288	324	360	384	432	480	540	576

Fig. 6.19 **The diatonic scale presented with *D* as the standard.**

If one uses another note such as *D*, 288 vps, as the standard, and multiplies this frequency by the 9/8, 5/4, 4/3, 3/2, 5/3, and 15/8 used in the diatonic scale, several new notes are produced, as displayed in Figure 6.19. The notes *F* and *C'* vary quite drastically from that value assigned previously using *C* as the keynote. For this reason, the semitones, called "sharps" and "flats," came to be and are provided by the black keys on the piano. They differ from the eight tones which are part of the major diatonic scale by a significant rate. If one checks the products of all notes within the octave by the factors previously mentioned, a complete octave is composed of 70 notes. This number of notes is impractical, to say the least; thus the typical scale called the chromatic scale is identical to the major diatonic scale to which is added five semitones. These twelve intervals are evenly distributed, differing from the previous tone by the product factor of $\sqrt[12]{2} = 1.05946$, taken into consideration in a scale referred to as the tempered scale. Thus, although accomplished musicians prefer the chromatic scale, the tempered scale is the accepted one used in playing most instruments. For practical purposes, then $C\sharp = D_b$ (*C*-sharp is the same as *D*-flat), $D\sharp = E_b$, $G\sharp = A_b$, and so on.

OTHER CHARACTERISTICS OF SOUND

Almost everyone has noticed the change in pitch of a train whistle as the train approaches him and the fact that it reaches its maximum pitch as the train is nearest to him. The pitch is then noted to change in the opposite direction as the train moves away from him. This phenomenon is known as the *Doppler effect* and is caused by a variation in the number of waves which reach the observer in successive seconds. As the train approaches him, the pitch gets continuously higher until the train is at its nearest point to the observer. As the train moves away, the pitch continues to decrease. The higher the velocity of the source of sound, the more obvious the change in pitch. That is, the speed of the source of the sound must be added to or subtracted from the velocity of sound

the science of sound and music **127**

depending on whether the object's motion is toward or away from the observer. The same effect is evident in other types of waves.

Another concept is responsible for causing the listener to lack proper appreciation of some musical performances. Much of the pleasantness produced by musical sounds can be readily lost by poor acoustics. An orchestra that performs in a concert hall is sometimes hindered by the unwanted reflection of the sound it previously produced. This reflection is known as an *echo* and can occur any time a surface in the hall is some 55 feet away from the source of sound. At approximately 1100 ft/sec, sound would return to its source in about one-tenth second, the length of time that must pass before the average listener can distinguish two separate sounds. One can stop the reflection of sound by installing curtains on the opposite walls and by placing acoustical tile on the ceiling. If one views the tile carefully, he will observe that the tile has small holes which individually serve to absorb some of the sound previously produced. A certain degree of reflection is advantageous to an orchestra. If the total amount of sound is not absorbed, it will reflect from all surfaces, producing an effect of having the orchestra surrounding the audience; thus the sound will continue on for a short time, a property known as *reverberation*. The best reverberation time for listening to an orchestra perform is about 1.5 seconds. When both echo and reverberation are adjusted for proper listening, the room is said to have good acoustics, another subjective measure encompassed in the quality of sound.

CHAPTER SUMMARY

Sound travels through various mediums at different velocities. Generally speaking, the velocity of sound increases with the density of the medium. The temperature of the medium also affects the velocity of sound. For each centigrade degree the air temperature increases, the velocity of sound in air increases 2 ft/sec (0.61 m/sec). As the air temperature decreases, the velocity of sound decreases accordingly.

There are two general types of waves—*transverse* and *longitudinal*. Transverse waves travel in the same direction as the disturbance, and longitudinal waves travel at right angles to the disturbance. Sound is generally considered a longitudinal wave.

Frequency, wave length, and *amplitude* are the fundamental characteristics of all waves. The *velocity* of a wave is calculated from the product of its frequency and wave length. In sound, amplitude is generally considered synonymous with loudness.

The general attributes of sound help classify the sound as *noise* or *music*. Since these inherent characteristics are subjective in nature, the given attributes of sound are *pitch, loudness,* and *quality.*

The rate by which a string vibrates is determined by the string's length, diameter, composition, density, and tension. The vibrating rate of a string is inversely related to length, diameter, and density of the string. In general, metallic strings vibrate at a lesser rate than do strings composed of plastic and other nonmetals. The vibrating rate of a string increases with the square root of the tension applied to the string.

The building of a large vibration by small impulses the frequency of which equals one of the natural frequencies of the resonating body is called resonance. If the impulses are slightly out of synchronization with one of the natural frequencies of the resonating body, a decrease in amplitude of the resonating body results.

The three most popular musical scales are the diatonic, the chromatic, and the tempered scales. Most common among the three scales is the diatonic scale. The tempered scale, however, is the accepted scale used in playing most instruments.

The Doppler effect applies to sound and to light. In sound, it is the change in pitch caused by the motion of the source relative to that of the listener. As the source moves toward the listener, the pitch increases. As the source moves away, the pitch decreases. Either the listener, the source, or both may be in motion for the Doppler effect to occur. A sound reflected from a surface is called an echo. A series of echoes of the same sound is known as reverberation.

QUESTIONS AND PROBLEMS

1. Discuss, in your own words, the similarities and differences between loudness, pitch, and quality of sound waves. (p.112)

2. A tuning fork sounded and held on a solid table appears louder than if the vibrating fork is suspended in air. Why is this observation true? (p.121)

3. Marchers who are remote from the band which is playing are generally out of step with the marchers nearer the band. Why is this observation true? (p.107)

4. Why does the pitch of a given factory whistle vary on a windy day? Why does the same whistle sound louder on certain days than on others? (p.108)

5. What factors determine the velocity with which a sound travels through a given medium? Why does sound travel faster in warm air than in cold air; faster in moist air than in dry air? (p.108)

6. If the velocity of sound decreases 1.1 ft/sec per °F, theoretically, at what temperature would the velocity of sound = 0 ft/sec? At lower temperatures than this value, what would the velocity of sound be like? (p.107)

7. In tuning a violin, how can the pitch of a string be raised? How can a violin produce so many tones with so few strings? (p.124)

8. A high-fidelity sound system often has frequency ranges above the range of human hearing. Of what use is this extended range? (p.121)

9. Some persons say that music exists in the brain rather than in the violin. React to this inference. (p.112)

10. Can sound be produced without having anything vibrate? If a metal pan falls to the kitchen floor but no one is around to hear it, is sound produced? (p.106)

the science of sound and music

11. Many different notes on a musical scale are produced by the same string on a musical instrument. How are the various notes produced? (p. 120)

12. Why do notes from plucked strings on a guitar sound different from those from a banjo or a viola, even if each note has identical frequencies? (p. 116)

13. In an organ do the long pipes emit tones of high or low pitch? Why is this? (p. 122)

14. What would be the effect on the music heard from a distant band if the different frequencies produced traveled at different speeds? (p. 107)

15. In what respect would the sounds emitted by a cornet be altered by taking the instrument from indoors where the temperature is 72°F to outdoors where the temperature is 40°F. Why? (p. 107)

16. A tuning fork will stop vibrating in a brief time. How can you explain why this observation is valid? (p. 114)

the science of sound and music

basic
electricity
and
magnetism
7

THE CHARGE ON THE ELECTRON
ELECTRICITY AT REST
ELECTRONS IN MOTION
HOW MAGNETISM DEVELOPED
THEORIES OF MAGNETISM
ELECTROMAGNETISM
ALTERNATING CURRENT
ELECTRICITY IN GENERAL

vocabulary

Alternating Current	Conductor
Direct Current	Electromagnet
Electricity	Electric Current
Electric Potential	Insulator

Man observed various electrical phenomena long before he discovered electricity. He had long been awed by lightning, the electric eel, St. Elmo's fire (the pale glow sometimes visible in stormy weather at prominent or pointed areas on a ship), and the shock he receives through his contact with electrically charged objects.

An additional phenomenon, by far the least spectacular, led to man's understanding of electricity. Some six centuries B.C., the Greeks realized that a piece of amber rubbed with cloth acquired peculiar properties. The amber would attract light objects such as animal hairs and small seeds and would keep them sticking against its surface. The effect became known as the *amber effect* since amber is one of the materials which exhibits the effect best. The word *electricity* is derived from the Greek term *elektron,* meaning amber.

Little attention was paid to this mysterious observation of attraction of one object for another until about 1600 A.D. when Queen Elizabeth's physician, William Gilbert, became interested in the unusual manner that various substances reacted when rubbed together. He observed that a molten mixture of sulfur and sealing wax hardened with a glossy surface, which, when rubbed with fur, attracted things to it. Gilbert, therefore, surmised that rubbed materials became filled with a light airy fluid which he named *electricity.* He could not explain the nature of the electrical fluid, but he noted that it appeared to flow from the cloth into the object being rubbed.

Further investigations by other experimenters revealed that some substances retained the electric charge placed on them, such as glass and well-polished wood. Other substances, such as copper and iron, were found to lose the electrical fluid instantly. The results of the combined investigation led to the classification of substances into two categories: those that stored the fluid or *insulators,* and those which permitted electricity to flow easily through them or *conductors.*

In 1672, German-born Otto Van Guericke discovered a method of generating a continuous supply of electricity. He used a ball of sulfur mounted on an axle. As the ball was rotated, it was rubbed with fur and thus was kept continuously charged. Later, F. Hawkbee improved Guericke's electrical machine by replacing the sulfur ball with a glass globe. The globe was rotated while in contact with a cloth surface; then the electric charge that had built up was discharged to a brass cylinder through a chain that connected the globe and cylinder. Hawkbee later succeeded in building a charge large enough to cause sparks as the electrical "fluid" jumped the gap between the cylinder and another conductor and "vanished" into the ground.

In 1746, an American investigator, Benjamin Franklin, switched his interests to the new field of electricity. He noted the similarity between the spark produced by the electrical machine and that produced by lightning. He concluded that lightning was a gigantic spark of electricity and wrote a paper on "The Sameness of Lightning with Electricity" and submitted it to the Royal Society of London. Franklin's ideas were met with general skepticism from the members of the Society with the exception of two French scientists who were inspired with Franklin's suggested experiments and confirmed his suspicions by performing them. The most famous experiment conducted by Franklin himself was nearly his last. He decided to test his theories concerning the similarity between lightning and electricity by flying a kite in a thunderstorm. He attached a metal key to one end of a ball of string and a kite to the other. In addition, he tied a silk string to the opposite end of the key by which to hold the kite while it was aloft. He found that he could cause sparks to leap from the key as he brought his knuckles near it and permitted the accumulated charge to escape through his body to the ground. Luckily, the accumulation of electricity was constantly lowered by its escaping via the moist air into the atmosphere. But for this leakage, Franklin would surely have met his death. From this time on, however, lightning was considered an electric spark and, more important, electricity was considered a natural and important component in our lives.

In 1773, Charles Du Fay conducted investigations which showed that two kinds of electricity existed. He caused an accumulation of electric charge on a glass rod by rubbing it with silk cloth. The silk thread which supported the rod also insulated it and prevented the charge from leaking off the rod into the ground. He then charged a globe made of sealing wax by rubbing it with fur and discovered that the glass rod was attracted toward the globe. He also noted that two suspended glass rods repelled each other when rubbed with a silk cloth. He then concluded that the charge on the glass rod was opposite that accumulated on the charged globe of sealing wax.

In 1800, Alessandro Volta, Professor of Natural Philosophy at the University of Pavia, noted that electricity flowed when chemical changes were occurring. He built a *pile* from pairs of silver and zinc discs, separated by moist cardboard sheets, and found that he could cause a continuous electrical discharge by connecting opposite ends of the pile with a wire. This arrangement was the first battery and was heralded as a great discovery with unlimited potential value. Later, Volta developed a battery of greater efficiency by using glass vessels filled with salt water in which were submerged strips of copper and zinc. He connected the copper in one vessel to the zinc strip in the next until all vessels were connected together. He discovered that a large spark was created when he lightly touched the opposite strips of the vessel at each end of the series by means of a wire conductor.

Further experiments with electricity by many investigators revealed that electric *current,* a name adopted from early theories of the fluid-like properties of electricity, caused thin wires to become hot enough to emit heat and even light. Current flowing through a wire displayed magnetic properties and led to the development of the principles involved in the electric motor.

Michael Faraday, British scientist of the 1830's, studied the electrolytic properties of liquids, that is, the effects caused by passing electric current through various solutions. He noted that the current caused solutions to separate into their component parts. Water, for example, was found to separate into hydrogen and oxygen. Observations such as this one strengthened the belief that electricity was indeed an ethereal fluid. Man had learned to accept this intangible phenomenon, but not to the extent that he accepted heat and light.

For electricity to become an integral part of man's life, more proof of its worth had to be found. Also, man had to develop a better understanding of its properties. During the nineteenth century, Dalton's *Atomic Theory* concerning matter was reorganizing the field of chemistry. Matter was no longer considered a continuous mass (from the days of Aristotle), but was thought to be composed of tiny atoms. The scientist conceived that electricity may also be composed of atoms, a proposal similar to one suggested by Franklin in which he formulated that "electric matter consists of particles extremely subtle since it can permeate common matter."

Further experimentation was concerned with the effect of electricity on solutions and on various gases. Michael Faraday experimented with the effects electricity had on various chemical solutions. He discovered that the number of atoms released as chemicals decomposed was dependent on the amount of electric current passed through them. Faraday's observation led him to the conclusion that there exists a set or constant amount of electricity that is required to release each atom.

Sir William Crookes experimented with the effects of electricity on gases. He conceived the idea that electricity existed in the form of particles. He also noted that a great electrical pressure was required to cause current to flow from wire to wire if the ends were separated by a gap of air. His continued investigations concerned the distance an electric spark would jump if the air pressure were decreased in a confined area and the electric current held constant. The ends of the wires carrying the current were sealed in a glass tube so equipped as to permit removal of air from it. He discovered that electricity flowed more easily between the two ends of the wires or *electrodes* as air was removed and that the snake-like appearance of the flash changed to a stream or ray of beautiful violet light. Further investigations by Crookes involved a paddle-wheel device placed in a tube from which the air had been evacuated. He conceived that the ray produced by an electric current passing through the tube caused the paddle wheel, riding on rails, to turn as a mill wheel is turned by water, a conjecture which lately has been disproved. He was correct in his conclusion, though, that electricity was composed of particles of some sort. Crookes' continued study revealed that this beam of particles was affected by a magnet, and that its color and brightness could be influenced by the insertion of various gases into the evacuated tube. The latter is the principle involved in our neon and mercury vapor lights of today. This concept of *cathode rays* implies that a stream of particles is emitted from the cathode, the point at which electric current enters the tube. The steam of particles flows through the gas to the conductor at the other end of the tube. (See Fig. 7.1.)

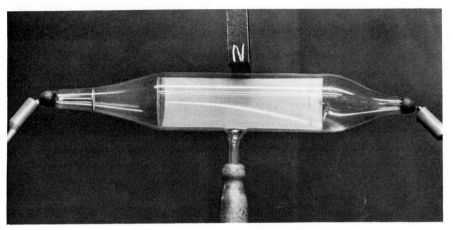

Fig. 7.1 **A model of Crooke's tube, more efficient than the model with which Crooke conceived the properties of the cathode ray. Note that the beam is deflected by the permanent magnet brought near its path.**

(Courtesy of Central Scientific Company)

Some time later, this discovery of cathode rays became more important as J. J. Thomson, Professor of Physics at Cavendish Laboratory in Cambridge, identified the tiny electric particles of the cathode rays as *electrons*. His measurements indicated that an electron had much less than a thousandth the mass of the hydrogen atom. Thomson's announcement to the Royal Society in 1897 of his discovery created great reactions, since this particle theory indicated that the atom was not elementary, but was itself composed of minute particles. The investigations which followed reinforced his belief until the electron became accepted as a building block of the modern atomic theory.

THE CHARGE ON THE ELECTRON

Dalton's theory of the atom was acceptable for less than one hundred years. Then Thomson and his contemporaries proved the atom was divisible and could be disrupted into electrical particles of varying properties. The connection revealed between electricity and matter was revolutionary in its effect on the thinking of man at the turn of the twentieth century.

What had man learned concerning electricity by the turn of the century? The work of such investigators as Gilbert, Franklin, Thomson, Crookes, and others led to the following conclusions:

1. There are numerous similarities and dissimilarities between electricity and magnetism. Like charges repel and unlike attract, just as like poles of magnets repel and unlike attract.

2. Various materials accept and hold a charge of electricity such as glass, sealing wax, silk, and dry wood and are known as insulators. Others do not hold an accumulated charge but permit it to dissipate. This category is known as conductors and includes primarily the metals.

basic electricity and magnetism

3. Electricity is produced in various materials by rubbing them with other given substances. These materials accumulate charges different from each other, and, hence, the oppositely charged materials attract each other. Objects made from the same material and charged by rubbing their surfaces with the same cloth repel each other.

4. Electricity is also a product of chemical change. Charges accumulate in reacting chemicals placed in containers to form a cell. Two or more cells connected together form a battery, a continuous source of electricity.

5. Electricity flows through wires and causes them to become hot. If the wires are placed a short distance apart, the current jumps across the gap in the form of a spark.

The summary of characteristics led to the conclusion that electricity exists as two separate charges of electricity, positive and negative. The names are arbitrary and have no real meaning except to point out that a relative difference exists between them. When one type of electrical charge is produced, the other is generated also and in identical amounts. As an electrically neutral amber rod is rubbed with a piece of electrically neutral animal fur, the rod acquires a "negative" charge and the fur a "positive" one. It is also conceivable that the charged objects could be again neutralized by merely permitting the objects to come in contact with each other once more.

As man continued his search for knowledge about electricity, he became aware that there were many similarities between the modern concepts involving the atom and those ideas concerning the nature of electricity. Man's discoveries revealed that many of the forces which hold atoms together and many of the chemical forces which hold molecules together are really electrical forces. Once this similarity was realized, many theories were corrected and new ones proposed. Some of the theories which concern the atom and its relationship to electricity follow:

1. All matter is made of atoms. Every atom is composed of a nucleus which is positively charged, around which are distributed a number of negatively charged electrons. The electrons of all atoms are identical; each has the same mass and the same amount of negative charge. The nucleus is composed of two different particles, the positively charged proton (whose charge is equal to but opposite from that of the electron) and the neutrally charged neutron (except the nucleus of common hydrogen).

2. An object which has an imbalance of electrons and protons is electrically charged. If it has more electrons than protons, the object is negatively charged. If the number of protons is larger than the number of electrons, the object is positively charged.

3. It is evident, since the proton is embedded in the nucleus and held there by great nuclear forces, that the charge of the atom varies according to the number of electrons it gains or loses. The electron is held in its orbital path with such a perfect balance of forces that a slight disturbance may cause it to leave

its orbit around the nucleus and be free to move about. A substance, then, is negatively charged when it has an excess of electrons and is positive when it has a deficiency of electrons. When the substance is neutrally charged (or has a net charge of zero), the number of electrons present equals the number of protons, and the net charge is zero.

4. All substances can be arranged in order of their ability to conduct electrical charges. Those placed at the top of the order, representing those substances which readily conduct electricity, are called conductors. Those placed at the lower level are classed as insulators. Almost all metals are conductors, and most nonmetals are insulators. A substance which is a good conductor is composed of atoms that hold their electrons in orbit very weakly and permit one electron or more per atom to move freely throughout the substance in all directions.

ELECTRICITY AT REST

The amount of data collected and the new principles conceived enabled the experimenter to create a vivid mental picture of electricity. He concluded that electricity is the flow of electrons, even when they flow through poor conductors such as air and water. Electrons *are* electricity. The name "amber," in reality a yellowish to brownish resin from which beads and necklaces were made in ancient times, is related to our modern term "electron" since with this material electricity was first produced by man.

The early experimenters noticed that electricity could be produced by *two* separate methods, an observation which led them to realize that there was more than one type of electricity. One type of electricity was found to be produced when two different nonmetallic substances were rubbed together and, therefore, was produced by friction. Electricity produced by friction was found to be capable of being stored on the surface of insulating substances such as rubber or glass. The charge could be totally removed by giving the electrons a path over which to escape such as that afforded by a wire attached to the storing insulator. This type of buildup of charge came to be known as *static* electricity since the charge did not move but remained at rest.

Early investigators also noticed that the charge which accumulated on the two kinds of metal plates of a battery did not instantly remove itself when furnished an escape route by means of a wire attached between the opposite metal plates. In fact, it was noted that the charge replenished itself after a time, and the investigators, therefore, concluded that this kind of electricity was continuously being produced. As investigations continued, this kind of electricity came to be known as *current* electricity, with the concept of current carried forth from past investigators' points-of-view. Electricity was identified as the flow of ethereal fluid and later as the flow of electrons. One conceptualized this kind of electricity as differing from static electricity not only in method of production, but also in that the electrons seemed to cycle in a confined container such as a battery and became available for use once again.

Knowledge about electricity was gained primarily through the study of static electricity, which has many mutual characteristics with current electricity. Discoveries concerning the properties of eletricity appeared very rapidly after Dalton's publicized works.

Each of us is aware of the electrical effects produced by scuffing our shoes across a nylon rug and then creating a spark by touching a door knob. The same thing results when we slide across the seat in an automobile and then touch the door to close it. We have also noticed the crackling noise and sparks produced by unfolding nylon clothing taken directly from the clothes dryer. These and numerous other examples indicate electrons are being transferred by the friction of contact between objects.

Charges can be accumulated on materials by transferring electrons from other materials. A charged rod brought in contact with a neutral object can transfer charge to the neutral object. If the object is a good conductor, such as a metal sphere, the charge spreads evenly over its entire surface. If the object is pointed at one end, the charge is found to be more concentrated at this region than at other areas on its surface. (See Fig. 7.2.) If the object is a poor conductor, the charge will stay concentrated in the area of contact. The method of accumulating charge on a second object is referred to as the *contact method*.

Fig. 7.2 The electrical charge accumulates on curved surfaces in accordance with the extent of curvature.

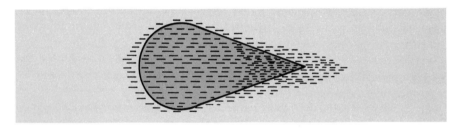

We have previously established that *like charges repel and unlike charges attract*. Therefore, how can both a negatively charged rod and a positively charged rod *attract* electrically neutral objects, such as bits of paper? The answer lies in the observation that the charges in the molecules which make up the bits of paper are capable of being arranged. The overall neutrality of the paper is lost as one end of each of the bits of paper has its charges so distributed as to take on a seemingly positive (deficient) charge while the opposite end becomes negative (excessive) in charge. This method of *polarizing* an object by bringing a charged object near it is known as charging by *induction*. For example, the lower half of clouds, such as those illustrated in Figure 7.3, is generally negatively charged. As clouds move through the atmosphere, they induce a positive charge on the air and on the surface of Earth. Lightning results as a means for the negative charge to become neutralized by yielding its excessive charge to Earth. Lightning rods make use of the principle of induction to permit the electric charges to be dissipated from the clouds through the rods to the ground; thus the lightning rods protect the structure to which they are attached from being "struck" by lightning.

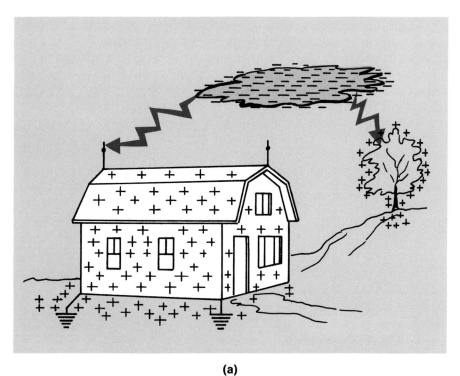

(a)

Fig. 7.3 (a) The typical manner in which a cloud dissipates its charge through the ground. (b) Some few clouds arrange charges oppositely and discharge by interaction with close-by clouds or the ground.

(b)

As man strove to learn the properties of electricity, one of the early concepts in which he became involved was the *magnitude of charge*. The investigator found that he could cause the magnitude or size to vary, a rationale based (1) on noticing the number of small bits of paper a charged object would attract after the object was rubbed with cloth and (2) on noting the extent to which a standard object, suspended by a silk thread and rubbed a set number of times with fur would be attracted or repelled by a second object, rubbed a varying number of times with fur or silk and brought near the suspended standard. Man adopted the latter technique to measure the amount of accumulated charge when he developed a detecting device for static electricity called an *electroscope,* illustrated in Figure 7.4. The amount that the sensitive needle is deflected by a charged object as it is brought in contact with the metal disc is some indication of the magnitude or quantity of charge on the object. A typical method of measuring the amount of electric charge on a given object is to compare the degree of deflection of the electroscope to that produced by a known or standard charge.

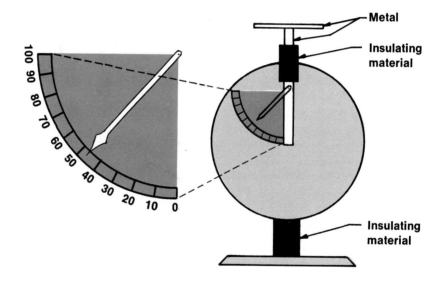

Fig. 7.4 **The electroscope generally is calibrated in relative units only; that is, from 0 units to 100 units which are not defined.**

ELECTRONS IN MOTION

Just as an object can have gravitational potential energy as a result of its position, an electron can have potential energy because of its position in an electric field. Similarly, just as work is required to move an object against Earth's gravitational field, work is necessary to move an electron against a positively charged electric field. The work done in the latter case increases the *electric potential energy* of the electron. As the electron again moves toward the positively charged field, the potential energy of the electron is converted into kinetic energy, the value of which depends on the change in potential energy as the electron moves about in the electric field. Consider an accumulation of

basic electricity and magnetism

electrons, a *charge,* that has potential energy because of its position in an electric field. The *electric potential,* the electric potential energy per charge, may be expressed:

$$\text{electric potential} = \frac{\text{electric potential energy}}{\text{charge}}.$$

The unit measure of electric potential is the *volt,* named after Allesandro Volta, an Italian physicist of the early nineteenth century. Electric potential and voltage are therefore interchangeable terms.

Electricity and heat have several attributes in common. If two conductors of heat, significantly different in temperature, are placed in contact with each other, heat "flows" from the conductor of higher temperature (because of greater particle energy) to the one of lower temperature. When both conductors reach the same temperature, the flow of heat ceases. In electricity, electric charge flows from the body of higher electric potential to the one of the lower electric potential. Charges flow only as long as the electric potential remains. Electric potential differs from electric potential energy in the same manner that temperature differs from heat.

There are numerous methods to create an electric potential between two objects. For example, if a rubber rod is rubbed with a piece of fur, an electric potential is produced between the rod and the fur. The electric potential cannot be maintained once the rod and fur are brought in contact with each other for an appreciable length of time or the rod and fur are otherwise neutralized. Chemical action can also develop an electric potential by upsetting the electrically neutral charge of atoms. Recall that an atom is electrically neutral; that is, the number of positive charges (protons) equals the number of negative charges (electrons). The atoms which lose the electrons are positively charged and form *positive ions.* The atoms which gain electrons are negatively charged and are *negative ions.* The positive ions and negative ions then can be concentrated on two dissimilar metals and an electric potential is developed. The accumulation is often accomplished in a device called a *voltaic cell.* The cell has the capability of maintaining the same electric potential and becomes a useful source of continuous electric charge. Two typical voltaic cells appear in Figure 7.5. A familiar

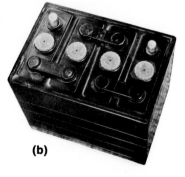

(a) **(b)**

Fig. 7.5 **(a) The common six-volt dry cell. (b) The six-volt battery used in some automobiles.**

(Courtesy of Consumer Products Division, Union Carbide Corporation)

basic electricity and magnetism

type of voltaic cell is the dry cell, used in our flashlights and portable radios (see Fig. 7.6). The container for the cell is commonly made from the metal zinc, which also serves as the negative electrode (cathode). A carbon rod is inserted in the center of the cell and serves as the positive electrode (anode). The inside of the zinc can is lined with chemically treated blotting paper and is further separated from the carbon rod by a paste containing various active ingredients. The paste actively dissolves the zinc and serves as a nonmetallic conductor to permit the cell to store electric charge and maintain its electric potential. This constant action is the reason that such dry cells have limited shelf life and that the battery in time develops leaks through its own container.

Fig. 7.6 Various views of the typical 1.5-volt dry cell.

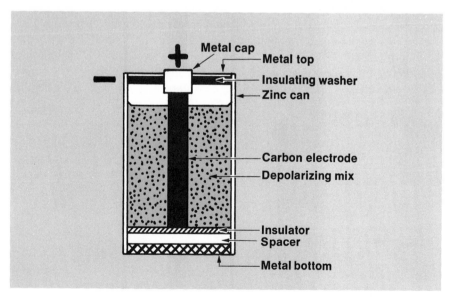

basic electricity and magnetism

Electric *current* is in reality the flow of electric charge. In conductors such as copper wire, it is the charge which flows through the *completed* circuit; that is, the path from the negative electrode to the positive electrode is a complete one, illustrated in Figure 7.7. In electrolytes, fluids which are nonmetallic such as water that contains an acid, a salt, or a base are commonly used. The charge which flows generally is composed of ions as well as electrons.

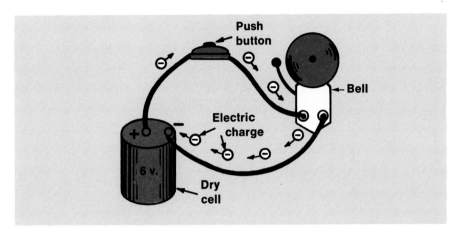

Fig. 7.7 A complete electrical circuit is necessary for electric current to flow.

The accepted unit of electrical charge is measured in terms of the *coulomb,* a value equal to the total electric charge carried by 6.25 billion billion electrons. The rate of electrical flow or intensity (I) is measured in the unit of the *ampere* (amp) and is equivalent to the flow of charge from 6.25 billion billion (6.25 $\times$ 10^{18}) electrons that pass a point in a completed circuit in a given period of time, the second. Five amperes, then, would mean that five coulombs of electric charge must pass through a detecting device or other point in the circuit in one second. Both the volt, the unit of electric potential, and the coulomb, the unit of electric charge are measures in the *mks* system. The *joule,* the *mks* unit of energy, then, can be expressed: 1 joule = 1 volt $\times$ 1 coulomb, or in terms of electric potential, 1 volt $= 1\dfrac{\text{joule}}{\text{coulomb}}$.

The quantity of electric charge that flows in a circuit per second depends on the opposition the charge encounters as it moves through the circuit—its resistance (R). The less resistance offered, the more current which flows. The unit of measure of electrical resistance is the ohm (Ω) named after George Simon Ohm. The symbol (Ω) is Greek Omega, since "O" would be misconstrued for "zero."

The resistance of a given conductor is dependent on several factors:

1. Substances vary in their ability to conduct the electric current, even among the metals, since different atoms vary in their ability to control their outermost electrons. For example, the resistance of iron wire is about seven times as high as copper wire of equal length and diameter.

2. Increasing the length of the conductor directly increases the amount of resistance to the flow of electric charge.

3. The cross-sectional area of the wire is also a factor since the resistance is inversely related to it. A conductor whose cross-sectional area is one square inch has twice the resistance of a conductor whose cross-sectional area is two square inches, assuming the two conductors are of like composition and of identical lengths.

4. The resistance of most conductors is affected by their temperature.

The filament in a light bulb has some ten times the resistance when glowing white hot as when it has not been turned on for some time. Other materials such as electrolytes and carbon conductors have a lower electrical resistance when they are heated than they do at lower temperatures.

HOW MAGNETISM DEVELOPED

Man has found it easy to accept the attractive forces that are always present in our environment. He is aware that various forces exist between objects which come in contact with each other, such as the way two smooth surfaces adhere to each other when pressed together or the manner in which small bits of paper stick to a comb that has been recently used. Man also accepts without questioning the way that gravity causes a falling object to plummet toward Earth, an event caused by the attractive force existing between them. Nor is attraction between objects which are not in contact with each other a new concept. All in all it is not difficult to comprehend that one electrically charged object should attract another which is oppositely charged even though the objects do not come in physical contact with each other.

However, when one discusses the forces of repulsion, there are no common experiences with which to make comparison. When one object pushes another, we expect to see physical contact between the objects when movement occurs. For without contact, there commonly is no motion involved.

But the awareness of repulsion without contact was discovered by ancient man. The Greeks long ago mined an unusual iron ore from the region of Magnesia in Thessaly. Chunks of this iron ore were observed to repel each other through some invisible force as well as to attract smelted iron to them in some magic manner. The ore became known as *lodestone,* from its ability to lode (lead) or attract small bits of iron. This unusual attractive or repulsive force was eventually noticed as capable of being transferred to pieces of iron which would in turn attract other bits of iron. The early Greeks, as well as the Chinese, also noted that a suspended piece of lodestone or iron which had been stroked with lodestone always rotated to a given position with respect to Earth. However, man considered this observation an isolated phenomenon with reference to other properties of magnetism he had identified, such as attraction, repulsion, and induction.

William Gilbert, court physician to Queen Elizabeth at the close of the sixteenth century, published *De Magnate,* which discussed the characteristics of magnetism realized at that time. Very little about natural magnetism is known today that was not part of man's storehouse of knowledge in Gilbert's time, including what magnetism really is, although many of its features have since been investigated in great detail.

Several important discoveries have been added to man's understanding of magnetism in the last 150 years. Hans Christian Oersted discovered in 1820 that an electric current in the vicinity of a compass needle would cause the needle, in reality a small balanced magnet, to be deflected. This new observation led to the development by William Sturgeon of the electromagnet, a device which was capable of lifting objects about twenty times its own weight. The electromagnet was the last of the outstanding discoveries in the field of magnetism until man was able to develop permanently magnetized materials. Recently, however, man has developed permanently magnetic materials, such as *alnico,* an alloy of aluminum, nickel, cobalt, and a small amount of copper, which are capable of attracting and lifting objects many times their own weight. The alnico magnet and other permanent magnets of various shapes and composition are illustrated in Figure 7.8.

THEORIES OF MAGNETISM

The process of making a magnet from a piece of steel by stroking it with another magnet is known as *magnetic induction,* the only rapid way of making a magnet before the discovery of electromagnetism (see Fig. 7.9). If a piece of soft iron or low carbon steel is magnetized through induction, the property of magnetism is found to be short-lived, whereas if hard steel is magnetized, the

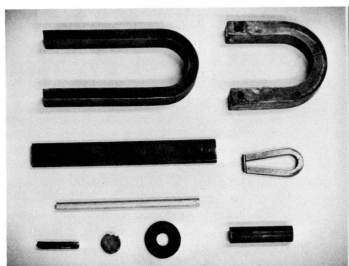

Fig. 7.8 **Permanent magnets differ in size, shape, and composition.**

Fig. 7.9 **One of the methods of inducing magnetism in a piece of steel to produce a permanent magnet.**

property is long retained and results in a permanent magnet being produced. The magnetism can be removed, however, by hammering, dropping, or heating the magnet. The soft iron which becomes a temporary magnet is very important and will be discussed in detail later.

Just as some materials retain magnetic properties longer than others, some vary in *permeability,* the capacity to afford a path for transmission of magnetism through them. This capacity is similar to that which materials have in their ability to transmit light. Iron, steel, cobalt, and nickel are said to have high permeability, whereas plastics, wood, copper, and aluminum have low permeability. In other words, permeability determines whether a material is magnetic or non-magnetic, but varies somewhat in degree. A comparison of some relative permeabilities of materials appears in Table 7.1.

TABLE 7.1

*Relative permeabilities of some
common materials*

Material	Relative Permeability
Bismuth	0.999833
Quartz	0.999985
Water	0.999991
Copper	0.999995
Air (S.T.P.)	1.0000004
Oxygen (S.T.P.)	1.0000018
Aluminum	1.0000214
Liquid Oxygen	1.00346
Nickel	400.0
Iron	5000.0
Permalloy	100,000.0
Supermalloy	800,000.0

Magnetic *flux* is a measure of the total amount of induction which has taken place as a magnet is produced. The measure is an indication of the strength of the magnet at various locations on or near the magnet and was described by Michael Faraday as magnetic *lines of force.* Such lines of force do not actually exist; rather the field of force is a *continuous* one.

If a bar magnet is placed beneath a sheet of paper and iron filings are sprinkled over the paper, illustrated in Figure 7.10, the small fragments of iron will align themselves and form continuous loops which follow the magnetic field of flux. The number of loops which fade out near the center of the magnet and reappear near the ends or *poles* of the magnet displays the variation in flux field or strength of the magnet toward the attraction or repulsion of another magnet.

If two magnets are placed adjacent to each other under the paper and iron filings are sprinkled on the paper, the results vary as to the arrangement of the magnets with respect to each other. If the magnets are arranged so that two opposite poles, a north and a south, are adjacent to each other (shown in Fig.

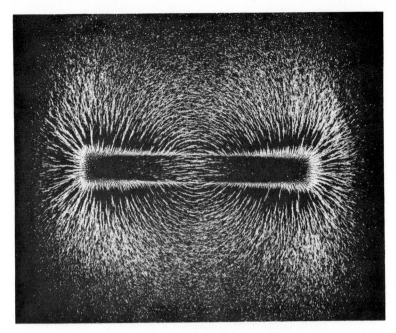

Fig. 7.10 A representation of the magnetic field about a bar magnet.

7.11a) an attractive force field exists. If like poles are placed adjacent to each other (shown in Fig. 7.11b), the force field is disturbed by the repulsive forces present. These illustrations disclose why like poles of magnets repel and unlike poles attract, the latter evidenced by the smooth continuation of force fields between the two magnets. An unusual arrangement of magnets is illustrated in Figure 7.11(c) in which four magnets create attractive and repulsive fields through the influence of one magnetic field on the other. If the pole at the top of the illustration is known to be a north pole, the identity of each pole can be determined by its effect on the magnetic force field of the known pole.

(a)   **(b)**   **(c)**

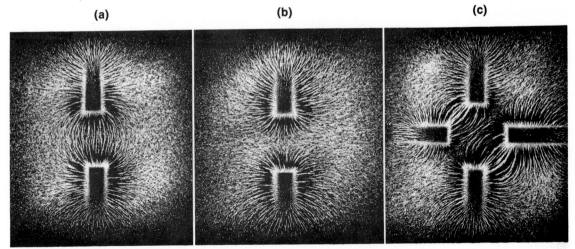

Fig. 7.11 (a) The magnetic field which interacts between opposite poles of two permanent magnets. (b) The magnetic field produced by two like poles of separate magnets. (c) The magnetic field produced by four poles from four separate magnets. The upper pole is considered a north pole. Can you identify the polarity of the other magnets?

Earth is considered a huge magnet and offers further evidence of the continuity of the field about a magnet by the manner in which a compass aligns with its magnetic field. The end of the compass that points to the north magnetic pole of Earth, regardless of the compass location on Earth's surface, is called the north-seeking pole. The end that points southward is thus the south-seeking pole. The names of the compass poles have been shortened to the *north* pole and the *south* pole, respectively. Since the north pole always points northward, the magnetic field of Earth or any magnet is continuous. The direction of the continuous magnetic field is conventionally considered to be from the north to the south pole on the exterior of the magnet, and from the south to the north pole in the magnet, illustrated in Figure 7.12(a). A spherical object, such as displayed in Figure 7.12(b), has a similar magnetic field to that of the bar magnet. Earth, represented in Figure 7.12(c), has a similar magnetic field. The field direction is the same as the external field of any magnet, and the internal field has a direction identical to that of either a bar magnet or a spherical magnet.

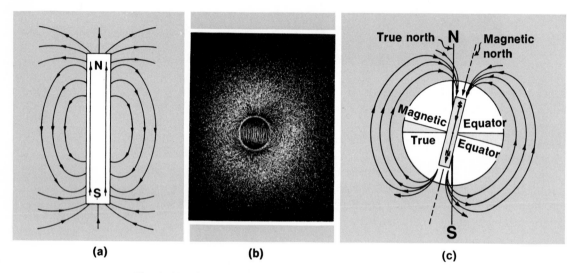

(a) (b) (c)

Fig. 7.12 **(a) The conventional direction of a bar magnet's magnetic field. (b) The magnetic field about a disk or sphere. (c) A representation of the magnetic field about Earth.**

The position of the north magnetic pole of Earth is thought to shift slightly from year to year, but the changes are quite small. The pole is considered to be approximately 12° west of geographic north when measured in New York, and 18° west of geographic north when measured in San Francisco. The pole is thus considered to be located in the region of Hudson Bay, some 1100 miles from the geographic north pole. The difference between true north and the compass reading at a given position on the surface of Earth is called *magnetic declination*. The degree of declination is illustrated in Figure 7.13(a) and 7.13(b).

basic electricity and magnetism

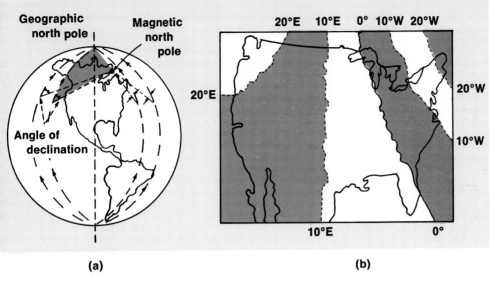

(a)   **(b)**

Scientists have long wondered about the exact nature of magnetism and recently have concluded that magnetically permeable materials have a crystal structure. The crystals can be aligned in such a manner that the individual magnetic properties become additive and cause the entire piece of material to become magnetized. In materials, these crystal structures, or *domains,* are arranged in no definite pattern, but actually change positions when they are exposed to an induced magnetic field so that the north pole of one domain falls in behind the south pole of the adjacent domain, which causes their individual magnetic field strength to become additive. (See Fig. 7.14.) When all domains are so aligned in the material, the material is said to have reached saturation and cannot become magnetically stronger. The fact that some materials can exist with domains aligned permanently has made the permanent magnet extremely useful in industry. Permanent magnets are found in electric motors, loudspeakers, telephones, timing devices, and in countless other devices.

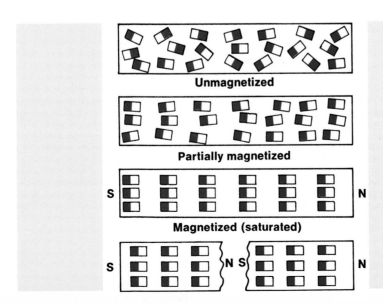

Unmagnetized

Partially magnetized

Magnetized (saturated)

Fig. 7.14 The domains and their suspended alignment in various degrees of magnetized substances. If the magnet is broken or cut, each piece is an equally strong magnet.

The theory concerning the presence of domains is supported by the fact that a physical blow on the magnet will diminish the magnet's strength; such a shock jars the domains and knocks them out of alignment. In contrast, a steel bar may become magnetized if struck with a hammer while the bar is being held parallel to a magnetic field. The blows cause the domains to shift so as to line up with the magnetic field in a manner similar to the process of magnetic induction. Further evidence in support of the *Domain Theory* is offered by the manner in which a magnet is entirely demagnetized when it is heated sufficiently. This demagnetization results from the tremendous amount of movement at the molecular level which occurs when heat or other forms of energy are added to the magnet. The motion rearranges the domains in a haphazard manner, a disarray which results in the magnet's loss of overall magnetic strength.

ELECTROMAGNETISM

Hans Christian Oersted's discovery of the relationship between electric current and magnetism led to a very important principle: when current flows in a wire or other conductor, the conductor is surrounded by a magnetic field. The field continuously encircles the wire, much the same as if the wire were to have an insulating tape wrapped around it, but the field extends into surrounding space in a limitless manner.

The strength of the magnetic field varies and is found to depend on one or both of two factors. First, the strength of the field is directly proportional to the amount of current flowing in the conductor. Second, as is often the case in practical application, the conductor is formed in the shape of a coil—a *helix,* as it is generally known. The magnetic strength of the wire is then related directly to the number of turns in the coil. Also, the polarity of the magnetic field, the direction of north and south, is dependent on the direction of current flowing in the conductor. Note the illustrations shown in Figure 7.15. One can apply a simple rule to determine the direction of magnetic field produced. The *left-hand rule,* as the relation is known, states that if the fingers on the left hand curl around the coil in the direction that the current flows ($-$ to $+$), the thumb will point to the magnetic north pole produced.

The magnetic field produced by an electric current can also be increased by the insertion of a soft iron core into the loops which form the helix. This arrangement is the usual form of the *electromagnet.* The iron core offers an easy path for the magnetic field inside the coil and thus provides a minimum of magnetic resistance. *Reluctance,* as this property is called, is related to magnetism as resistance is related to electricity. The reluctance of a core of iron is much less than a core of air or one of cardboard.

When electromagnets were first developed, man immediately realized many potential uses for them. For the first time, he could "turn" magnetism on or off at his command. Electromagnets were soon developed which could lift objects many times their weight, move them to a desired location, and release them. Others were developed that could readily separate permeable materials

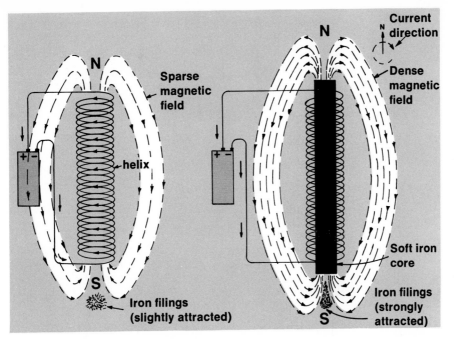

Fig. 7.15 A magnetic field is produced in a coiled wire by the electric current which flows from the battery. The magnetic field is enhanced by a soft iron core inserted in the helix. A magnetic field, though weak, is present in the straight sections of the wire, also.

from materials less permeable, such as fragments of iron from bits of aluminum or nickel. Industry has developed many other applications for the electromagnet, such as in the doorbell, the electric motor, the relay, and many other devices.

ALTERNATING CURRENT

We have seen how an electric current can produce a magnetic field or can create a magnet from a piece of nonmagnetized iron simply by sending a current through a wire wrapped around the iron. Now, let's look at the reverse: a magnet is capable of producing an electric current in a coil of wire. In Figure 7.16, we see a bar magnet being inserted and withdrawn from inside a coil of wire. The ends of the coil of wire are attached to a device which is capable of detecting an electric current. As the magnet is plunged through the coil, the needle deflects either to the left or to the right of center and returns to center (zero). As the magnet is removed, the needle deflects in the opposite direction and returns to center. In order for a current to be maintained, the magnet must be in constant motion. That is, the coil must be actively engaged in cutting through the magnetic field (some say cutting the magnetic lines of force) for current to be produced.

The fact that the needle is deflected in opposite directions leads the observer to deduce that the current must change directions. This change in current direction is the condition necessary to produce alternating current. Alternating current (AC) flows forward and backward in a conductor at definite time intervals. In contrast, the type of current produced by chemical action, discussed

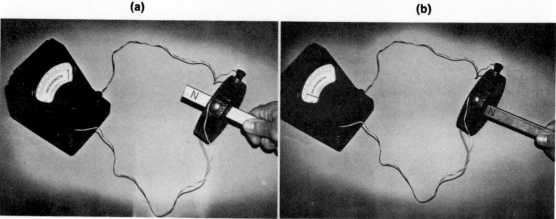

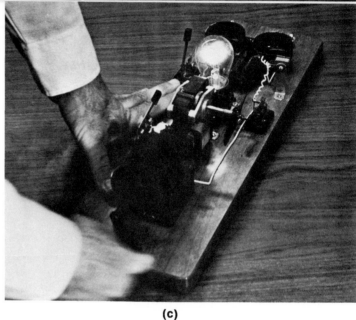

Fig. 7.16 (a) The magnetic field produced by the insertion of a magnet into a helix. (b) The direction of the magnetic field is reversed as the magnet is withdrawn from the helix. (c) Further evidence that electric current results from the disturbance of a magnetic field by a coil of a conducting wire.

(c)

earlier in the chapter, flows in one direction only and is thus considered direct current (DC). The AC which serves the home and industry typically alternates in direction 120 times per second or a total of 60 complete cycles per second (60 cps). In Europe, the AC completes 50 cycles per second and thus is said to have a *frequency* of 50 cps. Other areas may produce electricity with frequencies between 35 cps and 85 cps. In order for many appliances, particularly those which are motor-driven, to be useful in foreign countries, an adapting device must be used in conjunction with the typical connection or the appliance will be damaged instantly on insertion of the connection into the electrical outlet.

Alternating current is produced on a commercial scale in various types of electrical generating plants. Some plants make use of large rivers, waterfalls, or nuclear power generators to furnish the power necessary to turn gigantic coils of wire enclosed by a magnetic field and thus produce electric current by cutting the magnetic fields with the rotating coil. AC is generally produced at extremely high voltages, perhaps as high as 13,800 volts AC. As the AC is transmitted to distant cities, the voltage may be increased to as high as 500,000 volts to minimize current loss, then reduced to about 120 volts before being transmitted to the home to operate the various electrical appliances.

Alternating current is constantly changing both magnitude and direction with the result that AC produces a wave form which varies from a value of zero to some maximum value in one direction, then reverts back to zero and on to the same maximum value in the opposite direction. The variation in the amount of current produced is dependent on the number of "lines of force" actively being cut at any given time, illustrated in Figure 7.17(a). The greater the number of imaginary lines cut by the wire at any given time, the greater is the amount of current that will flow in the wire. The result of variations in the number of lines of force actively being cut causes the magnitude of current to vary according to Figure 7.17(b). In contrast, DC reaches its maximum current flow almost instantly and remains constant, as illustrated in Figure 7.17(c).

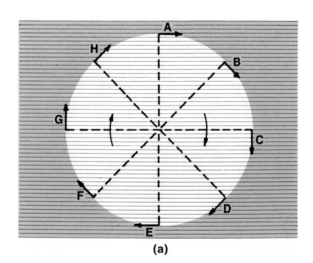

(a)

Fig 7.17 (a) The number of lines of force about a magnet which is actively cut by a conductor can vary. (b) The amount of current varies with the number of lines actively cut. (c) A comparison of direct current (DC) with alternating current (AC) in terms of variation in current intensity with time.

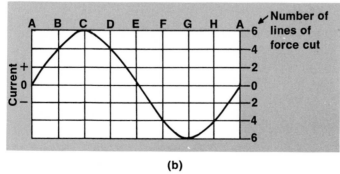

(b)

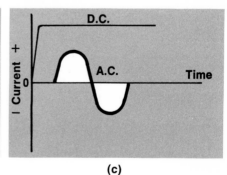

(c)

basic electricity and magnetism 155

A voltmeter specifically designed to measure DC shows the actual value of the potential difference. The AC voltmeter, altogether a different instrument from the DC voltmeter, reflects the *effective* value of the AC potential difference, an indication of approximately 0.7 to maximum value of the AC being transmitted. A 60-cycle, 120-volt alternating current actually reaches a maximum voltage of 170 volts twice each complete cycle, hence 120 times per second.

Alternating current is preferred over DC commercially for another important reason—economy. Not only is AC produced more cheaply than DC, but AC is transmitted long distances from the source without large heat loss in the wires, a property not evident with the transmission of direct current.

ELECTRICITY IN GENERAL

Electricity is not our latest controlled source of energy, but is probably our least understood. The properties it has are not clearly defined and are many times misconceived. A brief discussion of some phases of electricity, in addition to those previously discussed, points out some recent interpretation about its properties.

A sudden surge of energy is supplied by electrons to lamps the instant the switch is closed to complete the electrical circuit. An impossible task would be to turn off the lamp and to run from the room before the room becomes darkened. Electrical energy is conducted through a circuit at a speed which approaches the velocity of light, *but does not equal it.* During a telephone communication with a person in Europe, for example, a word is delayed in transmission long enough to permit the individual who is talking to complete the next word in the conversation before the previous word is received by the listener. The electron does not move faster than several million miles per hour whereas light travels over 600 million miles per hour. In addition, the electrons travel in all directions in a conductor and continue to move in random fashion but are accelerated by the presence of an external source of electrical energy toward the positive terminal of the source. The *electric field* so produced travels at a velocity which approaches the speed of light, *not the electrons.* In truth, then, electricity results from the movement of an electric field; electrons simply serve as the vehicle which carries it. When you receive an electrical shock, then, the electrons which produce the current are already present in your body and do not simply enter your body from some external source. The energy produced by the source causes electrons to transport energy in the form of an electrical field throughout your body and into the ground.

Another confusing property of electricity is closely related to the transmission of electrical energy. One can survive a shock produced by countless millions of volts or can succumb from shocks produced by sources of fifty volts or less. The damage from electrical shock is produced by current, not voltage. Currents of 0.05 amps can be fatal to man under certain conditions and twice this value is generally always fatal, causing irreparable damage to the heart and to the part of the nervous system which controls breathing. For man to receive a

basic electricity and magnetism

shock, a difference of potential must exist, however. Current will flow through the body between these two areas. An individual could hang onto a high-voltage electrical wire without receiving an electric shock as long as he did not come in contact with a wire, pole or other object which had a different potential. For this reason, birds can alight on wires without receiving a shock. But, if a piece of thread the bird is carrying touches another wire or object with a different potential difference, the bird will be instantly electrocuted.

CHAPTER SUMMARY

Electrons are tiny elementary particles that orbit the atom's nucleus. Their velocity approaches the velocity of light. Electricity has two separate charges, *positive* and *negative*. Since electrons are ejected in the form of rays from the negative electrode (cathode) of vacuum tubes, the electron's charge is considered negative. An atom is electrically *neutral;* however, if it has more electrons than protons, the atom assumes a negative charge. A deficiency of electrons causes an atom to be positively charged. *Insulators* offer great resistance to the flow of electric charge whereas *conductors* offer an easy path through which the electric charge may travel.

Electricity at rest is called *static* electricity. Electric charge in the form of electrons can be stored on insulators such as plastic or glass. Static electricity is produced by *friction* between two different insulating substances. One substance assumes a negative charge, the other a positive charge. Like electrical charges repel; opposite charges attract. A device to detect static electricity is called an electroscope.

An electric charge is an accumulation of electrons. Work done to move this charge against a positive electric field increases the electric *potential energy* of the charge. The electric potential energy per unit charge is known as *electric potential,* a value measured in volts. Electric potentials can be developed by friction between insulating substances, by chemical action, and by electric generators. The unit of charge is the *coulomb,* and the rate of charge flow is called *amperage.*

The quantity of charge that flows in a circuit depends on the electrical potential energy available; the magnitude of charge; and the opposition to flow the charge encounters, a concept known as *resistance.* Several factors regulate the electrical resistance in a circuit among which are the conductor's length, diameter, and composition.

The ancient Greeks are given credit for discovering that natural-occurring iron ores attracted bits of iron. Later, other investigators learned that iron bars stroked with lodestone would attract small bits of iron. Iron alloys poured in the form of bars and magnetized were found to lift iron objects many times their own weight. William Gilbert published his findings about *magnetism* in the sixteenth century. The relationship of electricity and magnetism was first conceived in the early part of the eighteenth century.

Permanent magnets may be made from iron and selected iron alloys by magnetic induction. The strength of the magnet produced depends on the alignment of the *domains* within the magnetic substance. *Permeability* is a measure of the time the domains stay aligned. Regardless of the length of shape of the magnet, the aligned domains produce a north and a south pole near the magnet's ends. Earth is a gigantic magnet whose magnetic poles slightly deviate from the geographic poles.

A conductor through which an electric current flows is surrounded by a magnetic field. If the conductor is wrapped around a soft iron core, the magnetic field is inducted in the core, and the looped conductor and the core form a temporary magnet. The instant the current ceases to flow, the core loses its magnetic field. The electromagnet, as the device is called, is very useful in industry and in home appliances.

A magnetic field surrounds a conductor through which an electric current flows. Conversely, an electric current flows in a conductor when the conductor is in motion in a magnetic field. If the conductor is moved back and forth in the magnetic field, the current produced alternates in the conductor, a phenomenon known as *alternating current,* the type of electricity typically used in the home. Alternating current can be transformed more readily than can direct current.

Current electricity results from the moving of an electric field in a conductor, *not* from the flow of electrons. The electric field moves through the conductor at a velocity that approaches the velocity of light. The danger from electric shock lies in the amount of current that flows, rather than in the electric potential, as is commonly thought. For current to flow, a difference of electric potential must exist between two conductors.

QUESTIONS AND PROBLEMS

1. If you charge a rubber balloon by rubbing it with wool, the balloon will stick to a wall or a wooden door but not to a metal door or appliance. Why is this statement true? (p.139)
2. Why are copper wires of large diameter rather than of small diameter used to conduct electric current throughout our homes and between cities? (p.146)
3. Each atom of iron may be alluded to as a tiny magnet. Why, then, are not all pieces of iron a magnet? (p.151)
4. A good conductor of heat is also a good conductor of electricity. Likewise, poor heat conductors are also used as electrical insulators. Can you think of reasons which explain these observations? (p.143)
5. Two rods which are identical in appearance are handed to you. One rod is known to be a magnet; the other is not a magnet. How can you determine which rod is the magnet, disregarding all observations other than those made of one rod's effect on the other? (p.148)
6. Trucks which transport tanks of gasoline and other highly combustible substances have a chain mounted on their undercarriages which comes in

basic electricity and magnetism

contact with the highway. What primary purpose does the chain serve? (p. 139)

7. How do birds that perch on high voltage wires remain safe without danger of being electrocuted? If the wires were close enough that a bird's wings touched two wires simultaneously, would the bird be in danger? Elaborate. (p.157)

8. How do alternating current, direct current, and static electricity differ in the following: (p.145)
   a. method of production
   b. magnetic field about each conductor
   c. uses
   d. flow of charge
   e. meaningful measures

9. Consider the various properties of electricity. List the various properties and cite examples of how each property is used by man. (p.137)

10. What factors might determine the strength of a permanent magnet? (p.146)

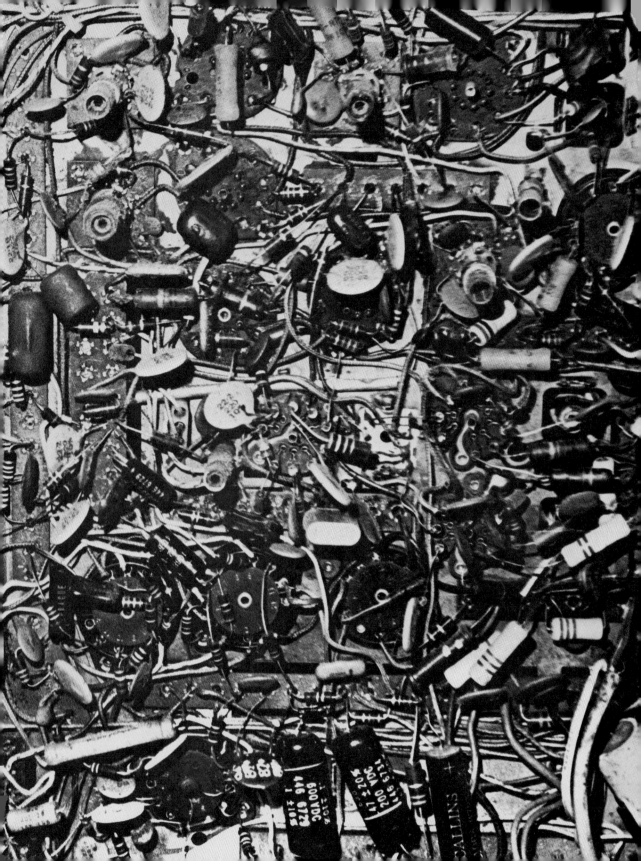

applied electricity and electronics
8

MEASUREMENTS IN ELECTRICITY
SERIES AND PARALLEL CIRCUITS
ALTERNATING CURRENT AND THE TRANSFORMER
HOUSEHOLD APPLIANCES
THE PRINCIPLES OF ELECTRONICS

vocabulary

Amplitude Modulation	Diode
Fluorescence	Frequency Modulation
Incandescence	Photoelectric Effect
Transformer	Transistor
Triode	

Electricity is the flow of electric charge in a circuit. The amount of charge that moves from the negative pole to the positive pole of the source in a given period of time is called the *current*. If the charge flows in one direction only, it is called *direct current* even though the amount of charge may fluctuate. If the charge continuously changes direction in the conductor, the flow is called *alternating current*.

The flow of electric charge is caused by some electric generating device which provides the work required to push the charge through the circuit. This charge travels from the negative pole of the generator through a path in the circuit and finally to the positive pole of the generator. The charge changes direction only as the polarity of the generator reverses.

There is a similarity between that which occurs when electric current flows in a circuit and when water flows in a pipe. The pressure of the water in the pipe results from the greater potential energy of the water at the reservoir than at the other end of the pipe. Charge flows in a conductor because there is a difference in electric potential across the ends of the conductor. The flow of water or of charge persists until both ends of the pipe or the conductor have a common potential. If the pipe is small, less water flows even though the difference in potential energy remains constant. In electricity, the amount of current which flows is less in a small wire than in one of larger diameter but of the same constituency even though the electric potential remains unchanged. If water in the pipe is forced to turn a paddle wheel placed in its path, less water also flows, since there is added resistance beyond that related to the size of the pipe. The amount of electric current that flows in a wire or the amount of water that flows in a pipe is directly related to the potential energy provided and is inversely related to the total resistance the flow of either meets.

MEASUREMENTS IN ELECTRICITY

Electricity has three basic quantities which are closely related to each other. For example, the *voltage* required in a circuit is equal to the product of the current and the total resistance. George Simon Ohm is credited with the original interpretation of their relationship in 1827. *Ohm's Law,* as the equation has come to be known, indicates that in a given circuit in which the only opposition to the flow of electric charge is resistance, there is a definite relationship between the values of voltage, current, and resistance. That is, $R = V/I$. Generally, the equation is restated in terms of I or V since these quantities are the two generally sought in practice, and thus: $I = V/R$, or $V = IR$. The unit of measure of opposition to the flow of electric charge is the unit called the

applied electricity and electronics

ohm (Ω). The current, *I*, is measured in amperes, and the electric potential, *V*, is measured in volts. One ohm of electrical resistance is present when a current of one ampere (from about six billion-billion electrons) flows past a point in the circuit in one second, caused to do so by an electric potential of one volt.

Example 8.1

How much current is required to operate an electric iron whose resistance is 20 ohms and which is plugged into an electrical outlet of 120 volts?

Solution: From Ohm's law:

$$I = V/R$$
$$= 120 \text{ volts}/20 \text{ ohms}$$
$$= 6 \text{ amperes.}$$

The three quantities are related in yet another way. A separate form of energy other than electrical is involved. This form of energy is heat, which, under given conditions, can be transformed into light energy. The electric charge flows through the conductor because of the voltage that forces it to move. The charge meets resistance to its motion in the form of friction, and as a result the conductor becomes thermally hot. The greater the opposition to the current flow, the more the conductor's temperature increases. In an electric iron, for instance, the heat emitted is not visible to us, but in an electric toaster or in a light bulb, the high heat produced creates visible light and permits us to see the glowing wire as one which is red-hot or white-hot, respectively.

The work done in overcoming friction assumes the form of heat and is equal to $V \times I \times t$, where t is the time in seconds that the current flows. The unit of work thus computed is the *joule*. If one joule of work is done in one second, the power required to perform the task is a *watt*, our basic power unit in electricity consumption. Power (p), produced in watts, is calculated by multiplying the voltage (V) times the current (I). If few electric charges flow through the circuit with a low voltage, then a low amount of power is consumed. Also, since $V = I \times R$ from Ohm's law, the power consumed can be expressed by $I \times I \times R$ or $p = I^2R$. From the last expression one can conclude that an increase in amperage causes greater energy consumption than does a proportional increase in resistance. This principle is kept in mind as engineers design various electrical appliances.

Example 8.2

An electric heater is connected to a 120-volt circuit which is protected by a 20-ampere circuit breaker. If the heater will cause the circuit breaker to "blow," (a) what must be the minimum resistance of the heater, and (b) what is the electrical power of the heater?

Solution:
(a) From Ohm's Law:

applied electricity and electronics   **163**

$$I = V/R \text{ or } R = V/I$$

$$\text{So: } R = 120 \text{ volts}/20 \text{ amperes}$$

$$= 6 \text{ ohms.}$$

(b) To establish the relation of electric potential and current to power, let's examine the units:

$$\text{power} = \frac{\text{energy}}{\text{charge}} \times \frac{\text{charge}}{\text{time}} = \frac{\text{energy}}{\text{time}}$$

$$\text{or power} = \frac{\text{joule}}{\text{coulomb}} \times \frac{\text{coulomb}}{\text{second}} = \frac{\text{joule}}{\text{second}}$$

$$\text{So: } p = V \times I = 120 \text{ volts} \times 20 \text{ amperes}$$

$$= 2400 \text{ watts.}$$

Experiments indicate that the energy expended in a wire is given by energy $= V \times I \times t$. Heat is produced by the dissipation of energy. Each unit of energy produces a definite amount of heat, or conversely, each unit of heat produced requires the expenditure of a definite amount of energy. If the relation is presented as an equation, $p = J \times H$, where J is a proportionality constant that equals 4.149×10^3 joules/kilocalorie. If the equation is solved for H, $H = p/J$ or $H = \dfrac{V \times I \times t}{J}$, then $H = 2.39 \times 10^{-4} \ V \times I \times t$. Then, to substitute Ohm's Law as $V = I \times R$, $H = 2.39 \times 10^{-4} \ I^2Rt$, a relation known as *Joule's Law*. Often the equation is expressed so that H is measured in calories, or $H = 0.24 \ I^2Rt$. For example, the amount of heat given off by the heater described in Example 8.2 is $H = 0.24 \times 2400$ watts $= 576$ calories per second. The wiring in a house has a given amount of resistance because of the innate characteristic being present in any conductor. If a short circuit occurs in which an extremely large amount of current flows, the heat that results may be enough to ignite the house and destroy it. Typical short circuits are caused by old, faulty wiring which lacks proper insulation and sometimes permits the conducting wires to touch. The current which flows through the wire is afforded an easier path through which to flow than an iron or a light bulb is designed to offer. Modern homes are better protected against this hazard by the use of advanced fuse systems and better insulated conductors.

For practical purposes, the unit of the watt is generally too small. The amount of electrical power consumed by man is tremendous, so he has devised a unit of measure of electrical power consumption called the *kilo-watt-hour* (kwh). The unit, as the name implies, is equivalent to the consumption of 1000 watts for one hour's duration.

Example 8.3

A television set that uses 600 watts (0.600 kilowatts) of electrical power, was left operating for ten hours. If the cost of electrical power consumption was 5¢/kwh, what would be the expense of operating the television set for this period?

Solution:

$$\text{Cost} = \text{watts} \times \text{hours}/1000 \text{ watts/kw} \times \text{cost/kwh}$$
$$= 600 \text{ w} \times 10 \text{ hrs}/1000 \text{ watts/kw} \times 5\text{¢/kwh}$$
$$= 30\text{¢}.$$

The company that supplies electricity to the consumer uses a meter similar to that illustrated in Figure 8.1. The meter is connected so that all electrical energy consumed must pass through it before electric current enters the home. The device measures the product of three variables—voltage, amperage, and time—and constantly adds the result to the dial readings. The difference in monthly dial readings, read from left to right, determines the cost of the electricity to the consumer.

Fig. 8.1 **The amount of power consumed by each household is measured by the kilowatt-hour meter.**

(Courtesy of General Electric Company)

SERIES AND PARALLEL CIRCUITS

A single dry cell, such as that used in a standard flashlight, has a potential difference of 1.5 volts whereas one cell of an automobile storage battery has a 2.0-volt potential. Batteries, made up of several single cell units, can deliver more voltage than the individual cell. The increase in voltage above 1.5 v or 2.0 v respectively indicates, then, that the cells are arranged and connected so that the voltage of each cell is additive, hence is part of a *series* connection. This arrangement calls for the negative terminal of one cell to be connected to the positive terminal of another. The total voltage produced thus is equal to the sum of the single cell voltages, a concept illustrated in Figure 8.2 (a) and (b). Mathematically, the result of connecting cells in series is given by the expression:
$V_t = V_1 + V_2 + V_3 + \ldots V_n.$

applied electricity and electronics　　　　　　　　**165**

Fig. 8.2 **(a) The method of connecting 1.5-volt dry cells to provide six volts. (b) A series of two 1.5-volt dry cells arranged to provide three volts, an arrangement typical in the common flashlight.**

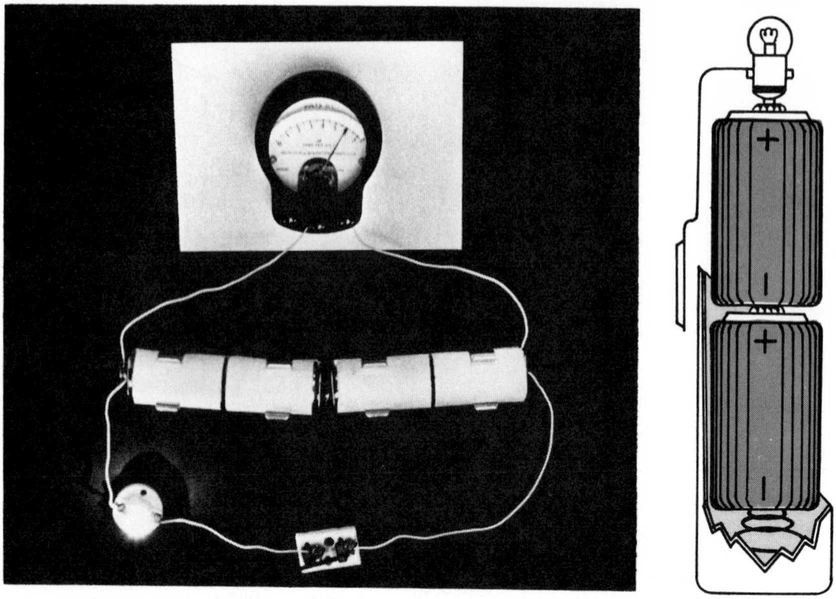

(a)

(b)

The electrical system of the automobile is generally the six-volt or the twelve-volt system. The six-volt system makes use of three cells of approximately two volts each, connected in series. The twelve-volt system is composed of a series of six cells which have an electric potential of two volts each. The arrangements are both illustrated in Figure 8.3.

Fig. 8.3 **Typical arrangements of cells to produce required voltages for automobile electrical circuits.**

applied electricity and electronics

Generally speaking, the dry cell has a relatively short useful life. The reason for this undesirable feature is the limited amount of electric charge which potentially can flow from the negative terminal through the circuit. The typical dry cell does not have the capability of replenishing its electric charge supply as in the case of the storage cell. However, chargeable dry cells are gaining in popularity and undoubtedly will replace the standard non-rechargeable dry cell to a great extent. The dry cell also differs in size and shape, though the standard cell delivers 1.5 volts. In Figure 8.4, all cells shown deliver 1.5 volts. Naturally, the larger cells will last longer since they can supply more current than the smaller cells.

Fig. 8.4 **The variation in size of dry cells of equal voltage.**

(Courtesy of Consumer Products Division, Union Carbide Corporation)

Sometimes the cells on hand are not large enough to offer the amount of current required by a given circuit. The solution to the situation is obtained by connecting a number of the 1.5-volt cells in *parallel*; that is, all positive terminals are connected as are all negative terminals. The method of connecting several dry cells in parallel is illustrated in Figure 8.5(a). Figure 8.5(b) offers a view of a large dry cell which is composed of several dry cells connected in parallel. Parallel circuits composed of numerous dry cells have a resultant voltage determined by the expression: $V_t = V_1 = V_2 = V_3 = V_n$, where V_t is the value of the largest cell in the group. If the largest cell is 1.5 volts, V_t equals 1.5 volts. The advantage to connecting dry cells of equal voltages together in parallel is that the current supplied by the cells is additive. If four cells, each with an electric potential of 1.5 volts were connected in parallel, the current made available to the circuit would be four times the amount available from the individual cell. The length of time the four batteries would last when they are connected in parallel is also significantly longer than the time the cells would last if used individually, due to several factors beyond the scope of the current discussion.

Cells are not the only device which the experimenter has sought to connect in series or in parallel. Resistances of various types are connected in both manners. Lights used on Christmas trees are ideal examples of both parallel and series circuits. The remaining lights on some sets will continue to "burn" even

applied electricity and electronics   **167**

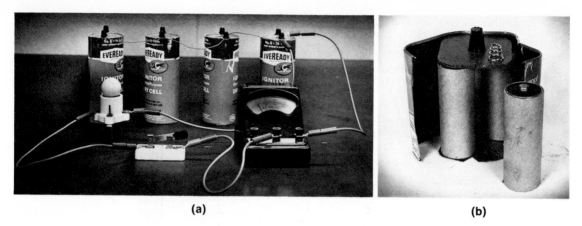

(a) (b)

Fig. 8.5   (a) The result of connecting individual dry cells in parallel. All positive poles are connected jointly as are all negative poles, a method which produces a potential difference of 1.5 volts. (b) This battery is composed of numerous 1.5-volt dry cells. Through parallel connections, the resultant total voltage of the battery is 1.5 volts.

though one or more bulbs have expired. In the case of other sets, when one bulb fails to light, the balance of bulbs in the set also fail to light. The former set of lights is arranged in parallel, the latter in series. Figure 8.6(a) illustrates a practical application of series and parallel circuitry as is used in the home. The circuitry of the modern automobile has become complicated, but necessary, as is illustrated in Figure 8.6(b).

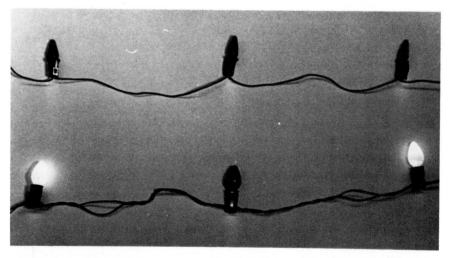

Fig. 8.6 (a)   The outmoded series circuit of Christmas tree lights shown at the top of the illustration has been replaced by the modern parallel circuit also shown. If one bulb in the series were to burn out, the entire set would fail to burn.

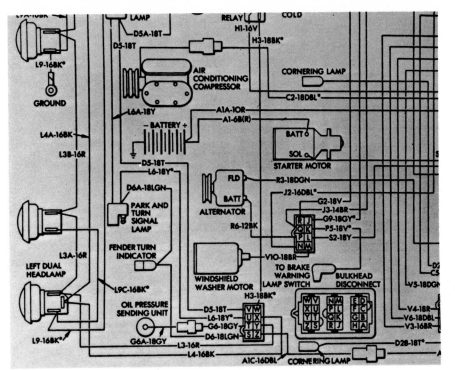

Fig. 8.6 **(b)** An automobile has an intricate electrical system composed of numerous parallel and series circuits as is illustrated in the schematic diagram.

(Courtesy of Chevrolet Motor Division, General Motors Corporation)

ALTERNATING CURRENT AND THE TRANSFORMER

Electricity has progressed to the degree that it now furnishes the energy for almost every task undertaken in the home and at work. It has furnished us energy with which to illuminate our homes and streets, to heat our homes, to communicate with each other, and to operate our household appliances.

Batteries remain our major source of direct current; however, an appreciable amount is produced by DC generators. The generator operates through the principle of cutting the imaginary lines of force of a permanent magnetic field by turning a wire coil through them, thus producing alternating current. Through innovative circuitry, the AC thus produced is then made unidirectional or *rectified* into DC (or more exactly into pulsating direct current, a concept beyond the scope of this book).

Alternating current, the type generally used in the home, is produced by huge generating plants such as the one illustrated in Figure 8.7. The energy which is used to turn the large turbines is produced by the mechanical energy of flowing or falling water and by the steam pressure produced through either atomic energy or the burning of coal.

The alternating current generator is sometimes referred to as an alternator, a term made common by Chrysler Corporation in the early 1960's. During this period the electrical systems of automobiles underwent renovation in which the alternator replaced the DC generator. The alternator is capable of delivering various voltages, other than simply twelve volts. The proper voltage for various components in the electrical system is obtained through the use of *transformers*. The transformer permits AC voltages to be increased or decreased to values other than those produced by the alternator.

Most homes make use of electricity which enters them at an electric potential of 120 volts. If the generator which produces the electricity is located very far from the area where it is to be used, the generator must produce a greater voltage than 120 volts, since the electric potential is readily lost in long-distance transmission of electricity and is the greater cause of expense to the consumer, not its production.

The generating plant actually produces electricity at a given electrical power value which may be measured in units as large as millions of watts or megawatts. Since power is the product of voltage times amperage, a decrease in one of the variables produces a proportional increase in the other. As has been previously discussed, heat produced by electricity is related to the square of the amperage, or $H = 0.24\ I^2Rt$. This mathematical statement indicates that heat is lost according to current flow primarily; thus electricity can be more profitably transmitted by lowering the current hence raising the voltage. Some generating plants increase the voltage to over 12,500 volts through the use of a transformer and decrease it through the use of a specific type of transformer at a substation near the community where the 120-volt electric potential is to be used.

The transformer has many applications in today's modern home. Its efficiency and simplicity are the reasons home appliances operate on AC rather than DC. In its simplest form, a transformer is composed of three integral parts: a primary coil of insulated wire, a secondary coil of insulated wire, and a laminated or layered soft iron core. The step-up transformer has more turns of wire around the core in its secondary coil than does its primary coil. Transformers are used to furnish high voltages for neon signs, radios, television sets, and X-ray units. Giant step-up transformers are employed to transmit voltages over long distances from the source. An illustration of a step-up transformer is shown in Figure 8.8(a). A step-down transformer, such as that illustrated in Figure 8.8(b), is applied to alternating current to reduce voltage with a corresponding proportional increase in current. The secondary coil has fewer turns about the core than does the primary coil, the coil which is identified as the one attached to the original voltage source—generally 120 volts. The step-down transformer supplies the low voltages required to operate electric trains, doorbells, kitchen appliances, and various components in television sets. Figure 8.8(c) exemplifies the component parts of a step-down transformer.

Some transformers have several secondary coils, some which step the voltage up, others down. For example, a typical transformer in a television set

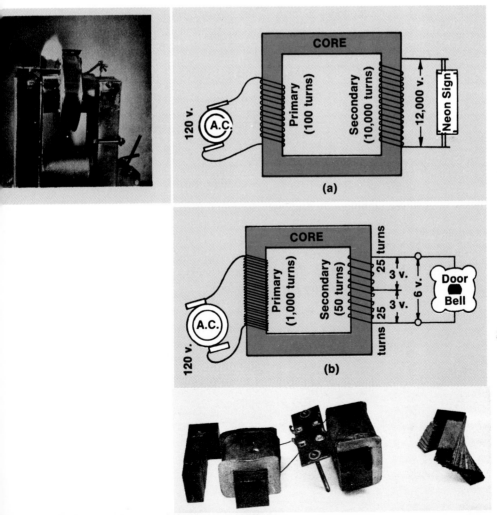

(a)

(b)

(c)

Fig. 8.8
(a) The step-up transformer has more turns of wire in the secondary coil than in the primary coil.

(b) The step-down transformer plays many important roles in household appliances. The primary coil has more turns than does the secondary coil.

(c) The major component parts in a step-down transformer. The soft, laminated iron core provides a path for the magnetic field produced to flow, a feature which increases the efficiency of the transformer.

supplies voltages that vary from about 6 volts to over 300 volts. Another furnishes up to 50,000 volts and more with other values available from the transformer connections as needed by the circuitry. Variable transformers have been developed which are used to operate toy racing autos by stepping the 120-volt AC down considerably to values which can be lowered even further to regulate the speed of the racers.

Various voltages are obtained through the relationships of the number of turns about the core in the secondary coil. The following proportion exists:

$$\frac{\text{Primary voltage (usually 120 volts)}}{\text{Secondary voltage}} = \frac{\text{Primary turns}}{\text{Secondary turns}}.$$

Example 8.4

A step-up transformer has its primary coil attached to 120 volts AC. The number of turns of wire in the primary coil is 200 and the secondary coil has 1400 turns. Calculate the secondary voltage.

Solution:

$$\frac{\text{Primary voltage}}{\text{Secondary voltage}} = \frac{\text{Primary turns}}{\text{Secondary turns}}$$

$$\frac{120 \text{ v}}{X} = \frac{200 \text{ turns}}{1400 \text{ turns}}$$

$$200\,X = 120 \times 1400$$

$$X = 840 \text{ volts.}$$

Transformers operate on a balance of power, so if voltage decreases, the current increases proportionally. The relationship follows:

$$\frac{\text{Secondary current}}{\text{Primary current}} = \frac{\text{Primary turns}}{\text{Secondary turns}}.$$

Heat loss, however, significantly decreases the actual efficiency of a transformer.

Example 8.5

A transformer has 500 turns in its primary coil and 100 turns in its secondary coil. If the current which flows into the primary coil is 5 amperes, determine the value of the secondary current produced.

Solution:

$$\frac{\text{Secondary current}}{\text{Primary current}} = \frac{\text{Primary turns}}{\text{Secondary turns}}$$

$$\frac{X}{5 \text{ amperes}} = \frac{500 \text{ turns}}{100 \text{ turns}}$$

$$100\,X = 500 \times 5$$

$$X = 25 \text{ amperes.}$$

applied electricity and electronics

The principle which makes the transformer function underlies the reason that there is great difficulty in raising or lowering the voltage of direct current by this process. The reader should recall that current flows in a coil of wire only when the imaginary lines of force of the magnetic field are in the process of being cut by the wire. In AC, the variation in voltage caused by its constant change of direction in the wire produces the same effect as when the magnetic lines of force are cut. Direct current, however, reaches its maximum potential almost instantaneously and stays constant so that a wire coil which conducts DC can interfere with the lines of force of a magnetic field only when either the coil or the magnetic field itself is being moved.

Transformers are integral parts of many home and industrial items. In addition to the previously mentioned applications, transformers play an important role in the telephone, the doorbell, the air-conditioner, the furnace, and many other electrical devices.

HOUSEHOLD APPLIANCES

Electricity serves us in the home in many ways. One of its earliest applications was that of man's source of artificial illumination. The invention of the carbon filament lamp by Thomas Edison in 1879 introduced electricity as a practical source of energy to serve man. The Chicago World's Fair of 1893 was illuminated by some 20,000 electric lamps and became the first commercial display of the illuminating potential of electricity.

The vacuum-type enclosure which contained a tungsten filament lamp, our most efficient source of artificial illumination, was placed in practical use at the New York World's Fair in 1939.

The incandescent lamp, as the tungsten-filament lamp was called, received its name due to the white-hot glow produced by the electrical energy which passed through the filament. An illustration of the development of filament bulbs appears in Figure 8.9 (a, b, and c). The filament offers high resistance to the current flow and reaches extremely high temperatures without melting. The

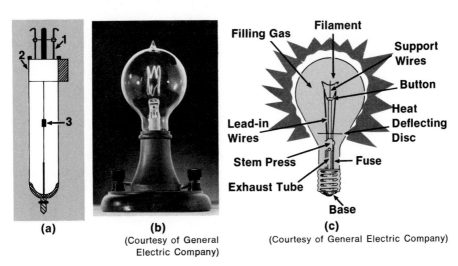

(a)

(b)
(Courtesy of General Electric Company)

(c)
(Courtesy of General Electric Company)

Fig. 8.9 (a) An illustration of the open arc lamp of about 1878. Obviously, the lifetime of this model was quite short and the lamp had little practical use. (b) Edison's first lamp, 1879. The lamp became one of the first practical applications of electricity. (c) The modern incandescent lamp has numerous components.

proper mixture of various inert gases causes the filament to last for thousands of hours without burning into two or more fragments, whereas Edison's first lamp lasted only some forty hours. The life of the filament is inversely related to the rate by which the filament combines with oxygen; thus the more expensive bulbs are carefully manufactured to prevent the leakage of air into the bulb. The incandescent bulb, by its nature, has a low efficiency and produces great amounts of heat. Some filaments reach temperatures as high as 4800°F even though they may be no thicker than a human hair.

The fluorescent lamp differs from the incandescent lamp in that light is emitted from it through the discharge of electrical energy from two points of significantly different electric potential. The filaments in a fluorescent lamp, such as the types illustrated in Figure 8.10 (a, b, and c), are slightly heated by the friction produced from the electric current which passes through them. The hot

Fig. 8.10 (a) The major components of the familiar fluorescent lamp. (b) The total components of the flourescent lamp. (c) The various shapes and sizes of common flourescent lamps.

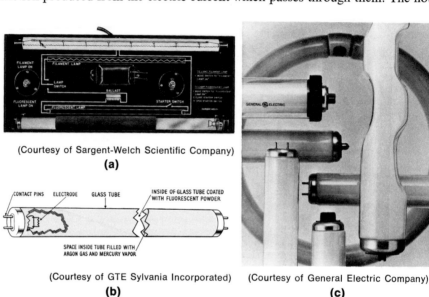

(Courtesy of Sargent-Welch Scientific Company)

(a)

(Courtesy of GTE Sylvania Incorporated)

(b)

(Courtesy of General Electric Company)

(c)

filaments emit electrons which collide with the particles of mercury vapor that are used to fill the tube, and the collisions produce a greater electric discharge that creates ultraviolet light. The invisible ultraviolet light in turn strikes the coating on the inside of the lamp and causes the coating to fluoresce or glow with light which is visible to us. Once the process is underway, the starter automatically disconnects the filaments from the electrical circuit, since the heated filaments are no longer necessary, and thus conserves their useful lifetime. (See Fig. 8.11.)

The fluorescent lamp has proved superior in many different ways to incandescent lighting. A given electrical power rating (in watts) produces over twice as much lighting efficiency for a fluorescent lamp as does an incandescent lamp of the same rated power. Also, the life of the flourescent lamp is about 8000 hours and is limited to this lifetime by the length of time the coating on

applied electricity and electronics

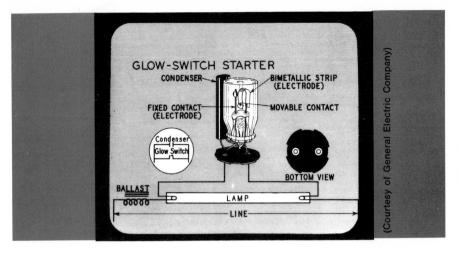

Fig. 8.11 **The internal elements of the starter used in conjunction with fluorescent lighting.**

the filament can last and thus furnish electrons to start the fluorescent process. The fluorescent lamp, though, is not suitable for use as a spotlight and does not typically function well in temperatures which vary far from a range of 60°F to 90°F; therefore, in many climates they are not very useful outdoors. Research by various lighting divisions has overcome this deficiency to some degree, and some additional applications have been accomplished. For example, Figure 8.12 displays a successful application of fluorescent lighting in tunnels. This illustration presents a true contrast of incandescent lighting, installed in the west-bound Tuscarora Tunnel on the Pennsylvania Turnpike in 1940, and the newly installed fluorescent lighting in the east-bound tunnel. Both installations were performed by Westinghouse Electric Corporation.

Fig. 8.12 **The east-bound tunnel pictured on the right is approximately a mile long. The tunnel is equipped with a new "tapered" fluorescent lighting system. Contrast the difference in brightness of incandescent lighting with that of the new fluorescent lighting.**

(Courtesy of Westinghouse Electric Corporation)

applied electricity and electronics   **175**

One of the latest lamps placed in use is the Metalarc lamp and its modified version, the Metalarc/C lamp. This type of lamp was introduced by General Telephone and Electronics' Sylvania Lighting Products Group and is a high intensity lamp with certain metallic iodides added to provide high light output and good color rendition. The lamp has an average rated life of 7500 hours and is used both indoors and outdoors. A close-up view of the Metalarc/C lamp is provided in Figure 8.13.

Fig. 8.13 **The metal arc/C lamp provides intense lighting wherever needed, inside or outside.**

(Courtesy of GTE Sylvania Incorporated)

A second application of electricity in the home involves its ability to heat. There are various types of heating appliances which use a resistance wire to produce heat such as toasters, irons, dryers, blankets, coffee makers, and stoves. The resistance wire is commonly composed of an alloy of nickel and chromium, called *nichrome*. The wire is wound into a coil or a screen-like grid pattern so that a long length of it can be installed in a small space. See Figure 8.14(a). The *element,* as the wire is often called, is insulated by ceramic materials or by mica sheets. Other elements, such as those illustrated in Figure 8.14(b) and (c) are enclosed in a metal tube to prevent damage to them from air and moisture.

The electric iron of today is capable of converting water into steam to aid the heat during the pressing of many fabrics (see Fig. 8.14 d). Temperature control of each of the appliances is accomplished by varying the resistance of the heating element. The heat regulator, called a *thermostat,* opens and closes the electrical circuit according to various settings of the calibrated temperature control. Thermostats, illustrated in Figure 8.15(a) and (b), generally are made of a metal strip composed of two different metals. Each of the metals is vastly different in the rate that it increases in length as its temperature is increased. The thermostat is heated by the filament sufficiently to cause it to bend away from the balance of the electrical circuit because of the variation in expansion of the metals, and the circuit is broken. As the element cools, the thermostat cools correspondingly and eventually returns to its original position, which again completes the electrical circuit. Therefore, the amount of heat given off by the hot filament is controlled by an adjustment of the tension which holds the metal strip in contact with the electrical circuit.

(Courtesy of Sargent-Welch Scientific Company, Skokie, Illinois)

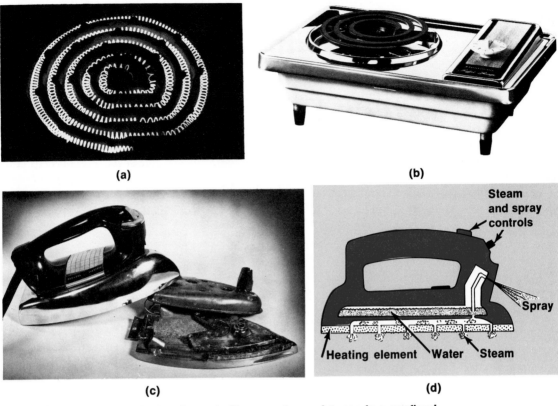

Fig. 8.14 **(a) The coiled wire element offers greater resistance in a confined space to current flow than would a straight wire element because of its greater length, a feature which produces greater heat. (b) The enclosed electrical element of a hot plate. (c) The element is often molded into the base of an electric iron. (d) Water is converted into steam by the heat of the element in the steam iron.**

(Courtesy of General Electric Company)   (Courtesy of General Electric Company)

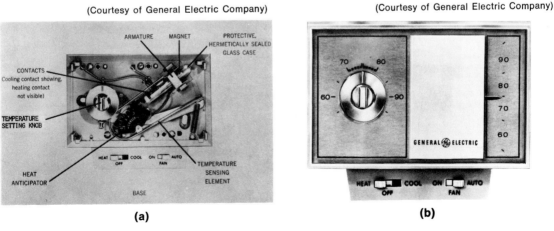

Fig. 8.15 **(a) A deluxe heating-cooling thermostat designed to provide close control of temperature in our homes. (b) The internal components in a heating-cooling thermostat as is typically used in the modern home.**

A third major application of electricity is that of converting electrical energy to mechanical energy through the electric motor. The DC motor is opposite in principle to the DC generator in that the generator converts mechanical energy to electrical energy. The generator produces electricity by virtue of rotating a coil of wire in a permanent magnetic field. If an electric current is connected to the coil of wire, known as the *armature,* the coil becomes an electromagnet and the generator is converted into a DC motor. The North pole of the armature is repelled by the North pole of the permanent magnet as is the South pole of the armature repelled by the South pole of the permanent magnet. The armature is caused to rotate by the repulsive magnetic fields, but the fields of the armature are reversed by design before the armature completes one full rotation. The reversal of fields of the armature forces the armature to continue to rotate, as is displayed in sequence in Figure 8.16 (a), (b), and (c). Then by simple conversion, the rotational energy is eventually transformed into more useful forms of mechanical energy or is used in its original form to do work for man.

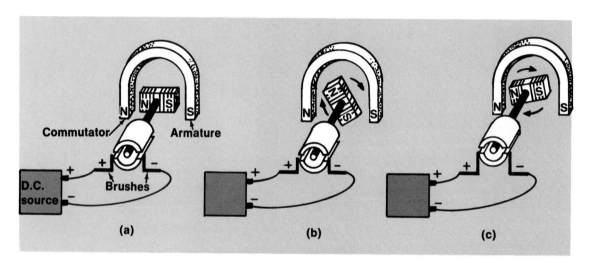

Fig. 8.16   (a) The repelling force of the fixed magnetic poles causes the armature to rotate. (b) The armature changes polarity and then is attracted by the fixed magnetic poles. (c) The polarity of the armature shifts again, so repulsive forces of like poles cause the armature to complete its rotation and the cycle starts over.

THE PRINCIPLES OF ELECTRONICS

The separation of electronics from the concepts of electricity is extremely difficult. The field of electricity is generally concerned with magnetism, lights, heating, and the production of electricity by generators and chemical action. Electronics is usually involved with the application of electricity in communication such as in radio, television, and in other devices where vacuum tubes and transistors are employed. The vacuum tube (or radio tube) is considered one of the most important inventions of recent times. It was discovered in principle by Edison shortly after he invented the incandescent bulb. He learned, by accident, that if an additional wire were attached to the outside of the bulb and then connected to the circuit, an electrical current was produced in it even though it was not connected to the filament. For this reason the phenomenon is known as "the Edison effect," although he did not realize its great capabilities. The vacuum tube and its contemporary, the transistor, are major components of the various electronic devices.

The simplest constructed vacuum tube is the *diode,* which contains two constituent parts (see Fig. 8.17). These two components generally consist of a *filament,* composed of tungsten wire, and of a metal cylinder which surrounds the filament, known as the *plate.* The components are sealed in an evacuated glass bulb to protect them from rapid oxidation. As a current is sent through the tube, the filament heats up sufficiently to cause electrons to "boil" from it. The

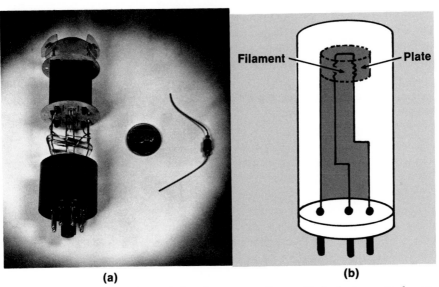

(a)   **(b)**

Fig. 8.17 **The diode, an electron tube which contains two major components is pictured with the rectifier that replaced it in modern circuitry. A stick diagram of a diode is also illustrated.**

plate is charged positively and thereby causes the emitted electrons to be attracted toward it at an accelerated rate. If the diode is connected to a source of alternating current, only the positive half of the cycle can cause electrons to move toward the plate; therefore, the current that flows closely resembles DC, and the AC has been *rectified.* Since the electronic device generally receives its electrical energy from an AC source, the diode is implanted in the circuit to convert the AC to DC since DC is required by the balance of the components in the electronic circuit.

The *triode,* a vacuum tube with an additional component, serves another function in the circuit. The third part permits the flow of electrons between the filament and plate to be regulated and is known as the *grid.* The triode, illustrated in Figure 8.18, can adjust the flow of electrical energy throughout the

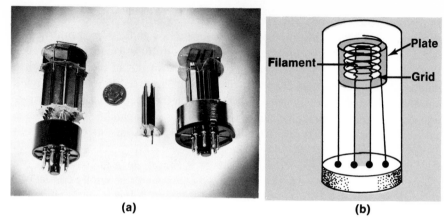

Fig. 8.18 The triode contains a third component part called the grid. The stick diagram reveals the typical arrangement of the three major components.

(a)

(b)

electrical circuit by repelling or attracting the electrons emitted from the filament. If the grid is positively charged, the electrons are attracted through it at an increased velocity in their route to the more positive plate; thus the triode is an *amplifier* as it magnifies the amount of electrical energy furnished the circuit. The voltage on the grid can be reversed rapidly by other electronic components and thus causes the triode to generate an alternating current of high frequency in which case it is known as an *oscillator*. The manner in which an individual tunes to a given radio station or TV channel involves his ability to adjust the frequency of the signal received by the set to the frequency of the signal the broadcasting station transmits.

The many disadvantages of vacuum tubes include high cost, large and bulky construction, high amount of operating current, limited life, and high operating temperature. However, the vacuum tube is ideal in many situations where electrical energy is readily available, where heat is no problem, and where simple replacement procedures are involved.

A more recent invention, the *transistor,* has replaced the vacuum tube in many situations. The transistor operates on a minimum amount of electrical energy, emits very little heat, and has a long life expectancy. The transistor is tiny compared to the size of the vacuum tube of equivalent energy output. This feature has permitted an electronic circuit to be so small that the electronic technician uses magnifying lenses to aid him in his circuit construction and repair.

Transistors function due to an unusual property of various elements. Two of the elements which have been associated with the development of transistors are silicon and germanium. The nature of the materials places them in a category between good electrical conductors and poor conductors or insulators. The *semiconductor,* as this class is called, is pressed against a metal plate or base. This pair of substances permits electrical energy to flow through them, but in only one direction. The transistor, then, can replace the diode in that an alternating current would have half of its cycle stopped and thus be rectified. The amount of electric current which flows through a circuit can be regulated by a transistor. Figure 8.19(a) illustrates the components of the transistor. The germanium "crystal" is attached at its base to a metal plate. The top surface is connected at two points of contact near each other to two connecting conductors known as leads. The current which flows through the contact point, referred to as the *emitter*, controls the amount of current which flows through the other contact point or the *collector*. A small difference in current in the emitter produces a much greater difference in the current available from the collector. The transistor serves as an amplifier, a rectifier, and an oscillator, and therefore has replaced the vacuum tube in many electronic circuits. Figure 8.19(b) illustrates two transistors and the typical vacuum tube each replaces in electronic circuitry.

Later research has revealed a device which has the potential of revolutionizing the field of transistorized circuitry. A Japanese investigator, Leo Esaki, is given credit for uncovering a simple semiconductor class of crystal that is even more amazing than the transistor. The *tunnel diode,* as it is called, operates on a negligible amount of electric current and is some few thousandths of an inch in diameter. With such an addition to the size of the transistor in common use, man will see telephones that are to be worn on wrists, pocket-size TV sets, ring radios, and many inventions not yet conceived.

Electronics serves as a means for transmission and reception of signals via radiated electromagnetic waves—radio waves. The radiations are produced by an alternating current of very high frequency and are transmitted in all directions. Sounds are converted into low-frequency electrical impulses (AC) by the *microphone* and then are magnified by the action of the *amplifier*. The *oscillator,* the third component of the circuit, generates high-frequency alternating current that is blended with the electrical impulse produced by the microphone in yet another component called the *mixer*. The resultant amplitude modulated (AM) signal is again amplified and is finally sent through the an-

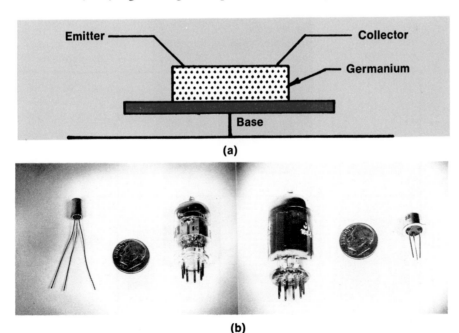

Fig. 8.19
(a) The components of a typical transistor as illustrated by a stick diagram.

(b) Transistors have replaced electronic tubes in many instances. Typical replacements are shown in the two illustrations.

applied electricity and electronics   **181**

tenna from which radiated waves are transmitted in all directions. The receiving set or radio is "tuned" to a transmitting station through the selection of the desired frequency from all those frequencies that reach the receiving antenna. The amplitude-modulated alternating current chosen has the sound waves extracted from it in the detector, the component which converts the carrier frequency AC into DC and discards it. The variations in the electrical impulse produced by the original sound are transmitted through the circuitry and are converted back into sound at the loudspeaker. Amplitude modulation has a disadvantage in that noises produced by storms, electrical appliances, and other disturbances are often found to accompany the carrier signal and appear as "static" to the listener. For this reason, frequency modulation (FM) has become increasingly popular. With FM, the frequency of the transmitter is varied by the sound from the microphone rather than the amplitude. Amplitude modulated and frequency modulated waves are illustrated in Figure 8.20. The FM receiver then converts the frequency differences back into sound much the same as AM waves are converted by AM receivers.

Not long after the transmission of sound over long distances was successfully accomplished, a concentrated effort was underway for a practical way to transmit light. Television, as the acceptable technique is called, converts light into electricity, transmits it through the air, and converts the electrical signal back into light. Light is converted into electricity by a phenomenon called *photoelectric effect*, the property present in some materials that causes them to emit electrons when their surfaces are struck with light. The television camera uses a rapidly moving beam of electrons which helps to divide the scene under

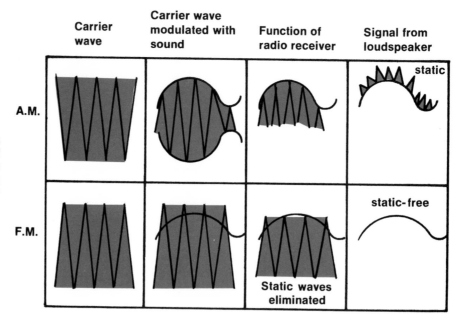

Fig. 8.20 **A comparison of amplitude modulation with frequency modulation.**

applied electricity and electronics

view into thousands of light and dark areas. The areas are converted into electrical impulses whose strength varies with the brightness of the area. These electrical signals are sent via radiated waves to the TV set, which reconverts the signals into areas of varying brightness and presents them on the television screen. The areas on the screen correspond exactly with the areas in the original scene and reproduce the picture as it was viewed by the camera. Only one area is transmitted at the same instant; however, all areas of the picture are received in less than $\frac{1}{30}$ second and appear to illuminate the screen simultaneously.

Several recent developments in the field of electronics have led to better ways to communicate efficiently over long distances. The *maser* (Microwave Amplification by Stimulated Emission of Radiation) and the *laser* (Light Amplification by Stimulated Emission of Radiation) permit man to beam extremely coherent and concentrated energy with almost no loss of intensity over great distances. Both devices are discussed in Chapter 9.

CHAPTER SUMMARY

Current, the flow of electric charge, moves in a circuit from the negative pole of the source of potential to its positive pole. *Ohm*'s law describes the relationship of current to electric potential and to resistance in a circuit. The current, measured in amperes, is equal to the electric potential, measured in volts, divided by the resistance, measured in *ohms*. The joule is a unit of work in electricity. The power consumed or provided by a circuit is measured in watts (joules/sec). According to Joule's law, $H = 0.24 \, I^2Rt$. The heat produced per unit of electrical energy consumed is measured in *calories* per unit time. The electrical power consumed by homes and industry is measured in kilowatt-hours.

Electrical circuits are constructed either in *parallel,* in *series,* or in combinations of parallel and series wiring. Batteries are composed of a series of cells connected so that the voltages of the cells are additive. House circuits are primarily wired in parallel; however, fuses are placed in series with lights and receptacles within a circuit. Sources of electric potential and resistances may be wired in parallel or in series.

Batteries produce *direct* current. Alternating current is produced by AC generators. Devices that increase or decrease the voltage of the AC produced by a generator are called *transformers.* Transformers are composed of at least two coils, the primary coil and the secondary coil. Step-down transformers decrease voltages, and step-up transformers increase voltages. The current that flows through the secondary coil of the transformer is inversely related to the change in primary voltage. Many household appliances contain transformers that provide proper voltages for the necessary electrical circuitry.

Electrical energy used in our homes is often converted into heat and light. The incandescent lamp and fluorescent lamp provide artificial lighting. Each type of lamp has advantages over the other type; therefore, most homes contain both types of lamps. Heat produced from electrical energy often heats our homes and our water, controls room temperature, and cooks our food. Household appliances convert electrical energy into *mechanical* energy. The conversion is accomplished by a motor, a device that contains a coil of wire that rotates in a magnetic field. Alternating current changes the direction of the magnetic field, and this changing field causes the armature to rotate by the law of repulsion and attraction of magnetic fields.

Electronics involves the use of electricity in radio, television, and other means of communication. The vacuum tube, a device which transforms electric potential or that changes AC to DC, is a major component in our communication devices. The *transistor,* however, has replaced the vacuum tube to some degree. Light is converted into electricity by the *photoelectric effect,* a phenomenon in which some materials emit electrons when exposed to light. Many electronic instruments operate on the principle of the photoelectric effect.

QUESTIONS AND PROBLEMS

1. What are the advantages of using AC rather than DC in our homes? (p.169)
2. Cite several examples of both series circuits and parallel circuits which are commonly used in our homes. (p.165)
3. Draw a diagram which shows the manner in which to connect dry cells to provide 7.5 volts. Draw a diagram which indicates the proper manner in which to connect eight 1.5-volt flashlight bulbs to a 12-volt storage battery. (p.165)
4. What electrical measurements are necessary to determine the electrical power in a circuit? Cite several examples. (p.163)
5. List the advantages of fluorescent lighting over incandescent lighting as either is used in a home. What are the disadvantages of fluorescent lighting in the home? (p.174)
6. List the advantages you can conceive in using transistors rather than electronic vacuum tubes in radios and television sets. (p.180)
7. How do radio waves differ from sound waves? How do audio and video signals arrive at our television sets simultaneously? (p.181)
8. Our bodies are damaged by electrical shock which results from the amount of current that flows through our body. The warning signs placed near sources of electricity read "Danger-High Voltage." Should the signs read "Danger-High Current"? Defend your answer. (p.158)
9. Why must the armature of an electric motor always be composed of a soft iron core rather than one of hardened steel? (p.178)
10. The transmission of electricity between cities is done at increased voltages (perhaps 10,000 volts) rather than at ordinary operating voltages. Considering the effect this voltage change would have on the current that flows, what might be the reason for this voltage adjustment? (p.170)

CHART of ELECTROMAGNETIC RADIATION

COPYRIGHT 1966 THE WELCH SCIENTIFIC COMPANY 7300 N. LINDER AVENUE, SKOKIE, ILLINOIS 60076

COMMON SPEED IN A VACUUM – 2.99793 × 10⁸ meters/second

electromagnetic waves

9

THE SPECTRUM
THE BEHAVIOR OF ELECTROMAGNETIC WAVES
REFLECTION, DIFFUSION, AND SCATTERING
REFRACTION AND DISPERSION
EFFECTS OF REFRACTION
POLARIZATION AND DIFFRACTION
SPECIAL APPLICATIONS OF ELECTROMAGNETIC RADIATION
A LOOK AT RELATIVITY

vocabulary

Critical Angle	Diffraction
Dispersion	Hologram
Normal	Real Image
Reflection	Refraction
Relativity	Spectrum
Virtual Image	

The discovery of electromagnetic waves is considered one of the most outstanding and major breakthroughs in the field of physics. The abstract idea was proposed in 1856 by James Clark Maxwell, a brilliant theoretical physicist from England. He suggested that electromagnetic waves, which consist of alternating electric and magnetic fields, do exist and that they travel through various mediums at the velocity of light, which, by the way, is also an electromagnetic wave.

Maxwell's hypothesis was proven in 1887 by Heinrich Hertz, a German experimental physicist. Hertz actually produced electromagnetic waves in his laboratory and confirmed Maxwell's predictions as to their properties.

The production of electromagnetic waves is possible in several ways. The various names which have been given to the waves refer to the method of their generation and detection, such as radio waves, infrared waves, and X-rays. Waves generated by oscillations in electric circuits are called radio waves. The usual source of infrared waves is a hot radiating object and X-rays are produced by the creation of disturbances in the energy level of orbiting electrons. There are other electromagnetic waves, yet all have methods of production which closely resemble those of one of these three basic waves. Generally speaking, most electromagnetic waves are produced by one common method, a fact which explains the accepted name for the entire spectrum.

Ordinarily, air is a very poor conductor for electricity, but when the potential difference between two wires is extremely large and the wires are relatively close to each other, a spark, which is a short burst of electricity, can span the gap. An induction coil, illustrated in Figure 9.1, produces a series of visible sparks which jump back and forth or oscillate between the terminals of the

Fig. 9.1 **The induction coil provides intermittent high voltages from a DC source of low voltage. The device contains a primary coil, a secondary coil, and an interrupter.**

electromagnetic waves

secondary coil. If a loop of wire such as that illustrated in Figure 9.2 is placed close to the induction coil, a spark can be observed to jump between terminals of the loop each time that a spark jumps across the terminals of the induction coil. Hertz correctly surmised that as the spark oscillates across the space between the terminals of the induction coil there is produced a magnetic field as well as an electric field. Both the electric and the magnetic fields change direction simultaneously. These direction changes, as proposed by Maxwell, are transmitted through space as electromagnetic waves. The frequency of the waves produced is identical to the number of oscillations per unit time of the spark. Therefore, waves of all frequencies can potentially be produced by this induction method.

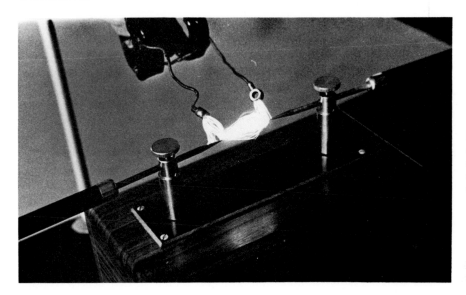

Fig. 9.2 The induction coil, along with the coil of wire, is a source of electromagnetic waves.

THE SPECTRUM

As was discussed in Chapter 6, a spring stretched between two rigid posts then compressed and released experiences simple harmonic motion. The frequency of the vibrating spring is the number of vibrations the spring completes per unit time, usually the second. For a series of waves, the frequency is the number of waves that pass a given point per second. Each wave has a crest and a trough, also discussed in Chapter 6. The distance between crests or other identical points on successive waves may be measured to determine the wave length of the vibrating body.

Electromagnetic waves also vibrate; hence the number of pulses of energy per second indicates their frequencies. Each type of electromagnetic wave travels at the same velocity in a vacuum. Each wave differs from the others in its frequency and wave length. The frequency of each electromagnetic wave which travels through space is the same as the frequency of the oscillating electric charge that generated it. If the electric charge oscillates between 4.3×10^{14} vps and 7×10^{14} vps, the electromagnetic wave which results is generally in the sensitivity region of the retina of the human eye. The wave length, since all electromagnetic waves travel at 186,000 miles per second (3×10^8 meters per second), is readily determined from the relation $v = f\lambda$ and in concept is similar to like measures made of sound energy. Thus an electric charge which oscillates once per second has a wave length of 186,000 miles whereas one that oscillates with a frequency of 1000 vibrations per second has a wave length of 186 miles. Other frequencies which we use quite often, such as those for AM and FM radio, have wave lengths ranging from 614 to 1785 feet and from nine to eleven feet respectively. Visible light ranges in wave length between fourteen and twenty-eight millionths of an inch in contrast to radio waves. Frequencies of several thousand cycles per second (kilocycles/sec or kilohertz) are employed as radio waves, and those of about fifty million cycles per second (50 megacycles/sec or megahertz) are used as the VHF television band. Other electromagnetic waves are presented in Figure 9.3.

In order to reach the high frequencies used for radio waves (550 kilohertz to 1500 kilohertz), an alternate method of production of electromagnetic waves had to be developed since the induction coil could not reverse the electric field produced rapidly enough to provide a major portion of the total energy spectrum. A small crystal was found to reverse the direction of the electric field more rapidly. Various crystals vibrate naturally at a constant and high frequency directly related to their respective size and composition much the same as a bell or chime has its own innate frequency. Each radio station has its own wafer of crystalline quartz about the size of a nickel which constantly vibrates at the assigned frequency of the station. Smaller crystals are designed and employed to produce the transmitting signal of FM radio and television stations. Other crystals, microscopic in size, vibrate at frequencies that are known as infrared or heat waves. The observation can be made, therefore, that heat waves are produced by vibrating atoms and molecules which carry an electric charge. Frequencies emitted in the range of visible light, 4.3–7.0×10^{14} vps, are obtained through the vibration of particles even smaller than the atom. These particles which must carry an electric charge are known as electrons, the orbiting particle of the atom. If the electron is caused to vibrate above 7×10^{14} vps. the wave produced may be the ultraviolet ray or the X-ray, depending on the exact frequency which results. Above the X-ray range lies the gamma ray region. The gamma ray is produced in the atom's nucleus through a disturbance of the nucleus' electric field. As was previously discussed, all electromagnetic waves travel at the same velocity, 186,000 miles per second in a vacuum, and differ in frequency and hence wave length. As the frequency of the wave increases, its wave length decreases since the velocity of the wave is constant (recall, $v = f\lambda$). Visible light frequencies travel through water at a velocity of about 140,000 miles per second and through glass at about 124,000 miles per second. In no known medium does light travel faster than in a vacuum. Several properties of electromagnetic waves are extremely interesting. First, as visible

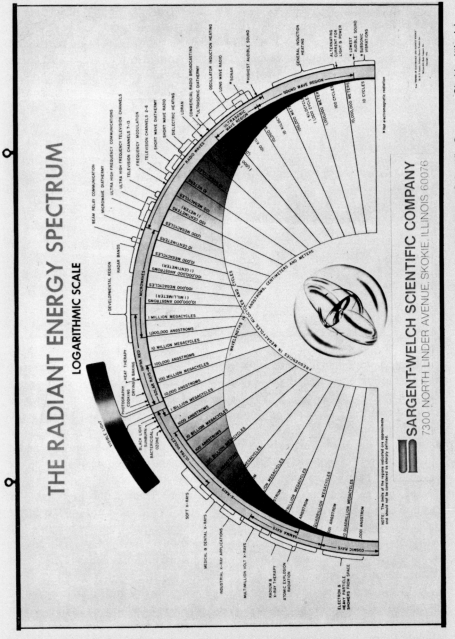

(Courtesy Sargent-Welch Scientific Company, Skokie, Illinois)

Fig. 9.3 **The electromagnetic wave spectrum includes the entire range of wave lengths of electromagnetic radiation which extends from gamma rays to alternating current used for power in our homes. Normally, cosmic rays and waves produced by vibrating matter are excluded from the spectrum but are included in the illustration for continuity. The latter wave lengths are marked with an asterisk.**

light travels through glass, it slows down instantly from 186,000 miles per second to 124,000 miles per second. But the light regains its original velocity as it leaves the glass; that is, the light immediately reassumes a velocity of 186,000 miles per second, a feature which causes the emitted ray to be parallel to the incident ray (see Fig. 9.4). Also, electromagnetic waves differ in their ability to

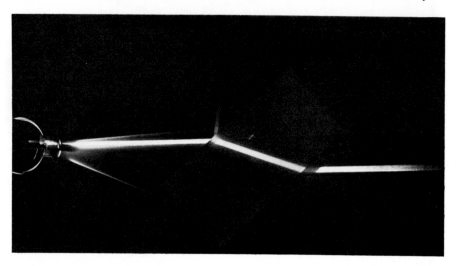

Fig. 9.4   The light ray which exits from a rectangular, transparent medium is parallel to the incident light ray.

penetrate matter. High frequency waves, such as radio waves, can penetrate the air but are absorbed by thin layers of metals. Low frequency radio waves are reflected from the ionosphere, a layer of our atmosphere composed of charged particles, whereas high frequency radio waves pass through the ionosphere into outer space. Radio reception is better at low frequencies at night than during the day because the ionosphere becomes more concentrated and is thus able to reflect radio waves back to Earth. Television frequencies, which are higher than the radio waves, penetrate the ionosphere and hence are not reflected. The TV frequencies (about 10^8 vps) can be received only in straight line distances from the transmitting station. Because of the curvature of Earth, the taller a transmitting antenna is, the greater is the distance from the antenna that a signal can be received. This relationship is expressed mathematically as $d = 1.23\sqrt{h}$ where d is the distance of transmission in miles and h is the height of the antenna in feet.

In addition, infrared or heat waves can penetrate dry air but are stopped by humid air. The human body also absorbs waves in the infrared region, a fact which accounts for the sun's being able to warm our bodies. The infrared rays, slightly longer than the red end of the visible band, also affect special photographic film, illustrated in Figure 9.5.

Electromagnetic waves shorter in wave length than the violet end of the visible band are unusual in many ways. Ultraviolet radiation (UV), like visible light, can cause photochemical reactions in which radiant energy is converted into chemical energy. UV produces ozone in the upper atmosphere and produces melanin in the human skin. The production of melanin results in a tanning effect of the skin: a "suntan." UV, however, cannot penetrate glass as do visible light and X-rays. X-rays penetrate more dense materials than do any electromagnetic rays of lesser frequencies; however, some of the X-rays are absorbed by matter. The latter point explains how man has learned to shield himself from the damaging effects of high-frequency radiation, including X-rays, gamma rays, and cosmic rays.

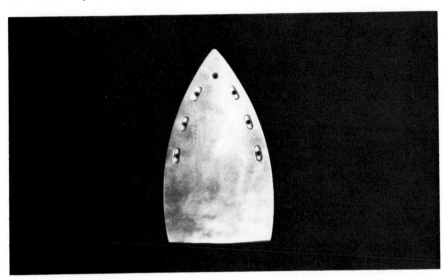

Fig. 9.5 Infrared photography permits photographing hot objects in total darkness as is illustrated by this photograph of a hot iron bottom.

THE BEHAVIOR OF ELECTROMAGNETIC WAVES

There are many properties that man has found useful in his application of electromagnetic waves. Some of the various properties apply for *all* frequencies, not only for visible light. The radio waves are reflected just as are visible light. Other waves are bent as they enter various media in the same manner that visible light is bent as it is transmitted from air into water or glass. The bending of electromagnetic waves is attributed to the change in velocity of the waves as they travel through media of differing densities. Most study of electromagnetic wave properties is done with frequencies within the visible region of the spectrum because this region is the most familiar to man.

In the study of electricity, the electron with an abnormally high velocity was presented as the responsible source which furnished the necessary energy to the atom to cause fluorescence. The atom can also be excited through its absorption of light energy. Ultraviolet light can deliver more energy than visible light, a factor which is the reason that many substances become excited when they are exposed to ultraviolet light and other high frequencies. Many such substances so excited emit visible light as they become de-excited. This action is a common method by which fluorescence is produced and the principle behind the action of substances colored with fluorescent paints. Detergents which

electromagnetic waves

supposedly clean clothes whiter than other detergents often contain fluorescent substances which cause the materials to absorb the ultraviolet light from the sun. The clothes then reflect more blue light than normal, and this characteristic causes them to appear whiter than ordinary to the observer.

Some substances emit certain frequencies of electromagnetic waves which are in the visible part of the spectrum. For example, the sodium vapor lamp emits yellow light, and the mercury vapor lamp emits various frequencies which appear blue-white. Neon lights appear red, and argon lights appear purple. Other substances do not emit light, yet they appear to have color because they reflect light rather than emit it. But, what is color? The scientist considers color a physiological phenomenon that depends on the frequency of light emitted or reflected by a substance. We perceive color by virtue of variances in frequency, the lowest visible frequency being red, the highest violet, and the multitude of colors and hues comes from frequencies in between. Some substances can reflect only that wave length which gives the object or material that certain color seen by an observer. For example, a cloth which appears red in sunlight appears red in red light. A blue cloth, as observed in sunlight, appears blue in blue light as well. On the other hand, the red cloth appears black in the blue light and the blue cloth appears black in the red light.

Other colors of light are reflected from the surfaces of the materials and still additional colors are absorbed. This phenomenon is the major reason that objects appear to be one color in the sunlight but other colors under incandescent lamps, mercury vapor lamps, or fluorescent lamps. Each source differs in the range and intensity of visible wave lengths that it provides. For example, the incandescent lamp provides greater intensity in the red region of the visible spectrum; thus red objects are more red in appearance in incandescent lighting than in fluorescent lighting. Fluorescent lighting, by the same token, enhances blue objects.

The visible part of the electromagnetic spectrum is that portion to which the human eye is sensitive, as is illustrated on Plate 1. The range is expressed in terms of *micrometers,* one millionth of a meter; in terms of *nanometers,* one billionth of a meter; or in terms of *microns,* one ten-billionth of a meter. Our range of sight encompasses wave lengths between approximately 400 nanometers and 700 nanometers (4000 to 7000 microns or 4×10^5 to 7×10^5 micrometers). The range of wave lengths and frequencies along with their assigned colors and hues appears in Figure 9.6.

Fig. 9.6 Wavelengths of visible light vary from 400 nanometers to 700 nanometers (4×10^{-5} to 7×10^{-5} cm).

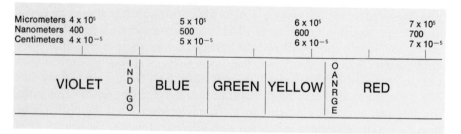

Micrometers 4×10^5			5×10^5		6×10^5		7×10^5
Nanometers 400			500		600		700
Centimeters 4×10^{-5}			5×10^{-5}		6×10^{-5}		7×10^{-5}

VIOLET | INDIGO | BLUE | GREEN | YELLOW | ORANGE | RED

electromagnetic waves

Ultraviolet radiation (200–400 nanometers) and infrared radiation (700–1200 nanometers) are included as portions of the spectrum which are normally associated with our light sources, though they are not visible to the eye.

Sunlight and some other sources of white light include practically all wave lengths between 400–700 nanometers. But another source of white light may be produced by mixing or superimposing red with green light, or even violet with orange light. Other combinations appear on Plate 2 as do the three colors which produce white light and thus are known as the primary colors—red, blue, and green. The various colors of light other than the primaries are obtained by varying the amount of each primary color used.

Elementary school students are aware that a mixture of blue pigment with yellow pigment produces the color green and that a mixture of red and yellow pigments yields the color orange. And, unlike the three colored lights of red, green, and blue which produce white light, the pigments of these colors produce the color brown when mixed together. The process of mixing pigments produces the opposite result from that of adding colored lights to produce white light. That is, the mixing of pigments is a *subtractive* process whereby the observer sees the color which is reflected, all other colors being absorbed. An object which appears blue has absorbed all visible wave lengths that struck it with the exception of those in the blue region. Closer examination of known wave lengths which strike the surface of the object reveals that colors on both sides of blue in the visible spectrum—green and violet—are also reflected. However, the colors red, orange, and yellow are totally absorbed by the object.

A piece of transparent colored glass, such as that viewed from inside a church, appears to have a certain color because of the wave lengths it transmits whereas all other wave lengths which form the white sunlight are absorbed. A piece of yellow glass transmits wave lengths in the yellow region and absorbs other wave lengths. The typical glass window used in our homes is transparent because it transmits practically all wave lengths in the white sunlight and thus appears colorless.

REFLECTION, DIFFUSION, AND SCATTERING

When electromagnetic waves strike a surface, some of them are reflected. A transparent medium reflects some few waves which strike it at an angle other than one perpendicular to its surface. A metallic surface reflects practically all wave lengths which strike it. For both types of surfaces, the angle which the incident beam of light makes with an imaginary line perpendicular to the surface is equal to the angle the reflected ray makes with the same perpendicular line. The angle of incidence, then, equals the angle of reflection, as is illustrated in Figure 9.7.

A surface which we view as shiny, such as one provided by a coat of high-gloss enamel on a wall, is very smooth. A pair of shoes which has a well-buffed surface is extremely smooth too and, like the enameled wall, reflects light exceedingly well. A rough surface such as that of suede shoes reflects little of the

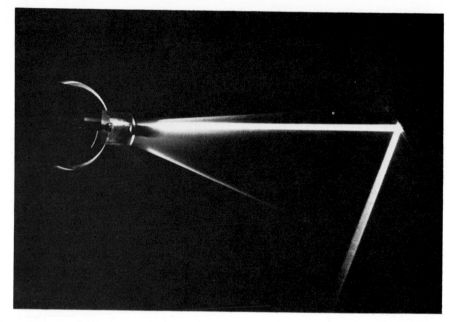

incident light. That light which is reflected is done so at angles other than that of the incident rays; therefore, the object does not have a high gloss. (see Fig. 9.8.) A "flat" enamel, then, causes the surface of the wall upon which it is applied to assume a rough, uneven surface and causes the incident ray to be either absorbed or reflected at an unusual angle. The light is said to be *diffused* in various directions. A newspaper is easy to read since the source of illumination of the paper is not reflected into our eyes and is, therefore, comfortable in terms of the slight amount of glare it yields.

Another similar phenomenon, *scattering,* occurs when a molecule absorbs the electromagnetic wave and releases it in another direction. The higher frequencies are more readily scattered than are the lower frequencies. The sky appears blue because the molecules of air in our atmosphere absorb the sun's wave lengths in the blue region and scatter them toward Earth, a feature which makes them visible to us. Without this scattering the sky would appear black and the clouds white. The thicker clouds, such as those which are responsible for bringing thunderstorms, appear much bluer or even black in contrast to the surrounding sky because of scattering of wave lengths in the blue region from the spectrum of electromagnetic waves.

REFRACTION AND DISPERSION

Sir Isaac Newton was extremely interested not only in the science of mechanics, for which he is well known, but also in the properties of light. Once he caused a narrow beam of white light to be transmitted through various shapes of glass, all of which had sides that were not parallel to each other. Some shapes that he tried include the sphere, the triangle, and the trapezoid. The results of his efforts have become quite famous and useful, but the spectrum of color that he produced from "white" light has been most outstanding. The most common display of such a spectrum found in nature is that which is produced by raindrops as the sunlight passes through them and the light is bent toward our view.

As a spectrum is produced by a prism such as is illustrated in Figure 9.9 (a) and (b), all the colors of the "rainbow" are formed, starting with red and changing through orange, yellow, green, blue, indigo, and finally violet. Newton recognized that white light is a mixture of all colors and that the prism simply serves as a mechanism to separate the colors from each other.

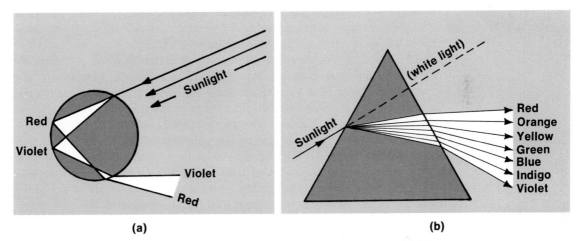

(a) **(b)**

Fig. 9.9 (a) A spectrum is formed by incident rays on raindrops, a condition that creates the rainbow. (b) A triangular-shaped transparent medium causes refraction of an incident beam of light at both surfaces. The amount of dispersion so created is sufficient to create a spectrum of color from the incident white light.

However, Newton did not realize the complexity of light, nor did Roemer, who successfully measured its velocity in the late seventeenth century. Roemer found that light was not instantaneous but traveled in a vacuum, such as that generally found in outer space, at a velocity of about 186,300 miles per second. This velocity has come to be known as the constant velocity of light, c. Since the work of Roemer, further investigation has revealed that the velocity of light varies in other media. That is, the velocity of light in water is $0.75c$; and in glass, about $0.67c$. Table 9.1 includes the velocity of light in other media.

The velocity of light is really dependent on the wave length of light being transmitted through the medium, with the exception of a vacuum in which all wave lengths travel at c. For example, the velocity of violet light in crown glass is 121,131 miles per second (195,057 km/sec), and red light is 122,565 miles per second (197,367 km/sec) in the same glass.

electromagnetic waves **197**

TABLE 9.1

Comparative velocity of light in various media

Medium	Velocity (miles/sec)	Velocity (km/sec)
Vacuum	186,300	300,000
Air	186,244	299,909
Diamond	76,983	123,966
Glass (light brown)	123,377	198,675
Glass (dense flint)	108,947	175,438
Ice	142,214	229,008

The strangest behavior of light, when it is compared to other types of energy known to man, has to do with its transmission through various media. A bullet, for example, slows down as it penetrates a block of wood due to its energy loss as it collides with the particles of wood in its path. As it emerges from the wood, it does so at a lesser velocity than that at which it entered the block. Light enters a second medium from air at velocity *c*, slows down as it seemingly penetrates the second medium, but, unlike the bullet, resumes its original velocity of *c* as it emerges. This behavior is explained most thoroughly if light is viewed as individual *photons* or bundles of energy rather than a solid beam as light is generally represented. As a photon collides, thus interacts, with the particles which constitute the second medium, probably the atom and its orbital electrons, the similarity of light with other electromagnetic waves is more readily recognized. The photon interacts with one or more of the orbital electrons and causes them to oscillate in the same manner as does the electric spark in an induction coil. The photon which strikes the surface of clear glass is absorbed by an atom and causes an electron of that atom to oscillate at a frequency equal to that of the photon. This vibration in turn causes the emission of a second photon of the same frequency as the incident photon. The second photon also travels at *c* until it is absorbed by another atom. The absorption causes the electrons to oscillate and a third photon to be released at the velocity of *c*. The process continues until a photon reaches the surface, where it continues through the original medium, air, at velocity *c*. All of the interactions and resultant photons released require time to occur and reveal why light travels at a velocity less than that of *c* in any medium other than a vacuum (or air, for practical purposes). Therefore, the photon of light energy which emerges from the second medium is not the same photon which entered it initially but one which originated from inside the medium.

The slowing down of light as it passes obliquely from one medium to another, such as from air into glass, causes, the light to bend as it passes from air into a transparent medium such as glass or water. When the entrance into the second medium is abrupt, the bending is abrupt. As the ray penetrates the second medium and passes through it, the ray is again abruptly refracted until it is parallel to its original path. When the change of medium is gradual, such as from

electromagnetic waves

warm to cool air or from cool to warm air, the bending is also gradual, a condition which creates mirages. When light is incident on the surface of transparent glass perpendicular or *normal* to the surface, that is, when the ray strikes the glass at an angle of 90° with the surface, no refraction occurs. If the light ray strikes the glass at an angle other than the normal, refraction becomes evident. Figure 9.10 represents rays of light incident on the surface of glass at various angles, their penetration, and their paths as they emerge. Note that some rays do not emerge but are reflected along the surface of the glass The angle which produces this effect is known as the *critical angle* and varies with the types of media involved.

Fig. 9.10 **As an incident ray of light strikes the surface of glass at the critical angle, the resultant ray is neither refracted nor totally reflected.**

The critical angle for diamonds is indeed small when its value is compared to that of glass and other seemingly transparent media. The critical angle for glass and water is about 43° and 48°, respectively. An incident ray of light which strikes the surface of a diamond at an angle greater than the critical angle, about 23°, is reflected completely. The incident ray is often reflected several times from the various faces of a diamond, a condition which seemingly causes the diamond to sparkle. This feature becomes more evident as the jeweler cuts the diamond's surfaces at proper angles to take full advantage of this property.

EFFECTS OF REFRACTION

There are many common instances of refraction as light passes from one medium through another. For example, consider a straight rod immersed in a tank of water. The side view, illustrated in Figure 9.11(a), reveals that as the rod is permitted to rest against the top of the tank, it appears to be bent or broken. A light ray, pictured in Figure 9.11(b), is bent as it passes through the

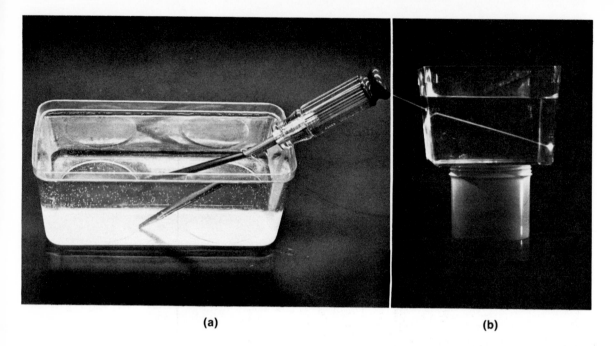

(a)   (b)

Fig. 9.11   (a) Refraction creates many phenomena when objects are viewed in two media in which the velocity of light differs in each. (b) Light rays incident on the surface of water other than perpendicular to the surface or at the critical angle are refracted until they can exit from the water and the transparent container.

tank at an angle other than the normal or perpendicular to the surface. Secondly, refraction is evident in air of significantly different temperature. As light passes from cold air above a chimney which is emitting the hot air and gases from a wood fire, the air appears to shimmer. The same phenomenon is observable above a hot highway on a summer day. The refracting of light as it penetrates through layers of air of varying temperatures causes mirages such as pools or lakes. On the highway the road often appears wet ahead of the auto, but the wetness disappears as the auto approaches it. As light travels from the cool air through the warmer air near the highway, the sunlight is bent upward into the observer's eye before it comes in contact with the hot pavement. The effect of this refracted light is similar to the effect of light which is reflected from water and thus fools the eye. Other mirages are caused by the refraction from the surface of a lake or river of light which is then refracted downward to the view of the observer. The observer sees the surface of the water but at an appreciable distance from where it actually is located.

A lens serves to aid the eye through the application of the property of refraction. This transparent material, usually glass, is the vital part of many optical devices such as eyeglasses, cameras, microscopes, telescopes, and projectors. Lenses are constructed in many different shapes and sizes ranging from that which is smaller than the lens in the human eye to that which is used in the gigantic telescopes, over sixteen feet in diameter.

The basic shape of lenses falls into two categories: convex and concave. the *convex* lens, illustrated in Figure 9.12(a), is thicker at its center than at its edges and thus causes parallel light rays striking it near the center to travel a greater distance in the glass than those rays nearer the edges. The convex lens then converges the rays to a point from whence they intersect the imaginary center of the lens. Figure 9.12(b) represents a *concave* lens which is thinner in the center than at the edges and diverges light rays. The concave lens, then, is opposite the convex lens in the manner in which it bends parallel rays of light. The rays that pass through the lens intersect at an imaginary point on the opposite side of the lens from that where they are diverged. Both types of lenses are used in optical instruments since man finds it necessary to converge and to diverge light to suit his many needs. He makes greater use of the convex lens since he is more interested in producing an image which is magnified and thus easily viewed. The size of the image and the distance the image is produced from the lens are directly related to the distance an object is located from the convex lens as is illustrated in Figure 9.13. The distance from the lens at which the parallel rays converge to a point (the focal point) is called the focal length (F).

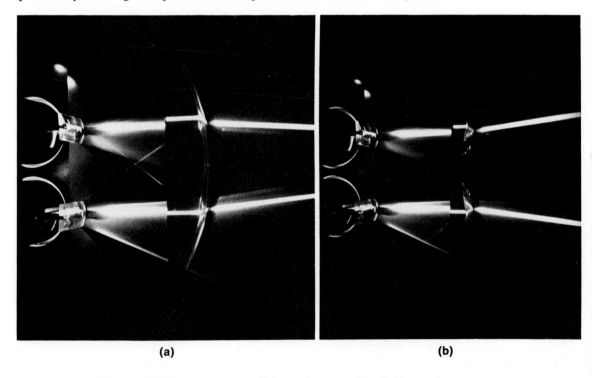

(a)   (b)

Fig. 9.12 (a) The convex lens converges light and causes the light rays to intersect. (b) The concave lens diverges light; that is, the parallel light rays which enter it are bent away from each other.

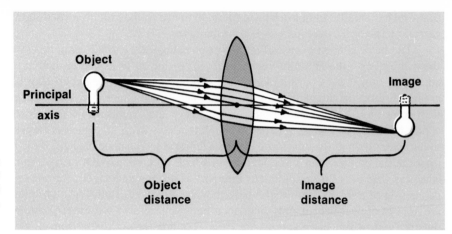

Fig. 9.13 **The image formed by a convex lens is always inverted.**

The focal point is the same distance from the center of the lens in each case. The value would differ if the curvature of the lens were different on either side of the lens. The more abrupt the curvature of the lens, the shorter is its focal length, illustrated in Figure 9.14.

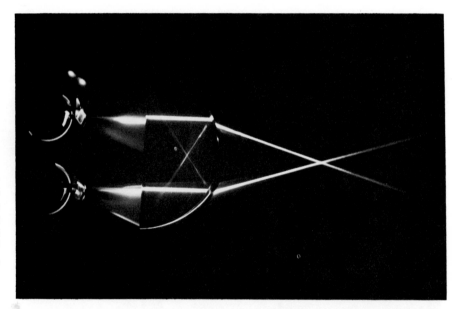

Fig. 9.14 **A convex lens refracts parallel light rays commensurate with the degree of curvature of the lens.**

The location and size of an image formed by a lens can be determined by simple construction as is illustrated in Figure 9.15. The drawing must be completed to scale, and some care must be used to be certain that the construction lines are straight and that they pass through the designated points as closely as possible. The object in the illustration is placed between the focal length and $2F$. An imaginary ray of light is shown being reflected from the tip of the arrow which serves as the object **parallel to the** principal axis, a line which passes through the center of the lens **as a viewer** would use it. The light ray is bent or refracted as it enters the lens, approaches the center, and continues through the lens. As it exists, the ray travels on a straight path through the point which terminates the length of the focal length, F, and continues on endlessly. A second ray, starting at the same point on the tip of the arrow as the first ray, is projected with no refraction involved through the point which marks the intersection of the two axes, the principal axis and a second axis perpendicular to the first about which one could also spin the lens as if it were a top. The intersection, in reality, is located at the center of gravity of the lens. The ray continues until it intersects the first ray, thus determining the location of the image. The distance the lens is located from the "center" of the lens is measured and the size is also readily determined, since both measures are converted to actual size and distance according to the original scale chosen. The construction also reveals that the image is inverted and therefore is deemed "real," thus capable of being projected. Regardless of the position of the object or its size, construction permits the location and the size of the image to be determined. Some diagrams of image positions reveal how the optical instruments make use of their various locations. Note the effect produced by the distance that the object is placed from the center of the lens in regard to the size and the location of the image. Of course, this observation explains why we must view large objects from great distances and how images of objects too small to be seen by the naked eye can be multiplied to such an extent that they become readily visible to us.

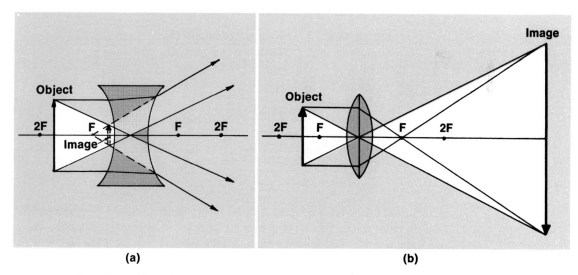

(a)   **(b)**

Fig. 9.15 **The size and position of the image formed by a convex lens can be determined by construction as can the image produced by a concave lens.**

Also, the location of the image and its size can be determined mathematically with little difficulty.

Example 9.1

A 6-inch ruler is placed 24 inches from a lens whose focal length is 8 inches. Determine (a) the location and (b) the size of the image.

Solution:

(a) Let D_o represent the distance the object is from the lens; D_i represent the distance the image is formed from the lens; and F represent the focal length. Then, the lens formula can be expressed so that the reciprocal sum of D_o and D_i is equal to F, or

$$\frac{1}{D_o} + \frac{1}{D_i} = \frac{1}{F} = \text{Substituting}$$

$$\frac{1}{24 \text{ in}} + \frac{1}{D_i} = \frac{1}{8 \text{ in}}$$

$$\frac{1}{D_i} = \frac{1}{8} - \frac{1}{24}$$

$$\frac{1}{D_i} = \frac{3}{24} - \frac{1}{24}$$

$$\frac{1}{D_i} = \frac{2}{24} \quad \text{Cross-multiplying,}$$

$$2\,D_i = 24$$

$$D_i = 12 \text{ in.}$$

(b) The size of the image is found from the relation:

$$\frac{\text{Size of image, } S_i}{\text{Size of object, } S_o} = \frac{\text{Distance of image, } D_i}{\text{Distance of object, } D_o}$$

$$\frac{S_i}{S_o} = \frac{D_i}{D_o}$$

$$\frac{S_i}{6 \text{ in}} = \frac{12 \text{ in}}{24 \text{ in}} \quad \text{Cross-multiplying,}$$

$$24\,S_i = 12 \times 6$$

$$S_i = \frac{72}{24} = 3 \text{ in.}$$

So, the image is found to be formed twelve inches from the opposite side of the lens from where the object is located (a conclusion reached since the value of D_i is positive). The image is three inches tall and is real. If the value of D_i had been negative, as can readily occur, the image would have been upright or *virtual* and hence not capable of being projected. The virtual image is formed on the same side of the lens as the object is located. Concave lenses always produce virtual images, since the light rays which pass through it are refracted away from the focal point and thus never intersect with another ray as did the light rays which penetrated the convex lens. The concave lens serves as a means to increase the focal length of a convex lens since two or more lenses can be used in combination. Myopia or nearsightedness and hyperopia or farsightedness are corrected by the use of a lens of a given shape in combination with the lens of the eye. Figure 9.16 (a) and (b) illustrates how the concave lens and the convex lens are applied to make the lens of the eye to function normally. Astigmatism, a condition of the eye caused by an uneven curvature of its lens, is corrected by special eyeglasses designed to overcome the condition.

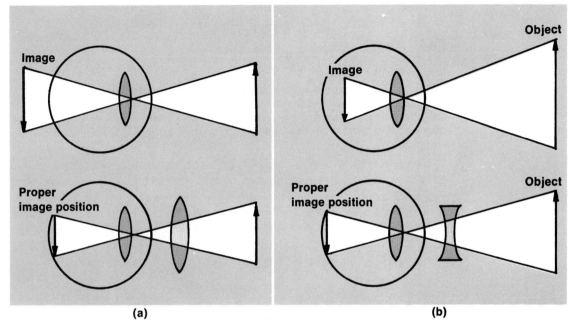

Fig. 9.16 (a) The manner in which a convex lens can correct farsighted-ness. (b) Nearsightedness is corrected by means of a concave lens.

POLARIZATION AND DIFFRACTION

The reader should recall from previous discussion that the electric field of an electromagnetic wave actually vibrates. In fact, an ordinary beam of light has its electric fields vibrating at all possible right angles to the direction of the beam. In Figure 9.17, such a beam is aimed at a crystalline substance such as tourmaline, herapathite or other polarizing crystal. The first disk, A, transmits that incident light which vibrates in the axis of the disk's crystalline alignment. Disk B is turned so that the crystalline alignment is parallel to that of disk A. As disk B is moved so as to overlap disk A, slightly less light is transmitted through the two disks. However, when disk B is oriented so that its crystalline alignment is normal (or perpendicular) to that of disk A, all light is absorbed and none is transmitted, for all practical purposes. An ideal substance will transmit fifty percent of the non-polarized light and absorb the rest.

electromagnetic waves **205**

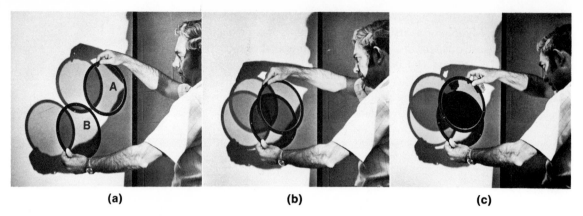

(a) (b) (c)

Fig. 9.17 Various crystalline substances have the capability to polarize light. Numerous automobile manufacturers place polarized windows in their newer models to minimize the glare from the sun and from the headlights of autos approaching them from the rear.

An analogy, shown in Figure 9.18, further demonstrates the phenomenon of light polarization. A rope, analogous to a non-polarized beam of light, can be caused to vibrate in any plane. If the rope is threaded through a picket fence, a child can cause the rope to vibrate along one axis only; thus the rope is *plane polarized*. If the rope is threaded through a second section of picket fence, the rope may vibrate as before if the sections are parallel or be prevented from vibrating altogether if the sections are perpendicular to each other.

**Fig. 9.18
Vibrations can pass through both slots when both are vertical. When the slots are turned at right angles to each other, the vibrations which pass through the first slot are stopped by the second slot.**

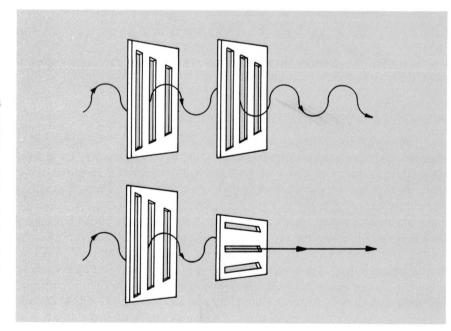

Plate 1   The part of the electro-magnetic spectrum visible to the eye contains many colors and hues.

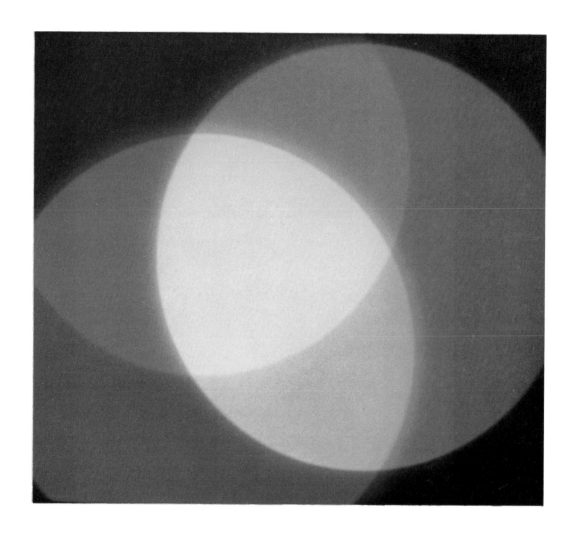

Plate 2 The color box provides evidence that the three primary colors of light are red, green, and blue-violet.

Many surfaces, including those of water and glass, have a tendency to polarize light as it is reflected from them. For this reason, Polaroid sunglasses are very effective in reducing the glare from highways, windshields, beaches, and lakes. A sportsman who enjoys fishing in clear, fast-moving streams should be aware of the benefit to him if he were to wear Polaroid sunglasses, since the depths of the stream become readily visible to him as the light is plane polarized by the glasses. His success in catching fish is enhanced tremendously if he can see them lurking in their lair.

When light is noted carefully as it passes close to the edge of any object, it is found to deviate slightly from its path. The deviation of light from its original direction due to the corners around which it is bent is called *diffraction*. This slight bending of light causes the edges of shadows produced to be blurred or diffused rather than sharp as are the sides of the object which produced the shadow. If a small source of light emanating from a pinhole in a screen (or collimated by lenses) is permitted to pass by an object and cast a shadow of the object on a screen, illustrated in Figure 9.19, the shadow is found to be bounded at the edges by narrow bands or fringes of light. The phenomenon called diffraction was explained by Newton who assumed that light is composed of small

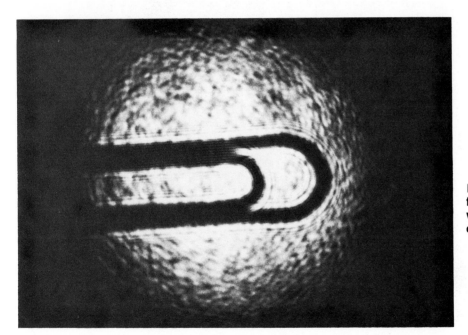

Fig. 9.19 Narrow fringe lines are visible around the object's shadow.

particles which obey the laws of mechanical motion and are radiated from a source of light at high velocities in all directions. The particle or corpuscular theory of light was replaced by the hypothesis that light was composed of many waves of extremely short wave lengths, a fact which explained such characteristics as reflection, refraction, polarization, and diffraction.

electromagnetic waves

The diffraction grating has become an optical device which partially replaces the prism in that it can also produce the spectrum and permit the measuring of the wave lengths of light. The gratings are produced by the use of a fine diamond point which rules grooves on a glass plate or a highly buffed piece of metal. The rulings on a glass plate produce a transmission grating and on a metal plate produce a reflection grating similar to the one produced by the grooves on a LP record. The transmission gratings, as illustrated in Figure 9.20(a) and a microscopic photo of a grating, as illustrated in Figure 9.20(b),

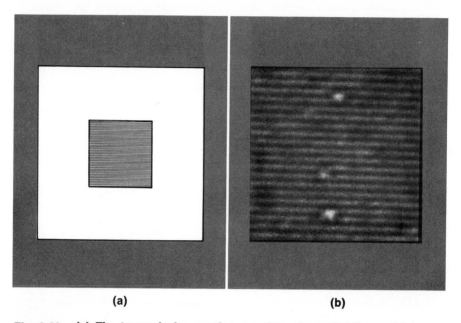

(a)   (b)

Fig. 9.20 (a) The transmission grating consists of parallel lines which produce fringe areas. (b) Microscopic views of diffraction gratings reveal that they are produced by scratching parallel lines on a glass surface or are made available through the innate characteristics of various crystalline substances.

are an inch or more wide and contain between 5,000 and 30,000 lines per inch (1970–11800 lines per cm). The grooves stop the light from penetrating the grating whereas light is transmitted through the areas between the grooves. Light from the point source is bent slightly, and all different wave lengths, corresponding to the various colors, form their own characteristic wave fronts and produce a complete and continuous spectrum. Through the use of diffraction gratings the wave lengths of visible light have been found to approach the values listed in Table 9.2. To further the reader's comprehension of how the scientist interprets the wave phenomena of light, Figure 9.21 displays various wave chains of the numerous visible colors in addition to a comparative relationship of ultraviolet rays, X-rays, infrared rays, and microwaves.

TABLE 9.2

*The wave lengths of visible light as determined
with a diffraction grating*

Color	Wave Lengths	
Red	0.000026 in.	0.000066 cm
Orange	0.000024 in.	0.000061 cm
Yellow	0.000023 in.	0.000058 cm
Green	0.000021 in.	0.000054 cm
Blue	0.000018 in.	0.000046 cm
Violet	0.000017 in.	0.000042 cm

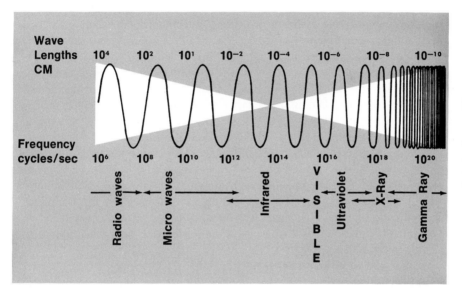

Fig. 9.21 The relative wave-lengths and frequencies of some common electromagnetic waves.

SPECIAL APPLICATIONS OF ELECTROMAGNETIC RADIATION

Man has long been able to use electromagnetic waves which he has produced to illuminate various surfaces in order to study them in closer detail. The distance that he must send these waves in order to permit closer observation of the object under study has brought about several problems. For example, if he did not have a method to collimate light or project it in the form of parallel rays such as is done by the lens of a flashlight, the light energy would be so dissipated in short order that he could not distinguish its presence on the surface of faraway objects. According to the inverse square law governing such projections of energy, the intensity of an uncollimated flashlight beam measured one foot from the source would be decreased to 1/100 the intensity at a distance

electromagnetic waves

of ten feet from the source. The same intensity which illuminates an area of one square foot at a distance of one foot from the source, then, illuminates an area of 100 square feet at a distance of ten feet from the source. Now, suppose that the object to be illuminated were twenty-five miles from the flashlight. The intensity measured one foot from the source would be decreased until each square foot illuminated on the object twenty-five miles away would receive but 1/17,424,000,000 as much intensity as that of the original measure at one foot. Few if any instruments would be capable of detecting the presence of such faint illumination on a surface twenty-five miles away due to the other sources of light which would also be reflecting from the same surface. Consider, however, how bright the moon would seem if the distance from Earth to the moon were much less. How long could man survive if the distance from the sun to Earth were halved since both heat and light energy would be tremendously increased?

Man has been successful in projecting detectable light beams great distances. A light beam was projected from Holmdel, New Jersey, and was seen by man without the aid of instrumentation at Murray Hill, some twenty-five miles away. The beam initially projected was so narrow that it only illuminated a circle of 200 feet in diameter at this distance. (Later models spread their beams less than 5 feet per mile.) Within a year after the successful experiment described above was performed, a beam was reflected from the surface of the moon and detected by instruments as it was reflected back at Earth.

The light sources which were employed to perform the remarkable feats were from two closely related devices—lasers and masers. Both instruments were invented in 1958 by scientists at Columbia University and Bell Laboratories. The term *laser*, which stands for *L*ight *A*mplification by *S*timulated *E*mission of *R*adiation, and *maser*, which stands for *M*icrowave *A*mplification by *S*timulated *E*mission of *R*adiation, actually describe how each source of radiation is produced.

Light which is given off by an electric lamp used in the home is incoherent; that is, the photons of light energy given off are present in many frequencies and phases of vibration. This feature is analogous to the frequency and phase relationship in which raindrops fall on a metal roof during a thunderstorm. Laser beams are coherent; that is, they are of one frequency and are in phase such as is a pair of railroad tracks, regardless of their deviations from a straight path. For this reason, a laser beam does not spread out appreciably over long distances as does the beam from a flashlight.

The laser, illustrated in Figure 9.22 (a) and (b), has been found to be very beneficial to man in many ways, and the list of applications is growing continuously. The path of the laser is so straight that it has been used to align long sections of pipe and telephone conduit where straightness is essential. As man developed better methods of timing events, he found that he could use the laser to study the surface of the moon to such detail that he could measure the depths of craters and the heights of lunar mountain ranges. The recent moon voyages have included experiments with lasers in which sensitive laser reflectors have been set up on the surface of the moon to permit studies of Earth's

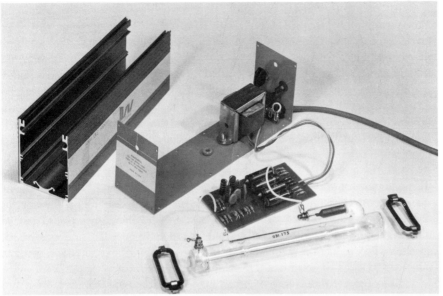

(Courtesy of Metrologic Instruments, Inc.)

(a)

(b)

Fig. 9.22
**(a) A small laser
assembly and its
container which
houses the power
supply and other
component parts
necessary to
produce the
coherent laser
light.**

**(b) Even the
powerful laser
beam is totally
reflected by the
lucite rod.**

atmospheric conditions and countless experiments concerning the moon and its distance from Earth as well as its rotational characteristics. The laser has found a place in medicine in that it maintains a pinhole-like beam which permits the device's total energy to be concentrated on a minute area. This feature permits the surgeon to fuse sensitive tissue such as the retina of the eye into place with no need for surgery.

Laser light can be used to transmit power of various types. In the case of radio signals, sound waves are transformed into electromagnetic waves then back again to sound waves as they reach their destination. Various forms of power such as electricity are transformed into laser light, sent to a given receiver, and then converted back to electrical power. Also, many more messages can be sent at the same time by laser beam than can be sent by radio waves.

electromagnetic waves   **211**

Another of the outstanding applications of the laser is its use in the making and viewing of *holograms*. A hologram is in reality a photographic plate that gives the observer a three-dimensional view of an object when the plate is illuminated, particularly with laser light. As the viewer looks obliquely into the hologram, he sees an image rather than a picture as is typically seen on a photographic plate. The three-dimensional effect is obtained in the interference pattern because each eye sees the hologram from a slightly different angle. As the observer shifts position slightly, he sees other parts of the image not visible at the first position, including around corners. The laser beam which produces the hologram is separated so that only part of it illuminates the object to be photographed. As this part of the beam is reflected from the object it recombines with the major part of the beam. This recombining of the original beam produces an interference pattern much the same as is seen on the surface of an oil film floating on water or as is present in multi-thicknesses of plate glass. As a laser beam or other light source is projected on the surface of the plate, the image is reconstructed. The hologram is invaluable in both mapping and photography and may lead to three-dimensional television in the near future.

One of the earliest types of lasers made use of a ruby cylinder about two inches long which was impregnated with chromium. Green light, when introduced into the ruby, excited the chromium atoms. The atoms "cooled off" by the emission of photons of red light, which in turn excited other atoms. This stimulation accounts for the production of a coherent beam of red light which is reflected internally and eventually exits through the more transparent end of the reflecting container. (See Fig. 9.23.) The more modern lasers make use of various inert gases and are more economical, intense, and efficient than was the early ruby laser. Note its application to drilling of small holes (square, triangular, or round) as is illustrated in Figure 9.24 (a) and (b).

Fig. 9.23 Reflected in the safety goggles of a researcher, this gas laser is being used to investigate various applications in industry.

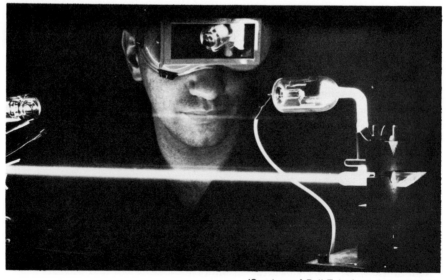

(Courtesy of Bell Telephone Laboratories)

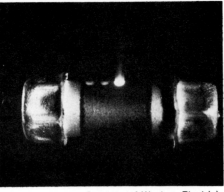

(Courtesy of Western Electric)

(a)

(Courtesy of Western Electric)

(b)

Fig. 9.24
(a) The laser is used to drill holes of various shapes and sizes in materials which are difficult to drill conventionally. The laser beam is piercing a hole in a diamond die. (b) A resistor is "trimmed" to a specified resistance by a laser beam.

Recently, Bell Telephone Laboratories received a patent for a carbon dioxide laser which can be used to burn holes through steel plates and to weld ceramic materials together—reportedly fifty times more powerful than other gas lasers. The CO_2 laser operates at a wave length of about ten microns (0.000039 in) and thus is in the invisible range of the infrared. With this frequency range, several applications are possible which are not with the other forms of gas lasers, including satellite to satellite communication and ground to ground communication such as has been successfully done using satellite Echo II (see Fig. 9.25).

Fig. 9.25
A researcher operates a gas flow laser which uses a mixture of carbon dioxide and nitrogen.

(Courtesy of Bell Telephone Laboratories)

electromagnetic waves

The existence of an X-ray laser in the future may not be impossible. The X-ray laser would permit holograms that would give three-dimensional pictures of internal objects, including man, crystals, and biological macromolecules. In addition, the X-ray laser would give the scientist a long-sought source of great energy in a concentrated form.

The *maser* is a device in which molecules emit energy because they are stimulated from an external source. The function of the maser is twofold: to energize the atoms and to stimulate their radiation. Light is emitted from atoms just as are microwaves; therefore, the laser is in reality an optical maser. The maser was found useful in microwave communication, but was discovered to be quite limited in this application due to the narrowness of the band of frequencies over which it would operate. The science of communication has progressed to such a degree that it is both possible and desirable to be able to send many telephone conversations, TV programs, and other communications over the same radio carrier simultaneously, thus minimizing the various costs involved.

The maser is very beneficial in the field of radio astronomy. With this device man has been able to determine the rotational properties of planets whose surfaces are not optically visible. He has been able to apply his observations to the Doppler Shift as it is related to astronomy, an application which has led to a greater comprehension of the universe.

An apparatus which has partially replaced the maser in its role as a communications device is the cooled parametric amplifier, abbreviated *paramp*. General Telephone and Electronics has successfully maintained ground stations for satellite communication throughout the world. In addition, the groundwork has been initiated for a satellite communication system which will provide worldwide coverage of telephone service along with live television broadcasting.

Another useful application of electromagnetic waves in the microwave region, that is, waves whose wave lengths are in the order of inches, is that of radar, *RA*dio *D*etection *A*nd *R*anging. The microwaves appear to travel in more nearly straight lines than do the longer radiowaves and are readily reflected by objects such as airplanes, mountains and the like. The direction of the beam is capable of being accurately measured as is the time which lapses between the instant the wave is transmitted and the instant it is returned through reflection. Thus an object can be tracked with extreme accuracy. A typical image of a transmitted signal and its return is represented in Figure 9.26 as might be seen on a radar screen. The application of microwaves in this sense has permitted air flights to become our safest mode of travel.

Microwaves are electromagnetic waves the same as are AM and FM radio, VHF and UHF, infrared, visible light, ultraviolet, X-rays, and gamma rays. We hear of radiation and identify it with X-rays and gamma rays, both potentially hazardous to living organisms. Color TV sets which lack proper shielding emit unwanted low-level X-radiation toward the observer and can damage biological tissue just as can other sources of X-rays. Therefore, government regulations set limits on permissive radiation levels to which TV manufacturers must comply. The frequency range of electromagnetic waves is capable of ionizing tissue and

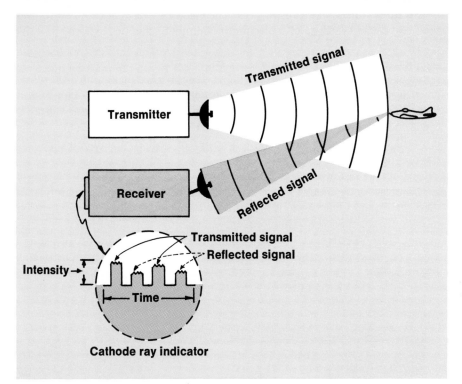

Fig. 9.26 The radar unit has served man well since its initiation in World War II. Airlines rely on the intricate system to establish flying conditions and to prevent midair collisions.

its effect is both cumulative and irreversible. Microwave radiation is not capable of ionizing the molecules of biological tissue because this frequency range, like infrared and radio waves, supposedly lacks sufficient energy. Research reveals that damage done to biological tissue by microwaves is in the form of thermal effects; however, governmental and independent researchers are continuing to study potential hazards from devices which emit microwaves, particularly to those patients who wear heart pacemakers.

Up to this point the character of microwave energy has been studied as to the manner it has been reflected, such as by metallic surfaces or other finishes which reflect light as do metals. Also, the study has included the way that electromagnetic waves, which encompass the microwave region, are transmitted through air, glass, and plastics. A third character involves the manner in which microwave energy is absorbed by matter, which in turn becomes thermally hot.

Some time ago a scientist for Raytheon Corporation discovered that his hand became extremely warm as he held it in a beam of microwave energy emitted from a radar device he was testing. His curiosity caused him to pursue this discovery and he is reported to have placed a bag of popping corn in the path of the beam. He found that the corn popped in seconds and the paper bag did not burn. This discovery led to the development of the microwave oven found in the modern home. The oven is an ingenious device that receives its energy from electrical energy (230 volts) which is converted into high frequency waves

electromagnetic waves

by electronic tubes and transformers. The energy travels back and forth in the food and causes a frenzy of molecular motion that generates extreme heat. The heat is, therefore, spread through the food, a feature which results in short cooking time. For example, hamburgers are cooked in one minute, bacon and baked potatoes in four minutes. The cooking is done in glass baking dishes, plastics or even paper plates. (Recall that metals reflect microwaves, so conventional pots and pans cannot be used effectively.) Also, the air inside and the walls of the range do not increase in temperature due to the microwave energy. The container is warmed slightly, but only by the transfer of heat by conduction from the cooked food itself.

Research has revealed that microwaves, at a frequency of 915 megacycles, are quite energetic enough for cooking; therefore this frequency, slightly above the UHF band of television, is assigned by the Federal Communications Commission for this purpose. This frequency, since it is contained inside the oven, cannot cause interference with television transmission or radio broadcasting as might be concluded. One of the popular microwave ovens is illustrated in Figure 9.27.

**Fig. 9.27
The microwave oven is rapidly finding its way into the modern home.**

Microwaves are still involved in contemporary research and are new enough in known character for one to conjecture that other uses will be developed for this range of electromagnetic radiation.

electromagnetic waves

A LOOK AT RELATIVITY

A mention of the term "relativity" brings to mind the name Albert Einstein. When he made public his special and his general theory of relativity and ultimately his field theory, the vast maze of mathematical equations and new concepts overwhelmed all men with their complexities. The theories proved extremely valuable in creating a new source of energy to aid man and eventually led to a new era—the Atomic Age.

Relativity is divisible into two categories. One part is called *special,* and the other *general.* The special theory, developed by Einstein near the turn of the twentieth century, has been proven by experimentation and is much easier to comprehend than the general theory that he developed in 1915, the latter of which deals with motions of bodies in accelerated frames of reference.

In any case, interpretations of Einstein's theories led to two amazing concepts. First, the observable length of an object is affected by its motion. Lorentz, a Dutch physicist hypothesized that the length of a moving object is changed according to the relation:

$$l = l_o \sqrt{1 - \frac{v^2}{c^2}}$$

where v is the velocity of the object through the air, l_o is its length at rest, and c is the velocity of light. For example, if a meter stick were to move lengthwise with 0.75 the velocity of light ($0.75c$), its observable length would be decreased to about 0.67 that of its length when at rest. Also, when $v = c$, the length of the meter stick is zero. The velocity of light, then, essentially is the upper limit for the velocity of any physical material. The mass of a moving object is also affected, but in a completely different manner. The relativistic correction for mass is given by the equation:

$$m = \frac{m_o}{\sqrt{1 - \frac{v^2}{c^2}}}$$

where m_o is the mass of an object at rest as the observer sees it and m is the mass of the object as it moves at velocity v. At $0.1c$, the mass of the object is only changed 0.5% of its mass at $v = 0$. At $0.5c$, the relativistic correction reveals an increase in rest mass of 15%. Some relativistic corrections for length and mass appear in Table 9.3.

Sound, heat, and light are various forms of energy with which man is familiar. Mass, according to Einstein's special theory, is also a form of energy. The relationship of mass to energy is given by the equation $E = m_o c^2$, where m_o represents the mass at rest and c is the velocity of light. The conversion of mass to energy has caused many drastic reviews of man's understanding of the laws of nature and has become man's most potent source of energy. The equation has been deemed valid through countless experiments that involve atomic nuclei. Its applications are limitless and no doubt will lead to the deletion of many of our present sources of energy as research continues.

TABLE 9.3

Comparison of relativistic mass and relativistic length at various velocities

Velocity ratio v/c in per cent	Velocity (miles/sec)	Velocity (km/sec)	Relativistic Mass (m/m_o)	Relativistic Length (l/l_o)
1	1,860	2,995	1.000	1.000
10	18,600	29,952	1.005	0.995
50	93,000	149,758	1.15	0.870
86	159,960	257,585	2.00	0.5
90	167,400	269,565	2.3	0.435
94	174,840	281,546	3.0	0.333
96	178,560	287,536	4.0	0.250
99.5	185,070	298.019	10.0	0.100
99.9	185,814	299,217	22.3	0.045
100	186,000	300,000	Infinity	0.000

CHAPTER SUMMARY

Electromagnetic waves are produced by a change in direction of electric and magnetic fields. All electromagnetic waves travel at approximately the same velocity through a medium, such as air. They may differ in frequency and in wave length. Electromagnetic waves are usually identified according to the manner in which they are generated.

Radio waves, visible light, and all other electromagnetic waves have common properties. All such waves may be reflected, absorbed, or transmitted by matter. Electromagnetic radiation is visible if its wave length is between 400 nanometers and 700 nanometers. White light is produced by the presence of practically all wave lengths in the visible band. Other combinations such as the primary colors, red, blue, and green, produce white light. Light is an additive process whereas the mixing of pigments is a subtractive process.

A substance that transmits light readily is considered *transparent;* however, if the waves strike the substance at an angle other than perpendicular, some waves are reflected. The angle of *reflection* equals the angle of *incidence*. Most waves are reflected from metals. If the surface is not smooth, the light is diffused. Some substances scatter an incident light beam, that is, they absorb it and release the wave at a lower frequency and in another direction. The amount of light reflected from an object, then, depends on several features: (a) the object's composition, (b) the object's surface, and (c) the object's position with respect to the incident beam.

Refraction is the bending of electromagnetic waves as they pass obliquely from one medium into another of different optical density. The velocity with which a given electromagnetic wave travels through a medium (other than a

electromagnetic waves

vacuum) depends on its wave length. Violet light travels more slowly than does red light in the same transparent medium. As light of any wave length passes from the second medium back into the original medium, the velocity of the light assumes its original velocity. That particular angle of incidence at which the refracted ray makes an angle of 90° with the normal is called the *critical* angle. Any angle of incidence greater than the critical angle produces a totally reflected ray.

Refraction of visible light occurs as it passes through air of different temperature, a phenomenon that creates mirages. Glass is a more highly refractive medium than is air; therefore, incident rays which pass through the glass are bent away from the normal. Lenses are constructed from glass and other transparent substances of similar optical density. *Convex* lenses are thicker at the center than at the edges. This type of lens converges parallel light rays to a point called a focal point. Images formed beyond this point are magnified and are real, that is, inverted and capable of being projected. *Concave* lenses bend parallel light rays away from each other and form virtual images, those not capable of being projected. Combinations of these two types of lenses permit correction of eye disorders and construction of such optical instruments as cameras, microscopes and telescopes.

The electric field of electromagnetic waves vibrates in many directions, all at right angles to the direction that the beam is traveling. Some crystalline substances have the inherent capability to permit light to vibrate in one plane only; thus the light is *plane polarized*. Polaroid sunglasses reduce glare through the polarization of light. Light is slightly deviated from its straight-line path as it passes through a narrow opening or between fine rulings in glass or other transparent media. The bending of light rays with the resultant spreading of the transmitted waves is known as *diffraction*. Diffraction gratings are designed to spread out the light rays and thus produce spectral colors.

Light emitted from a point source can be collimated by lenses and/or by mirrors. This special application of light has led to the development of *lasers* and *masers*. Light from a common incandescent lamp is incoherent; that is, the light emitted is composed of waves of many frequencies and of many phases of vibration. The electromagnetic waves from a laser are of the same frequency, are coherent, and in phase. Therefore, the laser light does not spread appreciably over long distances. The laser beams have been bounced off the surface of the moon and returned with sufficient intensity to be detected. Lasers have been used to transmit power, radio waves, and other waves used in communication. Laser photography has become an important asset to the geologist, the explorer, and industry. The maser is useful to astronomers and, in a limited fashion, is beneficial in communication. The operation of the laser and the maser employs such phenomena as excitation, fluorescence, and phosphorescence. *Microwaves*, as the name suggests, are electromagnetic waves of very short wave lengths. When microwaves are absorbed by matter, the matter becomes thermally hot. Microwave ovens apply this principle and shorten the cooking time of foods considerably as compared to conventional cooking methods.

The special theory of *relativity* represents several postulates. The application of a portion of this theory points out that length contracts as a body moves at relativistic speeds. Mass, on the other hand, increases with speed. At the velocity of light, *c*, the length of a body is zero, yet its mass becomes infinite. The most remarkable point of Einstein's special theory of relativity is the equivalence of mass and energy. A small mass at velocity *c* corresponds to an enormous quantity of energy. Through this application of the conversion of mass to energy, man has harnessed *atomic* energy.

QUESTIONS AND PROBLEMS

1. Before purchasing a piece of clothing you consider to be your favorite color, why should you view the article in direct sunlight rather than in artificial light? (p. 194)

2. Why does the mixture of all colors of light produce white light but the mixture of all colors of paint produce black? (p. 195)

3. Why does a glass prism separate white light into various colors of light? According to the additive primaries illustrated in this chapter, what colors of light would be produced by a prism that refracted orange light? What colors would be produced by a pale green light? (p. 197)

4. The velocity of light is said to be constant. Does this statement indicate that the velocity of light is the same in all materials? Defend your response. (p. 197)

5. Why does increasing the distance between a light source and the surface it illuminates lower the intensity of illumination with the square of the distance? (p. 210)

6. A convex lens has a focal length of 2 cm. Locate by construction the image of objects placed at the following distances from the lens: (a) 3 cm (b) 10 cm (c) 5 cm. (p. 202)

7. We can distinguish the color of an object only when the object is close to us; otherwise the object appears light or dark. Why? (p. 194)

8. We can see details better when objects are brought closer to the eye. Why? (p. 201)

9. Gasoline is a clear liquid, yet a trace of it on the surface of water creates a variety of colors. How are the colors produced? Why doesn't a piece of thin glass, such as a microscope slide floating on water produce the same effect? (p. 195)

10. Why do some paints cause a painted wall to appear "flat" while other paints produce a "glossy" effect? (p. 195)

(Courtesy of Brookhaven National Laboratory)

radioactivity and nuclear power

10

THE PARTICLES AND THE RAYS
NATURAL RADIOACTIVITY
HALF-LIFE AND ACTIVITY
FUSION AND FISSION
NATURAL AND ARTIFICIAL TRANSMUTATION
OF THE ELEMENTS
NUCLEAR POWER AND ARTIFICIAL RADIOACTIVITY
IONIZING RADIATION
RADIOISOTOPES, THEIR USES AND DETECTION
NUCLEAR ENERGY AND THE FUTURE

vocabulary

Activation	Activation Analysis
Alpha Particle	Atomic Mass
Atomic Number	Beta Particle
Binding Energy	Fission
Fusion	Gamma Ray
Half-life	Health Physics
Ionizing Radiation	Isotopes
Mass Defect	Nuclear Reactor
Nucleus	Plasma
Radioactivity	Transmutation

There are few, if any, children of our modern civilized world who have not heard of the atom. They are aware that the atom is composed of an extremely small, positively charged nucleus which is surrounded by orbiting electrons. However, there is little said which concerns the size of the atom, yet such information is available.

The diameter of the atom is of the order of 10^{-10} meters, and the diameter of the nucleus is of the order of 10^{-14} meters, a comparison which indicates the nucleus occupies about one trillionth the volume of the atom. The atom, like our Solar System, then, is composed mostly of empty space. Also, as in the case of the sun and the relation of its mass to the balance of the Solar System, most of the mass of the atom is found in its nucleus. If an atom of gold, for example, were suddenly magnified to the size of a modern automobile, the nucleus of this atom would be the size of a B-B. If this atom were to weigh 2000 pounds, the nucleus would account for 1,999 pounds of its weight. As was stated previously, most of the volume of an atom is void space.

The electron which orbits the nucleus of the atom is considered indivisible and is thus referred to as an *elementary particle*. The nucleus, however, is composed of two particles which are considered fundamental in nature or elementary, commonly known as the neutron and proton. Other particles are present as well, and some of them will be discussed in this chapter. (Presently, investigators are questioning the fact that protons are elementary. The particle may be found to have structure; and, therefore, is not really elementary.)

The discovery of radioactivity in 1896 permitted man to understand the nucleus of the atom in depth, in fact, to realize that the atom had a nucleus. The original study and the interpretation of radioactivity in terms of the structure of the atom are credited to the British physicist Ernest Rutherford. Other great discoveries that occurred almost simultaneously which concern the atom include the following: Germany's Wilhelm Roentgen and his discovery of X-rays, France's Henri Becquerel and his discovery of radioactivity, and England's J. J. Thompson's proof of the existence of the electron.

During the remainder of the nineteenth century, the curiosity which concerned the fundamental nature of radioactivity guided numerous contributions to the knowledge of the new science. The element thorium was found to be radioactive. Then Madame and Pierre Curie were successful in isolating two elements, polonium and radium, both of which were found to be radioactive. Eventually, three families of elements—uranium, thorium and actinium—were found to contain most of the 70 radioactive isotopes from the list of 340 isotopes which have been found in nature.

THE PARTICLES AND THE RAYS

The term *atom* means "cannot be broken into pieces," as interpreted from the Greek language. The atom, however, can be broken into pieces via chemical processes in which the atom is *ionized*; that is, the number of electrons which orbit the nucleus can be affected through the addition or subtraction of them during chemical reactions. Also, the nucleus of the atom can be broken apart, a phenomenon which, when it occurs spontaneously, is known as *radioactivity*.

The radioactive elements emit one or more distinct types of rays—alpha (α), beta (β), or gamma (γ)—the first three letters of the Greek alphabet. Alpha rays have been found to be positively charged, beta rays are negatively charged, and gamma rays have no charge whatsoever. Each ray responds to a magnetic or electric field in such a manner that either field can be used to separate one of the rays from the others. Note the separation of each type of ray by virtue of an electric field in Figure 10.1.

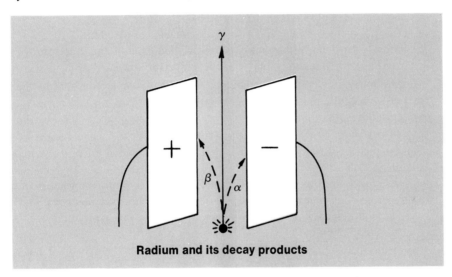

Radium and its decay products

Fig. 10.1 **An electric field permits separation of the three primary types of radiation: alpha, beta, and gamma.**

The total electrical charge of the nucleus is equal to the number of positively charged particles located in it, the sum of which is known as the atomic number, Z. This positively charged particle was discovered by Rutherford in 1920 during experiments with so-called positive rays and has been named the proton, Greek for "first" (protos). The proton has been recognized for some time as the nucleus of the hydrogen atom. A second particle, the neutrally charged neutron, is also located in the nucleus of the atom and was suggested by Rutherford in 1920 and ultimately identified by Chadwick in 1932. The neutron number, N, when added to the atomic number yields the mass number, A of the atom; that is

$$A = Z + N.$$

The electron, previously mentioned, was adapted as the unit of electricity in the latter part of the nineteenth century. The charge of the electron is equal in

radioactivity and nuclear power   **225**

amount but opposite in sign to the positively charged proton. The mass of the electron is often used as a unit of mass in measures which concern atomic particles and is valued at 9.106×10^{-28} grams. Thus, the mass of the proton is 1,836 electron masses, and that of the neutron is equivalent to 1,837 electron masses. Similarly, the charge of the electron is accepted as the unit of charge and is considered to be -1. The proton's charge, therefore, is $+1$. The neutron, as the name implies, is neutral and has a net charge of 0. The three particles, the electron, the neutron, and the proton, constitute the elementary particles of each atom. Other particles which are present in different atoms are divided into four groups because of their variations in mass. Still other particles have been identified which have identical masses but differ in other properties, perhaps in charge. The list of the number of particles found in the atom is constantly being lengthened, but some are better understood such as those briefly discussed below and arranged according to mass.

1. Leptons, particles whose mass is less than protons.
   a. neutrino—a lepton which accompanies beta decay.
   b. electrons—the particle which orbits the nucleus of the atom.
   c. muons—particles whose mass is larger than that of the electron; they travel at 99.5% c and result from nuclear disturbances.
2. Mesons, particles whose mass is greater than that of the lepton but less than that of the proton.
   a. typical mesons which differ in various properties are represented by such symbols as $+$, o, $-$, K^+, K^o, K^-, and others.
3. Baryons, particles found in the atom that are very massive in comparison to all others.
   a. nucleons—particles found in the nucleus, such as protons and neutrons.
   b. hyperons—particles with masses greater than those of neutrons which have characteristics different enough that there are six which can be separately identified.

The closely compacted nucleus, then, is primarily composed of positively charged protons and neutrons with no net charge. The repulsive force between protons is a strong one indeed and should cause the nucleus to burst apart. The nucleus is so tightly bound that there is no other mass on Earth as dense. Estimates indicate that one cubic centimeter of mass composed of nothing but nuclei would approach 10^{14} grams (100 trillion grams) whereas a like volume of water or gold would have a mass of one gram and approximately twenty grams respectively. Particles in the nucleus are bound together by a force many times stronger than any other known force in existence—the nuclear force. Little is known about this strange attractive force, but it more than counters the repulsive electrical force in some atoms; therefore, nature contains numerous stable atoms of the various elements. Several properties of this nuclear force are known, however, one of which reveals an important limitation. Electrical forces decrease with the square of the distance between the charged particles; nuclear forces decrease significantly more with distance. For instance, if the distance between two charged particles is doubled, the electrical force is decreased to one-fourth its original magnitude. Nuclear forces decrease to less than one-sixtieth the original value when the distance is doubled between particles, whether neutrons or protons. In the lighter elements the nucleons are closely packed, and electrical forces are readily overcome by the larger nuclear forces. Thus the nuclei of the lighter elements are generally stable. Exceptions to this statement will be pointed out as the discussion proceeds. In the heavier elements, the distance between protons on opposite sides of the large nucleus is great enough that the electrical forces are larger than the nuclear forces; thus the nucleus of heavier elements is less stable.

The neutrons help to stabilize the nucleus since the force which they exert is attractional (gravitational) only. This attractional force is aided by the attractional nuclear force between protons and between protons and neutrons. The total attractional force which exists in the nucleus, then, balances somewhat the electrical repulsive force which exists between the protons. The heavier elements require a greater number of neutrons than protons to assure stability. For instance, $^{238}_{92}$U has 92 protons and 146 neutrons in its nucleus and still is unstable (radioactive). A lesser number of neutrons in the uranium atom's nucleus lowers the balance of forces to the point that the nucleus splits apart since the repulsive forces are significantly larger than the attractive force.

There seems to be a limit as to the maximum number of protons a nucleus can contain regardless of the number of neutrons present. All nuclei which contain 84 or more protons are unstable. That is, from $Z = 84$ to $Z = 105$, there are no stable nuclei; all are radioactive. So that one may summarize the topic at hand, generalizations regarding stability in terms of Z number and N number follow.

Generally speaking, various combinations of Z number and N number produce stable nuclei whereas other combinations lack stability. The Z/N ratio is equal to or approximately equal to unity for the light elements. In the case of helium-4, lithium-6, boron-10, carbon-12, nitrogen-14, oxygen-16; $Z = N$, and the ratio of Z/N is 1.0 or unity. Above calcium-40, where $Z = 20$ and $N = 20$, no stable nuclei exist with equal numbers of protons and neutrons. As Z increases, the repulsive electrical forces of the positive nuclei become much greater than the attractive forces in the nucleus. The loss of stability of acting forces creates the potential for particle emission; hence a radioactive nucleus is created. The increase of additional neutrons in the nucleus serves to increase the distance between protons and thus to decrease the amount of repulsive force which tends to rip the nucleus apart. The stable nuclei of the heavier elements reveal the N number equal to about 1.5 times the value of the Z number.

The elements then are radioactive due to their emission of mass and energy in an attempt to reach stability. The most important types of radiation, of course, are alpha, beta, and gamma radiation.

Alpha radiation is actually a stream of alpha particles. The alpha particle is identical with the nucleus of the helium atom most abundant in nature. That is, the particle is an atom of helium with the two orbital electrons missing. A quick look at the periodic chart reveals the helium atom has a Z number of 2 and a mass number, A, of 4. Therefore, the helium nucleus is composed of 2 protons and A–Z or 2 neutrons. As the alpha particle (α) loses velocity and thus energy, it plummets through the atmosphere and attracts two free electrons that ultimately are drawn toward the particle and orbit it in a manner identical to that of other orbital electrons.

Beta radiation is in reality a flow of high speed electrons which are emitted from the nucleus of an atom. They are of two types: the negative electron or negatron (e^-) and the positive electron or positron (e^+). The masses of the negatron and positron are equal, but both particles are opposite in charge. The positron is in effect the antiparticle of the negatron. As the negatron loses energy in its trip through space, it has the capability of becoming the orbital electron for an atom which has been positively ionized previously. The positron generally collides with matter and often comes in contact with a negatron with a result that both are totally annihilated, that is, converted into energy.

If a nucleus is too rich in N number, it is unstable. Stability is reached through the conversion of a neutron into a proton, an electron, and an antineutrino within the nucleus:

$$n \rightarrow p^+ + e^- + \bar{\nu} \text{ (anti-neutrino)}.$$

The proton, p^+, which remains in the nucleus, adds to the Z number one unit of positive charge whereas the neutron number, N, is reduced one unit. The A number remains constant, but the element changes to the next higher element on the periodic chart in all physical and chemical properties. The electron, e^-, is emitted from the nucleus and is commonly referred to as beta radiation. The anti-neutrino, $\bar{\nu}$, represents unaccountable energy.

If a nucleus is too rich in Z number, stability is reached through the reaction:

$$p^+ \rightarrow n + e^+ + \nu \text{ (neutrino)}.$$

The proton is converted into a neutron and the positron is ejected. Additional energy loss which is unaccounted for is attributed to the neutrino.

One should realize that the above discussion should not be taken to mean that a neutron is composed of a proton and an electron nor that a proton is made up of a neutron and a positive electron. The neutron disintegrates into a proton and an electron in the same manner that a tree can be reduced to toothpicks. The tree, however, is not composed of toothpicks.

Gamma radiation is a stream of photons from electromagnetic radiation, generally higher in frequency than the X-ray. The photon is particulate in the manner in which it acts as a result of collisions with matter, but it has no rest mass. Therefore, this bundle of energy is basically different from alpha and beta radiation. The velocity of the gamma radiation, like the balance of the electromagnetic spectrum, is a constant, c, 186,000 miles per second (3.0×10^{10}

cm/sec or 3.0×10^8 m/sec). X-rays differ from gamma rays in that X-radiation is electromagnetic radiation released by energy disturbances of orbital electrons whereas gamma radiation is energy that dissipates from disturbances in the nucleus. The frequencies may overlap, and, therefore, both types of radiation are indistinguishable but for their respective origin. There are numerous examples of this slight difference between waves of various energies which comprise the electromagnetic spectrum. The amount of energy which is released as gamma, X-ray, radio, television, ultraviolet, or other types of electromagnetic radiation is directly related to the frequency of the oscillating object which produces it. This phenomenon, discovered by Max Planck in 1900, can be expressed as

$$E = nhf$$

where E represents energy, n is a small whole number which involves the quantum state (a factor beyond the scope of this book but necessarily mentioned for clarity), and f represents the frequency of the radiation. The factor h is a constant and therefore does not change. The only way then that energy from an electromagnetic wave can be increased, since n and h do not change, is by increasing the frequency of the wave. The energy unit of *all* types of radiation, including alpha, beta, and gamma, is the *electron volt* (*e.V.*), the energy an electron would acquire in accelerating through a potential difference of 1 volt.

Some years ago, Louis deBroglie suggested that since light and other electromagnetic radiation seemed to have properties causing it to act like waves as well as like particles (photons or quanta), there was logic in the conclusion that the elementary particles have associated wave length. The hypothesis was strengthened by increased knowledge of atomic structure, and resultant experimental measurements led to the collection of ample data to afford proof of the theory. For instance, an electron which is accelerated through a potential of 100 volts, and so has an energy gain of 100 *e.V.*, reaches a velocity of 5.9×10^8 cm/sec with a frequency of 4.84×10^{16} vibrations per second. An alpha particle with energy of 100 *e.V.*, however, reaches only a velocity of 6.9×10^6 cm/sec and a frequency of 4.79×10^{16} vibrations per second. An alpha particle of energy equal to 1,000,000 *e.V.* (1 *MeV*) has a velocity of 6.9×10^8 cm/sec and a frequency of 4.79×10^{20} vibrations per second, a figure which is representative of frequencies of electromagnetic waves which are classed as X- and gamma radiation.

About fifty years ago, scientists became aware that atomic nuclei contained vast stores of energy. The ability to detect the number of protons and neutrons an atom had in its nucleus in conjunction with the development of the mass spectrograph led investigators to a startling conclusion. The total weight of the individual nucleons, the protons and neutrons in the nucleus, amounted to more than the actual weight of the nucleus. With the exception of the hydrogen nucleus where A = 1, all nuclei weigh *less* than the sum of the particle weights. On the modern atomic scales the weight of various particles expressed in atomic mass units (a.m.u.) is presented in Table 10.1.

radioactivity and nuclear power

TABLE 10.1

*Comparative weights of various particles
in Atomic Mass Units*

Particle	Atomic Mass Units
alpha	4.001506
electron	0.000548
neutron	1.008665
proton	1.007277

Let us look at the *mass defect* in the oxygen atom whose A number is 16. The chart of the Nuclides, published by Knolls Atomic Power Laboratory, lists the atomic mass of $^{16}_{8}O$ as 15.00401. The Z number and N number are both equal to 8. The mass defect is revealed by the following calculation:

$$\text{Mass defect} = 8(1.007277) + 8(1.008665) + 8(0.000548) - 15.99491$$
$$= 0.137010 \text{ atomic mass units (a.m.u.)}.$$

The energy in electron volts for each atomic mass unit has been calculated from the relationship of mass to energy and found to be

$$1 \text{ a.m.u.} = 931.48 \text{ MeV}.$$

Therefore, the amount of mass of the $^{16}_{8}O$ atom converted into energy is 0.137010 a.m.u. $\times$ 931.48 = 127.62 MeV. This enormous amount of energy is known as the *binding energy* and is, in effect, the energy with which the particles are held in the nucleus.

NATURAL RADIOACTIVITY

The growth of knowledge in new frontiers of science is usually a chain reaction, since one great discovery generally leads to numerous others. For example, Roentgen discovered the X-ray in 1895; then Becquerel discovered radioactivity the following year. Becquerel revealed that radiations come from uranium salts which create silhouettes of the individual crystals on photographic plates. He also found that uranium compounds can cause electrically charged bodies to discharge, a discovery which led to a quantitative method of measure of radiation and the development of the electroscope and the ionization chamber. Later Rutherford used the discoveries of Becquerel to study the penetrating power of radiation from the uranium salts and found that the radiation consisted of two types: alpha rays, which are readily absorbed by matter, and a more penetrating type, the beta rays.

Through the ionization method Pierre and Marie Curie showed that the activity of the uranium salt is directly related to the mass of the uranium in the salt. Through this effort, they demonstrated the atomic nature of radioactivity. They also discovered two other naturally occurring radioactive elements, radium and polonium, for which they were awarded the Nobel prize. Further research by the Curies proved that radium has over a million times the activity as the same mass of uranium.

Many more radioactive elements which occur in nature have since been discovered. From the list of about 2000 isotopes that have been identified, approximately 1700 have been found to be radioactive, about 50 of which were created by nature. The exploits of man have artificially created the remainder of the known radioactive isotopes. The reader should note that there are approximately 300 stable isotopes in Earth's interior and her atmosphere, about 250 of which are available in a concentrated amount.

HALF-LIFE AND ACTIVITY

The nucleons, which comprise the nucleus of an atom, contain large amounts of energy as a result of electrical and nuclear forces. Some of the nuclei have excessive energy, a feature which causes them to exist in an excited state. If the amount of excess energy is high, alpha particles are ejected. Excited nuclei with insufficient energy to expel an alpha particle may reach stability through beta or gamma emission.

For any given nucleus, there is no way known to predict when the particle or energy will be emitted. The nucleus may disintegrate in the next second or at some future time. Yet in a large mass of the isotope, a constant rate of disintegration can be determined. This rate is commonly expressed in terms of *activity,* the number of disintegrations per second. So that activity could be expressed in a more beneficial manner, radium-226 was chosen as the standard. Close examination revealed that one gram of radium-226 emits 3.7×10^{10} alpha particles per second. The standard was called the *curie* (Ci) in honor of the discoverers of radium. A curie of iodine-131 has a mass of about one-millionth gram, and a curie of uranium-238 has a mass of 1000 kilograms. So that the unit can be adapted to any radioactive isotope, the unit now represents 3.7×10^{10} disintegrations per second of any given nuclei and therefore encompasses all types of radiation.

The rate of disintegration of a given mass of a radioactive isotope is expressed in yet another term—*half-life.* Half-life is the time required for a given mass of an isotope to disintegrate to half its original activity. Half-lives of the naturally occurring isotopes are in thousands or millions of years by necessity. That is, a shorter-lived isotope would have decayed to activities well below detection long ago. Those isotopes produced by man have much shorter half-lives that vary from a few thousand years to millionths of seconds. Some radioactive isotopes and their respective half-lives appear in Table 10.2.

TABLE 10.2
Half-lives of some commonly used radioisotopes

Element	Atomic Number (Z)	Mass Number (A)	Particle (s) Emitted	Half-life
Calcium	20	45	Beta	163 days
Carbon	6	14	Beta	5,730 years
Chlorine	17	36	Beta	300,000 years
Cobalt	27	60	Beta, Gamma	5.24 years
Hydrogen	1	3	Beta	12.26 years
Indium	49	114	Beta	72 seconds
Iodine	53	131	Beta, Gamma	8.05 days
Phosphorus	15	32	Beta	14.3 days
Potassium	19	40	Beta, Gamma	1,300,000,000 years
Uranium	92	238	Alpha, Gamma	4,510,000,000 years

FUSION AND FISSION

The work of early twentieth century scientists showed generally if two light nuclei combine to form a heavier one, the new nucleus weighs *less* than the sum of the original ones, an observation which holds true for elements lighter than iron or nickel. If two nuclei which are heavier than iron are fused together to form the heavy elements, the new nucleus weighs *more* than the sum of the two nuclei which formed it. According to this latter condition, if heavy nuclei are divided into parts, energy is released and the sum of the fragments' weight would be less than that of the original nucleus. Thus a given amount of matter is lost—converted to energy. The application of this source of energy is *atomic energy*. Combining nuclei and splitting nuclei are more commonly known as *fusion* and *fission*, respectively.

Fusion generally involves the combination of light nuclei to form a heavier nucleus. For fusion to occur, the various nuclei must be brought into contact with each other by overcoming the repulsive forces between protons. The mechanism which is commonly used involves the projection of one nucleus toward the target nucleus at a tremendous velocity. The impact of the two nuclei is such that the nuclei fuse and release tremendous amounts of energy.

Typical examples of fusion involve some of the mechanisms by which the sun "burns":

$$^2H + {}^2H \rightarrow {}^3He + {}^1n + 3.25 \text{ MeV}$$

$$\text{or } {}^2H + {}^2H \rightarrow {}^3H + {}^1p + 4.0 \text{ MeV}$$

$$\text{or } {}^2H + {}^2H \rightarrow {}^3He + {}^1p + 18.3 \text{ MeV}$$

A gallon of ordinary water contains a total of twenty drops which are composed of hydrogen-2 (^2H) or "deuterium". The amount of energy which could be obtained in theory from the fusion of the deuterium in a gallon of water is equivalent to the energy that is released by combustion of 300 gallons of gasoline. The energy is indeed much greater than that obtainable from any conventional source. Additionally, the vast amount of water available on Earth suggests an infinite source of potential energy.

Fission is the process in which the nuclei of various large atoms split into two approximately equal nuclei. The basic materials of the fission process are generally uranium or thorium minerals, both fairly abundant. The process was discovered in 1939 by two German scientists, Otto Hahn and Fritz Strassman, as they were bombarding a uranium target with neutrons in an attempt to produce heavier elements. They found that barium, an element about half the size of uranium was produced by the neutron bombardment. Later, the process was called fission after its similarity to biological cell division.

The potential for heavy nuclei to split into equal pieces depends on the delicate balance of various forces within the nucleus. The repulsive electrical forces present there cause the nucleus to expand from the typical spherical shape into an elongated shape, similar to that of a peanut shell. The forces of attraction are overcome by the repulsive forces, and the nucleus separates with a fantastic amount of energy—over six million times that obtained from the explosion of a molecule of TNT.

Fission also generally produces smaller fragments—free neutrons. These neutrons have the potential of causing other atoms to split with the consequence that more neutrons are released to cause additional atoms to fission. The total process is then a *chain reaction* that can continue to release neutrons to split other atoms if the mass is of proper or *critical* size. If small, segregated masses of the fissionable material are suddenly brought together, great energy is released. The concept is familiar to us all and is the primary mechanism of the atomic bomb. The critical mass of the bomb used at Hiroshima was approximately the size of a baseball. Other models have used even larger masses of the fissionable uranium.

A chain reaction does not occur with naturally occurring uranium, since the particular nucleus (^{235}U or U-235) which fissions is only 0.7% abundant in all uranium. The common nucleus (^{238}U or U-238) absorbs neutrons without splitting and thus actually prevents the fission process from occurring by its presence. The separation of the two nuclei is accomplished at various and strategic plants within the continental United States as well as in several foreign countries.

NATURAL AND ARTIFICIAL TRANSMUTATION OF THE ELEMENTS

Transmutation is a process in which one element is changed into another. The process is brought about by the ejection of an alpha, beta, or other charged particle from the nucleus of the initial element. For instance, uranium, as we find it in nature, emits an alpha particle. The instant the alpha particle leaves the uranium atom's nucleus, the atom is no longer uranium; its physical and chemical properties are that of the element thorium. Note the following reaction:

$$\frac{238}{92}U \rightarrow \frac{234}{90}Th + \frac{4}{2}\alpha.$$

In this reaction, the mass numbers (A) on both sides balance as do the atomic numbers (Z). Since the reaction is nuclear in character, the number of orbiting electrons is ignored. The thorium atom produced is also radioactive and undergoes transmutation according to the following reaction:

$$\frac{234}{90}Th \rightarrow \frac{234}{91}Pa + \frac{0}{-1}\beta.$$

Note that both the A and Z numbers balanced. The transmutation of the uranium atom continues through the emission of charged particles until the resulting element is lead, Z = 82, A = 206. Elements which are also produced through the natural transmutation of uranium include bismuth, polonium, protactinium, radium, radon, and thallium. The radioactive decay scheme of uranium is but one of numerous transmutation series that occur in nature.

The ancient alchemists and their predecessors spent two thousand years in an attempt to change common elements into gold, silver, and other valuable elements. Of course, none succeeded, for their most determined efforts were all centered around chemical reactions. The transmutation of one element into another requires a nuclear reaction. Regardless of the violence of the chemical reaction, the nucleus of an atom is protected from its immediate environment by the protective shield the atom's orbiting electrons produce. To change an element into another, recall, the number of charged particles in the nucleus must be altered. For example, to change lead into gold, three positive charges must be removed from the lead nucleus since the Z number of lead is eighty-two and that of gold is seventy-nine.

The answer to the alchemists' plight was at hand, yet their knowledge of nature was as yet too incomplete to realize the solution. The constant decay of radium, uranium, and other available elements would have given them a mechanism, for transmutation is possible by the addition of charged particles to a nucleus as well as by the emission of charged particles from a nucleus. Transmutation can also be caused by the addition of uncharged particles such as neutrons to some elements as well, a technique that is discussed later in the chapter. The alpha particle ejected from uranium would have been an excellent particle with which to create new elements. This particle could have caused transmutation to occur, since its energy is sufficient to pass through the orbiting electrons' protective shield with the ease that a bullet passes between the rotating blades of an electric fan.

In 1919, Rutherford bombarded nitrogen nuclei with alpha particles from a natural-occurring radioactive source and successfully transmuted nitrogen into oxygen. The nuclear reaction was as follows:

$$\frac{14}{7}N + \frac{4}{2}\alpha \rightarrow \frac{17}{8}O + \frac{1}{1}p.$$

His continued research led to the development of artificial transmutation. In the years that followed, the endeavors of many scientists culminated in the discovery of fission, a process previously discussed.

Artificial transmutation has been expanded considerably through the development of devices which accelerate charged particles to high velocities. Today, scientists bombard target nuclei with the more energetic particles such as high speed electrons, protons, and the positively-charged nuclei of the lighter elements. New elements thus have been created which have atomic numbers from 93 to 105. Some of the names assigned to the new elements honor the scientists involved in early research in radioactivity or in the actual discovery of the new element. Other names reflect their source. These new elements, listed in order of their increasing Z numbers, are neptunium, plutonium, americium, curium, berkelium, californium, einsteinium, fermium, mendelevium, nobelium, lawrencium, rutherfordium, and hannium. Each new element has a relatively short half-life; therefore, if they existed naturally on Earth any time during her history, they have decayed to a level that is undetectable.

NUCLEAR POWER AND ARTIFICIAL RADIOACTIVITY

At the present time, the chief source of the enormous amount of energy used in our modern world is fossil fuels—coal, petroleum, and natural gas. The amount of this source of energy is greatly limited, however. For that reason there seems to be little doubt that our next major energy source will be provided by the nuclear reactor. The reactor permits man to control chain reactions and to release energy from the tremendous store of fissionable materials as the need arises. The use of nuclear power to propel giant ships is a matter of reality. Also, many large industrial areas rely on nuclear power as their source of electrical energy. The cost and reliability of this nuclear-produced electricity are becoming competitive with the fossil fuel source.

Reactors are in reality devices for starting and governing chain reactions which are self-sustaining. The application of the reactor fits into one of the following categories: (1) to furnish vast numbers of neutrons for scientific studies; (2) to provide new elements through the process of neutron irradiation; (3) to provide heat for electrical power generation, propulsion, and other industrial uses (see Fig. 10.2).

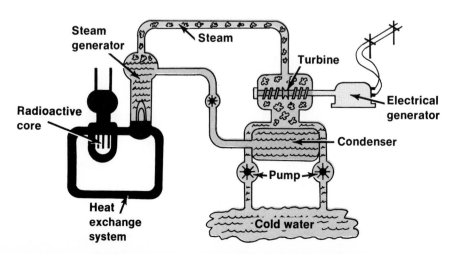

Fig. 10.2 Electricity is provided by nuclear power as presented in the schematic diagram.

Fig. 10.3
A cutaway
schematic diagram
of a high-
temperature, gas-
cooled reactor.

The typical nuclear reactor is composed of five primary parts: (1) a core which furnishes the fuel; (2) a moderator that enhances the fission process by slowing the neutrons' velocity; (3) a regulator which adjusts the number of free neutrons, the control which directly affects the fission rate; (4) a device that transfers the heat from the core; and (5) a type of shielding to provide areas for the operator and experimenter which are free from secondary radiation (see Fig. 10.3).

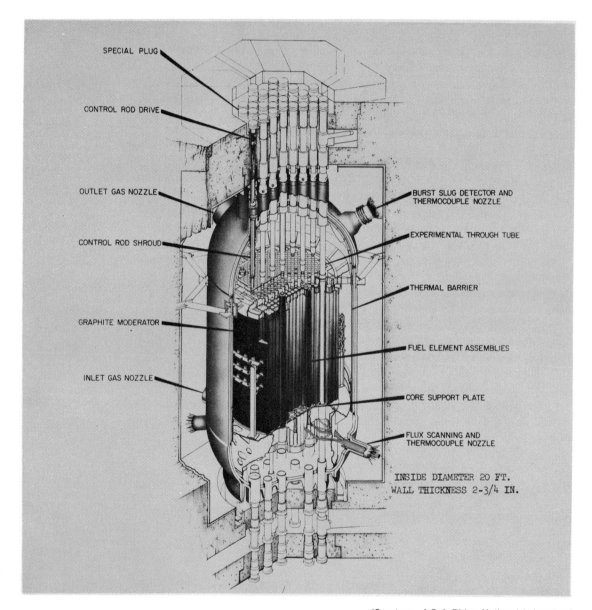

SPECIAL PLUG

CONTROL ROD DRIVE

OUTLET GAS NOZZLE

CONTROL ROD SHROUD

GRAPHITE MODERATOR

INLET GAS NOZZLE

BURST SLUG DETECTOR AND THERMOCOUPLE NOZZLE

EXPERIMENTAL THROUGH TUBE

THERMAL BARRIER

FUEL ELEMENT ASSEMBLIES

CORE SUPPORT PLATE

FLUX SCANNING AND THERMOCOUPLE NOZZLE

INSIDE DIAMETER 20 FT.
WALL THICKNESS 2-3/4 IN.

(Courtesy of Oak Ridge National Laboratory)

radioactivity and nuclear power

The material which fissions when struck by slow-moving neutrons provides the fuel for the reactor. Uranium-235 commonly furnishes the fission energy for many reactors. Other reactors make use of enriched uranium which has a higher percentage composition of uranium-235 than is ordinarily provided in nature. The fuel can be maintained in a liquid state, but more often is used as solid metallic uranium or uranium oxide, both of which are shaped like pellets, plates, or cylindrical rods, the latter illustrated in Figure 10.4.

(Courtesy of Oak Ridge National Laboratory)   (Courtesy of Oak Ridge National Laboratory)

(a)   **(b)**

Fig. 10.4 (a) The fuel area and control rods are constantly monitored by technicians for safety purposes. The water shielding produces reflections which lead to the observed complexity of the nuclear reactor assembly. (b) The fuel area releases tremendous amounts of heat in addition to a beautiful blue glow known as Cherenkov radiation.

During the fission process, the neutrons which are released travel at high velocity. The probability of the neutron's colliding with other atoms in order to continue the chain reaction is greatly enhanced if the neutron's velocity is lowered. A substance called a moderator permits the free neutrons to collide with its atoms without the neutrons' being absorbed and thus slows the elementary particles through numerous collisions (see Fig. 10.5).

radioactivity and nuclear power   **237**

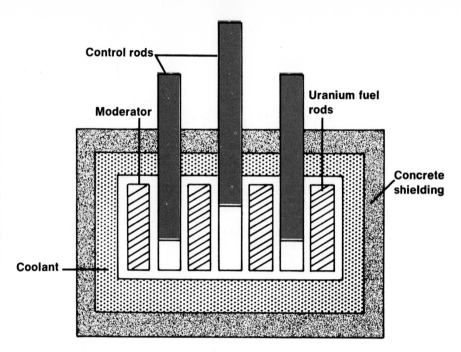

Fig. 10.5
The moderator increases the efficiency of nuclear reactions by slowing the neutron's velocity so that fission and fusion reactions are enhanced. The cadmium control rods readily absorb the unused slow neutrons and further control nuclear reaction rates.

Control rods

Moderator

Uranium fuel rods

Concrete shielding

Coolant

The nuclear reactor is generally controlled by adjusting the number of free neutrons which are present in the core at a given time. Substances such as cadmium or boron control the number of free neutrons by absorbing a high percentage of those which collide with their atoms. The two elements are manufactured and molded into the form of control rods. The length of the rod which is introduced into the core adjusts the number of free neutrons left to interact with other matter.

The neutrons and the fission products they produce collide with the moderating material or the highly absorbent control rods. The kinetic energy the particles thus lose is converted to heat energy. In order to prevent the total amount of heat energy from becoming excessive, various coolants are circulated to carry the heat away from the central core. Coolants such as air, helium, carbon dioxide, water, "heavy" water, and liquid sodium or lithium are typically used. The research reactor operates at less than 200°F; however, the reactor designed to generate power operates at temperatures in excess of 500°F. The vast amount of heat is converted into electrical energy or mechanical energy as the case may be.

Not all of the energy produced by fission is released as heat energy. Much of the total energy takes the form of radiation. Therefore, the interior walls of the reactor are protected by a heavy steel lining to prevent radiation damage. Also, the outer regions of the reactor are shielded with concrete, water, or some other suitable protective barrier designed to protect the personnel in the immediate vicinity of the reactor. Further shielding may take the form of massive superstructures, structures built to prevent extensive damage to surrounding areas if

radioactivity and nuclear power

by remote chance an accident occurs. Modern technical developments and detailed safety standards have made the latter type of shielding hardly necessary, though federal regulations force its installation in even the most remote areas.

Some reactors are built in a manner which takes advantage of a product of fission other than the heat energy provided by the process. The research reactor is designed to create a beam of neutrons incident on a given area, or a region of high neutron density into which various ports permit researchers to introduce various study materials. The physicist may use the reactor to study crystal structures or nuclear reactions. The chemist may use the reactor to analyze various substances for their exact composition or to determine the effects of radiation on chemical compounds. The biologist may use the reactor to produce genetic effects in plants and animals as well as to develop species which are more beneficial to man. The engineer, the medical doctor, and other professionals have also found many uses for the reactor and have helped to prove its worth to mankind.

The application of the nuclear reactor as a source of power has had a dramatic effect in the United States as well as in remote regions which lack coal, oil, and water of sufficient quantity to furnish needed energy. In 1960, the electrical power unit was provided only for demonstration of its potential. By 1963, large units were under construction which could provide about 600,000 kilowatts of electrical power from the heat energy of reactors as a concentrated effort. Five years later the total capacity of nuclear powered electrical energy under construction was in excess of 50,000,000 kilowatts—more than the total electrical energy required by the United States at the close of World War II. The future electrical power consumption of industry and home is expected to reach some 600,000,000 kilowatts by the end of the decade. The nuclear reactor will provide one-fourth of the total energy required.

The nuclear reactor is aiding the economy through its application in reducing electrical energy costs and in conserving the diminishing supply of fossil fuels. For instance, one gram of fissionable material provides 17,000 kilowatt hours of electricity, assuming seventy-five percent efficiency for the conversion of heat energy to electrical energy. In comparison, one ton of completely fissionable material is equivalent in power release to 3,000,000 tons of coal or 12,000,000 barrels of oil.

The heat energy provided by the reactor is sometimes used directly by industry. The steam provided by the transfer of heat from the core to the water coolant is used to dry, evaporate, and distill industrial products at temperatures which approach 400°F. The future plans include the use of reactors to provide temperatures above 3000°F. Such a source of heat would replace the typical industrial furnace that is used to gasify coal and to furnish intense heat for other industrial processes.

The heat from nuclear reactors is also readily converted to mechanical energy and thus is immediately available as a source of power for locomotion. Such a power reactor was built into the hull of the submarine USS Nautilus and was put to test in 1955. The United States now has more than 115 ships that are nuclear powered which have been propelled by pressurized water reactors over 15,000,000 miles. The various advantages of nuclear powered ships over conventionally powered ships include the elimination of areas to store fuel and the greater efficiency with which the engines operate. The advantages, though numerous, are counteracted by the cost which is substantially more than the cost of conventional engines. But, the routes, because they do not include necessary fueling stops, make travel more direct, and, therefore, make the additional expense of nuclear power less prohibitive.

Nuclear power is also applied in rocket propulsion. In the nuclear rocket, the element hydrogen is forced into the reactor, heated to extremely high temperatures, and thrust outward with tremendous force as gaseous hydrogen.

Small power reactors have been put to use in several categories of research application. The small unit has been used to furnish electrical power for a whole series of space probes and has been used as a power source for exploration of the moon. In addition, the power reactor has furnished necessary power to operate lighthouses and unmanned weather stations and radar bases. Other uses of the small unit are under consideration.

IONIZING RADIATION

Considerable concern has arisen from the increased application of nuclear power and the resulting level of radiation to which the public is exposed. Many people feel that the radiation safety standards are set too high and fear the results of levels they deem excessive to both present and future generations. Monitoring of streams, soils, and air is a constant task for professionals in the field of radioactive waste disposal. Their vigil is a costly endeavor but a necessary one (see Fig. 10.6).

Fig. 10.6
Air monitoring systems help maintain the rigid safety control of radioactive waste released into the atmosphere.

(Courtesy of NECO)

radioactivity and nuclear power

Natural external radiation exposures reach each inhabitant of Earth from such sources as cosmic rays from outer space; radioactive thorium present in concrete; radium that concentrates in granite, an important building material; and various radioactive elements, including uranium, which are present in the soil. Internally, radioactive potassium is a naturally occurring radiation hazard since the radioisotope is present in most body tissue, our food, and our water. Other radioisotopes have been placed in our environment as a result of man's application of nuclear energy. These man-made radiation sources have added measurable levels of radioactivity to our surroundings.

Human exposure to radiation is measured in *rems,* an accepted abbreviation which stands for "roentgen equivalent man." The unit refers to any type of ionizing radiation that produces a biological effect equal to the absorption of one roentgen of gamma or X-radiation. Commonly, exposures to radiation are extremely low and thus are measured in millirems (1 rem = 1000 mrems).

Gamma and X-radiation, like heat and light, are electromagnetic waves. Thus they travel through space at a velocity of light, *c*. As the radiation strikes body tissue or other matter, it transfers part or all of its energy to that with which it collides. Molecular structures are disrupted, and free ions are produced from broken chemical bonds. The free ions recombine in various manners and produce new and unusual compounds which create potential genetic changes. Alpha and beta particles ionize tissue just as does electromagnetic energy, but gamma and X-rays are often more energetic and thus potentially can free more ions.

Since the establishment of the Atomic Energy Commission, a new profession called *health physics* has become immensely important. The health physicist is devoted to the task of protecting man and his environment from unwarranted exposure to radiation. The Environmental Protection Agency, in conjunction with the Federal Radiation Council, has set maximum exposure levels for the general public. One of the health physicists' major tasks is to see that these levels are adhered to by the public. They must see, for instance, that individuals are not exposed to over 500 mrems per year of whole body exposure. A group of individuals who comprise an appropriate or statistical sample of the population are not to be exposed to over an average of 170 mrems per year of whole body radiation. Therefore, if one individual receives more, then the health physicist is to see that another individual receives proportionally less. In this manner, genetic changes which might be brought about in offsprings are held to a minimum. Genetic effects from radiation may not be the area of great concern that early investigations indicated. The survivors of Hiroshima, the Japanese city leveled by an atomic blast near the close of World War II, have been under constant observation. According to a published study twenty-five years after the nuclear detonation, virtually no genetic effects have been discovered among the survivors or their offsprings that resulted from the high levels of radiation present in the atomic blast or in the fallout that hovered over the city and surrounding countryside.

Although our greatest concern relative to exposure from radioactivity lies with low-level values over a long period of time, intense exposure to radiation should also be considered. The lethal dose of radiation has been determined through studies of individuals accidentally exposed to intense radiation areas; people in the immediate vicinity of the nuclear blasts at Hiroshima and Nagasaki, Japan; and animals placed at strategic points during nuclear detonation tests. Most studies reveal that an absorbed dose of 450 rems over the entire body in one day's time would be fatal in thirty days to half the people exposed. The value, therefore, is known as LD-50, a lethal dose of radiation to fifty percent of those people exposed.

If a total of 600 to 700 rems were accumulated by a population in one day, practically all of the exposed individuals would expire in a short time; therefore, this range is known as LD-100. If the dose of radiation were accumulated over a long length of time such as weeks or months, an individual could withstand much greater amounts, a feature which permits the use of intense radiation to destroy or to retard the growth of cancerous body tissue.

Presently, radioactive materials released into our air, soil, and water are diluted by natural actions. Thus the risk to the general public is almost immeasureable and might well remain so. However, safety standards must be maintained and surveillance continued to prevent an increase in radiation level and the possible risk to health that would follow. Once long-lived radioisotopes are added to our environment in quantity, their removal would be virtually impossible.

RADIOISOTOPES, THEIR USES AND DETECTION

The second major application of controlled nuclear reactions is producing radioactive atoms of various elements, called *radioisotopes*. Radioisotopes are produced in several ways. First, many radioactive atoms and some stable ones are produced as by-products of the fission process, since the nuclei of the fissionable material split into fragments which are in reality lighter nuclei, some of which are radioactive. Secondly, the neutrons produced by the fission process collide with and are absorbed by some of the stable atoms introduced into their path as a "target." Some of the new nuclei produced by absorbing a neutron are unstable, thus radioactive. Many of the radioisotopes used by man are prepared by this technique of neutron bombardment (see Fig. 10.7).

Every element known by man has at least one isotope which is radioactive. The radioisotope does not differ chemically from stable isotopes of the same element. All isotopes of an element, stable or radioactive, have the same Z number (the same number of protons) but a different A number. The element tin, for example, consists of twenty-five isotopes or *A* numbers—108 to 132—each identical in all chemical properties, but different in the number of neutrons in the individual nucleus. Some isotopes are stable, others radioactive. But how can man decide which atoms are stable? He has learned that the radioactive isotope can be detected by various instruments in an indirect manner. Atoms are

Fig. 10.7 **A small radioactive sample is removed from one of the pneumatic tubes which provides access to the source of neutrons in the reactor and rapid sample removal. After irradiation, the now radioactive sample is removed for analysis.**

(Courtesy of Brookhaven National Laboratory)

constantly colliding with others in gases or liquids, but not enough energy is involved in the typical collision to disturb the atoms so that they lose some of their orbiting electrons as a result of the collisions. The alpha and beta particles, however, are energetic enough to cause one or more atoms with which they collide to lose their orbiting electrons and thus their electrical balance. An entire avalanche of electrons often results from collisions with the atoms in gases and the cumbersome energetic alpha particle. The atoms which lose the electrons are said to be *ionized* or in effect to become positively charged particles rather than electrically neutral atoms. The ionization which results becomes the mechanism by which radioactivity can be detected.

The Geiger-Müller counter is commonly used to detect both beta and gamma radiation. This device, displayed in Figure 10.8, uses a Geiger tube which detects the presence of radiation. The tube (see Fig. 10.9) consists of a central wire mounted into a hollow cylinder which is made of metal or is metallically coated. The tube is filled with one of several available gases, and the radioactive substance is placed beneath the vertically mounted tube. The various particles enter the tube and ionize the gas. The ionized gas in turn is attracted to the charged central wire or the metal in the tube, according to the charge on the components involved. A pulse of electrical current results and is interpreted as an "event" by the electronic counter. The Geiger counter varies in sensitivity to the three elementary particles and is most sensitive to beta radiation. No measure of particle energy is directly available from the Geiger system; however, the degree by which aluminum, lead, or paper sheets placed between the

radioactivity and nuclear power

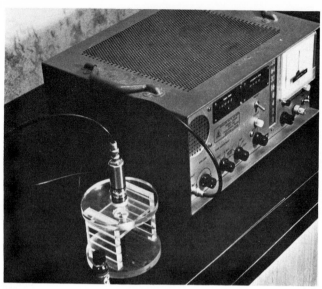

Figure 10.8

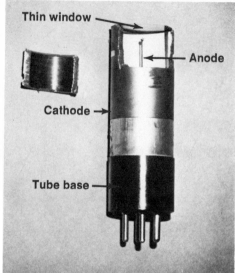

Figure 10.9

Fig. 10.8 Beta and gamma radiation are detected by the Geiger-Müller counter with varying efficiency.

Fig. 10.9 The Geiger-Müller tube is generally composed of metal by design and is filled with a mixture of gases. The radiation enters the tube through a thin window where the particle from radioactive decay interacts with the gas molecules present and is detected.

radioactive source and the Geiger tube absorb the radiation can be interpreted as a measure of particle energy by a trained observer. Gamma radiation is best studied by an instrument which uses various crystals as the detector. As gamma radiation enters the crystal, the alignment of its atoms is disrupted as the atoms absorb energy, and in turn the excited atoms emit a flash of light as they move back to their original energy state. The flash of light which results is called a *scintillation*. The brilliancy of the scintillation is directly related to the energy of the incident gamma ray. The balance of the instrumentation is capable of measuring the brilliancy of the flash and of interpreting the energy of the gamma ray accordingly. The interpretation of data from a given isotope permits the identification of the isotope with accuracy that is many times more sensitive than any method of chemical analysis.

A third instrument used to detect the presence of radioactivity is the cloud chamber. As a charged particle is introduced into a closed chamber where a "fog" has been created, the particle creates a visible path similar to that produced by a high-flying jet plane we see in our skies. The length of path charged particles produce as well as the degree of deflection caused by an electric or magnetic field brought near the chamber permits the scientist to determine the

radioactivity and nuclear power

particle's mass, charge, and energy. Many of the previously discussed particles were first discovered during the use of the cloud chamber. Ionizing radiation is also detected in the cloud chamber, but the mechanisms involved are beyond the scope of this book.

The bubble chamber, a fourth type of detecting and measuring device, permits particle trails in its liquid to be observed. The trails are actually gas bubbles created by particles in the liquid. Often the liquid used is hydrogen that has been heated in a pressurized chamber to a point near boiling. The pressure is suddenly released in the chamber, and an incoming particle causes the hydrogen to boil along the particle's path. The path disseminates almost instantly; therefore, time photography is the only practical means of observing properties of the particle. Mass, charge, and energy of the incident particle can be measured with the bubble chamber.

The detecting and measuring devices discussed are similar in principle to other devices which are currently in use. As investigations continue, more properties of the elementary particles will be revealed. Instrumentation will undoubtedly become more sophisticated and efficient as each investigator offers his contributions to the storehouse of available knowledge that concerns elementary particles.

The development and sophistication of the process of bombarding stable elements with neutrons and thus inducing radioactive isotopes in many cases has led to numberless applications which involve minute amounts. All the hairs on an individual's head contain the same amounts of metallic elements. The constant quantity can be determined with such accuracy that a single hair can be traced to a given individual. Also, the analysis of a hair from Napoleon's head and one from the head of King Eric XIV of Sweden reportedly has proved that the bodies of both men contained excessive amounts of arsenic. Both statesmen may well have been victims of arsenic poisoning.

In addition, specks of dirt and grease too small for the naked eye to see can be identified and related to similar spots. This type of measure, along with a determination of the amount of gun powder on a suspect's hands, has led to the solution of many crimes.

Similar applications of this technique, known in scientific fields as *activation analysis,* have produced great value in lunar studies, animal and human physiology, geology, and biology. Industry has applied the principle in many different manners such as the amount of automobile piston wear under various operating conditions, the optimum mixing time of paints, and countless characteristics of flow studies.

Radioisotopes and their particle energy, particularly the gamma ray, have replaced the costly X-ray devices and the rare and naturally occurring element radium. In the medical field, the energy yielded from the radioisotope has been used to retard the growth of unhealthy tissue and to sterilize surgical instruments. The energy of the radioisotope's particle has been used to preserve agricultural products and to sterilize seeds and thus prevent their germination. The radioisotopes which emit gamma radiation are potential sources of energy with which to make radiographs, the sequel to X-ray photographs. Industry has also found use of the radioisotopes in thickness gauging and in producing structural changes in wood, plastics, and other materials, thus making them more beneficial to man.

Another phenomenal application of the radioisotope is that in which it becomes an invisible tracer. Minute amounts can be mixed unnoticeably with varying proportions of a stable isotope of the same element and traced throughout a closed system. For instance, the length of time required for a chemical reaction to occur can be measured with amazing accuracy. On the other hand, a small amount of a radioisotope placed on a small, inert device can be followed through a pipe to determine where the pipe goes, or if poured directly into the liquid carried in the pipe, a concentration of the radioactivity readily marks the area of small leaks which might otherwise be undetected. The same technique has been invaluable in the fields of medicine and agriculture.

NUCLEAR ENERGY AND THE FUTURE

Man has not forgotten the devastating power unleashed to bring World War II to a rapid close. He has realized, though, that the sudden release of nuclear energy has advantages over conventional methods of energy release now used by man (See Chapter 24). Operation Plowshare, an application of instant nuclear power release in which Earth's surface is altered for the benefit of man, is a prime example of peaceful application of harnessed nuclear fission. Fission, however, results in the release of large quantities of various unwanted secondary radioactive isotopes that create decontamination problems and cause the area near the explosion to be hazardous for extensive periods of time after the blast. However, research is well underway to develop a "clean" explosion, one with fallout composed of little or no long-lived radioactive isotopes. Once the radioactive fallout produced by nuclear fission is successfully curbed, the construction of highways, waterways, airports, and level sites for large cities will be considerably expedited.

A new fuel and a new cycle are being presently investigated. It involves the fusion of hydrogen nuclei (protons) and of boron nuclei. Then the new nuclei fissions by dividing into stable helium nuclei. This reaction is free of radioactive by-products and therefore represents a valuable source of energy.

The process of fusion has been going on in nature for billions of years, since fusion is the mechanism through which stars obtain their energy. Fusion has been accomplished in the laboratory numerous times; however, the process requires temperatures which exceed 100 million degrees Fahrenheit. This fantastically high temperature must be maintained for periods long enough for a sufficient amount of fusion of light atoms (obviously, gaseous) to have occurred. The atoms in the mixture of gases are torn apart into nuclei and electrons (at cooler temperatures the molecules of gas undergo some degree of ionization). This mixture of fragments (or ionized gases) has come to be known as *plasma*. A container for the plasma created problems for scientists long after the theory was conceived. No conventional container can be used, since the plasma would either disintegrate it or be cooled below the plasmatic state. The answer was determined from a characteristic of the plasma itself—the fact that the plasma is electrically charged and thus would respond to the force of a magnetic field. Therefore, the plasma by which man has uncovered numerous secrets of the atom and of the universe can be contained in the form of a properly shaped magnetic field. Through the slow, controlled manipulation of the magnetic field, the scientist can place extreme conditions on the plasma itself. Many scientists feel that full control of thermonuclear energy, though some time away, will lead man to a source of energy which will be inexhaustible.

CHAPTER SUMMARY

The atom is composed primarily of elementary particles: *electrons. neutrons,* and *protons.* Various isotopes of each element are unstable, and stability is reached by spontaneous ejection of elementary particles and/or energy, a phenomenon known as *radioactivity.* Primary radiation released by unstable atoms is *alpha, beta,* and *gamma* rays. Alpha and beta rays are in reality streams of elementary particles. Other particles released are classified into various families among which are the leptons, the mesons, and the baryons. The ratio of protons to neutrons in an atom's nucleus seems to govern the *stability* of an atom. Upon the emission of a charged particle the nucleus of the unstable atom assumes a new identity instantly. Alpha particles are composed of two protons and two neutrons; therefore, the Z number of an atom decreases two units, and the A number decreases four units upon the ejection of an alpha particle. A beta particle is negatively charged and has negligible mass; therefore the Z number of the radioactive atom increases one unit and the A number remains the same. Gamma rays do not change the identity of the unstable atom since they are uncharged and have undetectable mass. The energy of all radiation is the *electron volt.* The nucleus of an atom is held together by mass that has been converted to energy, a measure of which is called *binding energy.*

Becquerel and others discovered that about fifty isotopes of the elements found in nature are radioactive. Most natural-occurring radioisotopes are among the heavier elements such as radium and uranium.

The rate by which an unstable isotope disintegrates is called *activity,* a measure generally calculated in disintegrations per second. The *curie* is the activity of one gram of radium, 3.7×10^{10} dis/sec, and is the accepted unit of measure of activity for all types of radiation (More precise measures indicate that one curie of activity is released from 0.97 gram of radium rather than one gram as previously calculated).

Half-life is the time required for the activity of a given mass of a radioisotope to decrease to one-half its original value. The loss of mass from radioactive decay among the heavier elements is negligible.

Fusion is the combining of light nuclei to form heavier nuclei. The resultant atom assumes an identity according to its new Z and A numbers. Fission is the splitting of a nucleus into two nuclei of approximately equal Z and A numbers. The mechanism by which stars "burn" is fusion, primarily of hydrogen atoms to form helium atoms. The process of *fission* is responsible for the tremendous energy supplied by nuclear reactors.

Elements are transformed into other elements both by man and by nature. Uranium, a natural-occurring radioactive element, eventually is transmuted to lead through a series of alpha and beta particle emissions. *Artificial transmutation* of one element into another has been accomplished for several decades. All elements from $Z = 93$ to $Z = 105$ have been transmuted from lighter elements. The stability of a target atom is upset by the bombardment of its nucleus by particles or by electromagnetic energy.

Nuclear reactors are devices designed to initiate and to control chain reactions. The reactor is fast becoming a major source of heat and electricity. The reactor is also a plentiful source of neutrons. As these neutrons are absorbed by target elements placed in their path, the unstable atom that is produced undergoes radioactive decay and transmutes into a new element.

Earth is constantly exposed to cosmic radiation, gamma radiation, and charged particles. Radiation in the form of high frequency electromagnetic waves has the potential to ionize tissue with which it collides. In addition to cosmic radiation and gamma radiation, tissue is exposed to X-radiation during diagnosis and treatment of numerous diseases. The primary unit by which ionizing radiation is measured is the *rem*. A total absorbed dose of ionizing radiation equal to 600 rems over a short period of time may be fatal to man.

Radioisotopes are unstable isotopes of the elements. Many radioisotopes occur in nature. All others are artificially produced during fission or by bombardment of a stable element with particles or energy. Every element has at least one radioisotope. The radioisotope does not differ chemically or physically from a stable isotope of the same element. The charged particles and energy emitted during radioactive decay are detected by *Geiger-Müller counters, scintillation detectors, cloud chambers,* and *bubble chambers.* In addition, other instruments similar in principle have been developed. Radioisotopes are used as tracers in many scientific fields. Gamma radiation from some radioisotopes has partially replaced the X-ray devices. The particles and energy from radioisotopes have been found to affect various materials in a manner beneficial to man.

In our growing world, energy requirements are increasing more rapidly than our population. Projected energy needs for the future reveal that a source of energy other than that provided by our depleting fossil fuels must be developed. Both conservation and economic considerations demand that some source of energy must furnish an increasingly larger amount of heat and electricity. Undoubtedly nuclear energy, as provided by controlled fission of the elements, will furnish a large portion of the required energy worldwide. The fuel for the reactors may be produced from the fusion of nuclei and charged particles as matter exists in a plasma state, perhaps our fourth state of matter.

QUESTIONS AND PROBLEMS

1. During certain nuclear reactions, mass is lost. Where does this mass go? In other nuclear reactions, mass is gained. What is the source of this additional mass? (p. 232)

2. How are X-rays and gamma rays similar? How do they differ? (p. 229)

3. List and briefly discuss the quantities which must be conserved when an atom of a radioisotope decays. What quantities, if any, are not conserved? (p. 234)

4. How many atomic mass units do atoms lose when the following types of radiation are released? (p. 230)
   a. alpha emission
   b. beta emission
   c. gamma emission
   d. X-ray emission
   e. neutron emission

5. What forces tend to cause an atom to maintain its stability? What forces cause an atom to lose its stability? (p. 227)

6. How can an atom weigh less than the weight of its individual components? Other atoms weigh more than the sum of the weight of their individual components. How is this situation possible? (p. 230)

7. Estimates indicate that the weight of the energy which Earth receives from the sun is about 3 pounds/mile^2 in a given unit of time. How could this value be determined if the sun's energy which reaches Earth has no mass? (p. 229)

8. Both fusion and fission are sources of tremendous amounts of energy. Cite numerous examples of how man applies each source for his benefit. (p. 232)

9. The radioactive atom has played an important role in the conquest of disease and in the development of better diagnostic procedures. Can you list other applications of radioisotopes in which their use is equal or superior to other available techniques? (p. 242)

10. Relate the hydrogen bomb to nuclear fusion as the star mechanism. (p. 232)

11. Is the neutron a combined proton and electron and the proton a combined neutron and positive electron? Support your answer. (p. 228)

12. Why is the major process which goes on in the sun known as fusion? Is fission commonly found anywhere in nature or is the fission process totally man-made? (p. 232)

13. If a stray neutron were to strike one of Earth's uranium mines, is there cause for concern that Earth may be destroyed? Defend your answer. (p. 233)

14. Neutrons are more commonly used to bombard targets to produce radioactivity than are alpha or beta particles. Why is this statement valid? (p. 227)

15. Consider the elements from Z= 93 to Z =105. What is the probable origin for each name chosen? What major contributions may be involved that led to the choice of name for each element? (p. 235)

radioactivity and nuclear power

atoms, molecules, and chemical change
11

A CHART OF PERIODIC PROPERTIES
SIMILAR ELECTRON ARRANGEMENTS—
SIMILAR PROPERTIES
THE MODERN PERIODIC CHART
WHY ATOMS REACT
MOLECULES AND EQUATIONS
HOW TO "READ" A CHEMICAL EQUATION
ANOTHER WAY THAT ATOMS REACT

vocabulary

Covalent Bond | Electronegativity
Ionic Bond

In Chapter 4 we found that all matter consists of submicroscopic particles called atoms and that atoms contain a very dense nucleus of protons and, in all atoms but the protium isotope of hydrogen, one or more neutrons. We also know that the electrons of atoms are arranged into various shells and subshells.

In this chapter we shall learn the important consequences of how many electrons are in the outermost electron shell of atoms. In general, it is the electrons in the outermost shell that determine the chemical properties of the atoms of different elements.

A CHART OF PERIODIC PROPERTIES

If one lists the elements in order of increasing atomic number, such as:

H	He	Li	Be	B	C	N	O	F	Ne	Na	Mg	Al	Si	P
1	2	3	4	5	6	7	8	9	10	11	12	13	14	15

S	Cl	Ar	K	Ca	etc.,
16	17	18	19	20	

a curious periodic repetition of physical and chemical properties can be noticed. Thus, helium ($_2$He), neon ($_{10}$Ne), and Argon ($_{18}$Ar) are all gases and are all chemically very unreactive. Note that $_{10}$Ne is the ninth element after $_2$He and that $_{18}$Ar is the ninth element after $_{10}$Ne. Lithium ($_3$Li), sodium ($_{11}$Na), and potassium ($_{19}$K) are all silver-colored solids and extremely reactive elements. Potassium is the ninth element after sodium, which is the ninth element after lithium. From the preceding, one would expect then that if any two elements are picked from the list of elements above that are separated by seven other elements, the two elements would have similar chemical and physical properties. This is indeed the case. It should be pointed out that hydrogen is an exception to this rule.

It is convenient to list together the elements with similar properties and still maintain the order of increasing atomic number. This can be done in the following manner:

(a)   H   He

(b)   Li   Be   B   C   N   O   F   Ne

(c)   Na   Mg   Al   Si   P   S   Cl   Ar

(d)   K   Ca

Thus, all the elements in row (c) have properties similar to the elements directly above them in row (b). The same relationship is true for rows (d) and (c). Hydrogen is again the exception.

In about 1869, Dimitri Mendeleev, a Russian chemistry professor at the University of St. Petersburg (today called Leningrad), noted the periodic repetition of the properties of each ninth element and arranged those elements known in his day in the manner shown above. In 1872, Mendeleev published his Periodic Classification of the Elements, part of which is shown below:

Li	Be	B	C	N	O	F
Na	Mg	Al	Si	P	S	Cl
K	Ca	☐	Ti	V	Cr	Mn
Cu	Zn	☐	☐	As	Se	Br

A portion of Mendeleev's original Periodic Chart.

The three blank spaces shown were left empty because at that time no elements existed that had properties similar to the properties that Mendeleev felt should be exhibited by elements in the blank positions. Now it is not sufficient for a scientific theory or concept to merely correlate known data. The real test of the validity of a new concept is to correctly predict the outcome of future tests or discoveries. Convinced that his classification of elements was a valid one, Mendeleev predicted that the three elements missing in his chart would be discovered and that they would fit into his chart in the blank spaces. He even forecasted the physical and chemical properties that the unknown elements would have. Within twelve years all three elements had been found, and their predicted properties turned out to be extremely close to the values determined experimentally.

SIMILAR ELECTRON ARRANGEMENTS— SIMILAR PROPERTIES

Why should every ninth element have similar chemical properties? The explanation lies in the number of electrons in the outermost shell of the atoms of each of the elements. To illustrate, consider the electron arrangement in the atoms of the Li, Na, and K column of a periodic chart.

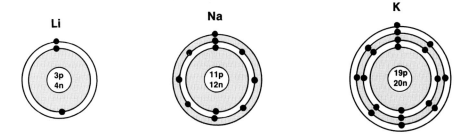

It is seen that Li, Na, and K all have only one electron in their outer electron shell.

On the other hand, F and Cl each have seven electrons in their outer electron shell. Similarly, Be, Mg, and Ca all have two electrons in their outer shell; B and Al both have three electrons in their outer shell, etc.

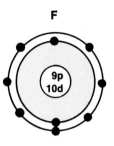

F

9p
10d

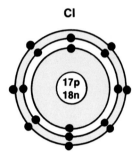

Cl

17p
18n

It would be helpful then to indicate the number of electrons in the outer shell of all the atoms in a particular column by the use of Roman numerals.

I	II	III	IV	V	VI	VII	VIII
H							He
Li	Be	B	C	N	O	F	Ne
Na	Mg	Al	Si	P	S	Cl	Ar
K	Ca						

A portion of the Periodic Chart. The Roman numerals above each column indicate the number of electrons in the outer shell of each element listed.

It is obvious then that atoms of those elements that have the same number of electrons in their outer shell have similar properties. Since sodium and potassium, for instance, each have one electron in their outer shell, they both have only one electron that can be involved in electron interaction with other atoms. Consequently, they both react in a similar manner.

As pointed out earlier, Ne and Ar are chemically rather unreactive; at the same time they have eight electrons in their outer shell. Indeed, all atoms with eight electrons in their outer shell are chemically very stable and unreactive.

THE MODERN PERIODIC CHART

Scientists now believe that man has already discovered all the elements that exist naturally on earth. Man has found on our planet a total of 92 elements and has synthesized in the laboratory 14 more elements, giving a total of 106 elements.

Whereas Mendeleev arranged the elements in his chart according to increasing atomic weight, modern charts have the elements arranged according to increasing atomic number. A modern periodic chart is shown in Figure 11.1.

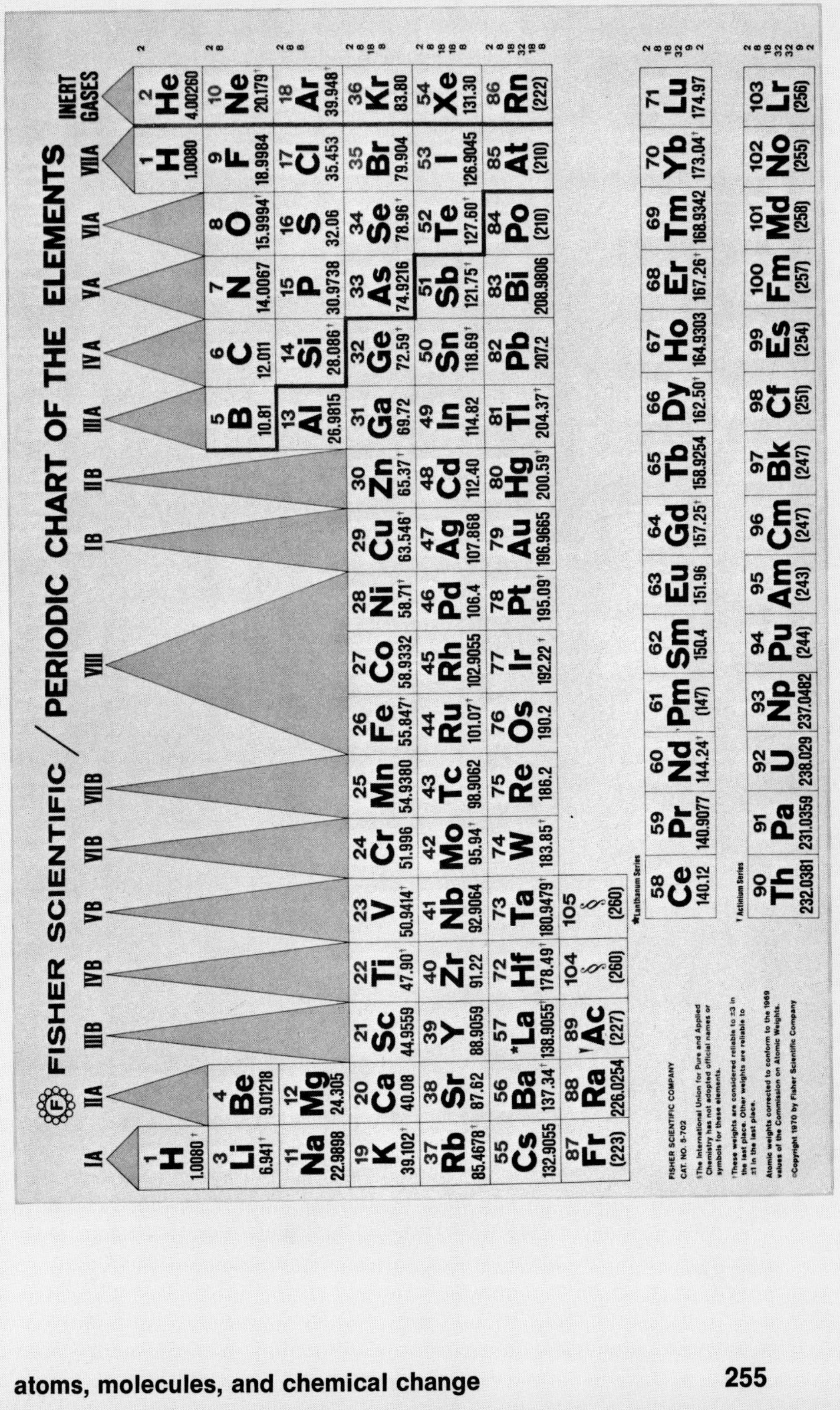

Fig. 11.1 A modern periodic chart.

The great advantage of the Periodic Chart lies in the considerable amount of knowledge that the chart makes available to one who knows how to "read" it. The chart (see Fig. 11.1) is arranged so that the following information is supplied:

1. All elements to the left of the dark staircase line are metals, and all the elements to the right are non-metals. Note that a majority of the elements that make up our universe is metals (84 of the 106 known elements).
2. The number above the symbol for each element represents the number of protons in the nucleus of the atoms of that element and is called the atomic number of the element. This number is also the number of electrons surrounding the nucleus of all atoms of this element.
3. The number below the symbol for each element represents the atomic weight of the element.
4. The number of neutrons present in atoms of an element can be found by subtracting the atomic number from the rounded-off atomic weight. For example, the *atomic number* for bromine (Br) is 35, and the atomic weight is 79.909. Therefore, the number of neutrons in an atom of bromine is 80 − 35 = 45. There are a few exceptions to this procedure that will be discussed later.
5. All elements in a particular column, called a family, have similar chemical properties. Thus, the elements of the alkali metal family—Li, Na, K, Rb, Cs, and Fr—are all silver-colored metals in the solid state and are all very reactive elements. The elements of the halogen family—F, Cl, Br, I, and At —are all reactive elements that react in the same manner as alkali metals to form salts that melt only at very high temperatures. The elements of the noble gas family—He, Ne, Ar, Kr, Xe, and Rn— are all gases and are very unreactive. In fact, it is due to their lack of reaction with the "common" elements that they are called noble gases.
6. The size of the atoms in the chart decrease from the left side to the right side and from the bottom to the top of the chart.
7. Metallic properties of the elements decrease from the left side to the right side and from the bottom to the top of the chart.
8. The electronegativity (a measure of the attraction an atom has for an extra electron) of the elements increases from the left side to the right side and from the bottom to the top of the chart.

WHY ATOMS REACT

Strangely, some atoms have an affinity for one or more electrons in spite of the fact that the atoms of an element are all electrically neutral; that is they have the same number of positive protons as negative electrons. This attraction for extra electrons varies with different elements and is highest for elements in the upper right hand corner of the Periodic Chart, and lowest for elements in the

atoms, molecules, and chemical change

lower left hand corner of the chart. The attraction of the atoms of an element for one or more extra electrons is called the electronegativity of the element. As a consequence of the electronegativity of fluorine, for example, the following reaction takes place. The fluorine atom, which has the highest electronegativity of all elements, attracts one electron and, as a result, then has ten electrons in its electron shells but still has only the original nine protons in its nucleus. Consequently, the structure then has a -1 charge associated with it. An atom with

Fluorine atom (F)   **Fluoride ion (F^{-1})**

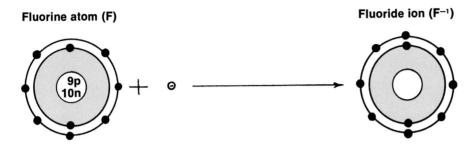

a charge, either positive or negative, is called an ion, and the charge present on the atom is designated as a right-hand superscript (F^{-1}). Also, the suffix *ide* is used with the name of the element to indicate a negative ion. Positive ions have the same name as the atoms the ions were made from. Thus, fluorine forms the fluoride ion, oxygen forms the oxide ion, and sulfur forms the sulfide ion, but sodium forms the sodium ion which has a $+1$ charge.

In the case of oxygen, two electrons are captured. Note that the fluorine atom and the oxygen atom both formed ions with eight electrons in their outermost electron shell. We will find that atoms that capture electrons will capture just the right number to result in an ion with eight electrons in its outer shell.

Oxygen atom (0)   **Oxide ion (0^{-2})**

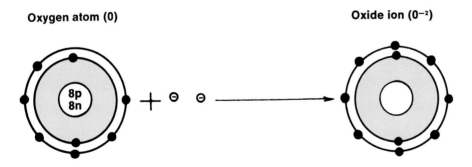

In general, atoms on the right side of the Periodic Chart (columns V A, VI A, and VII A) are sufficiently electronegative to attract electrons, whereas atoms on the left side of the chart (columns I A, II A, and III A) have such low electronegativity that they actually lose electrons to the more electronegative atoms on the right side of the chart.

atoms, molecules, and chemical change   **257**

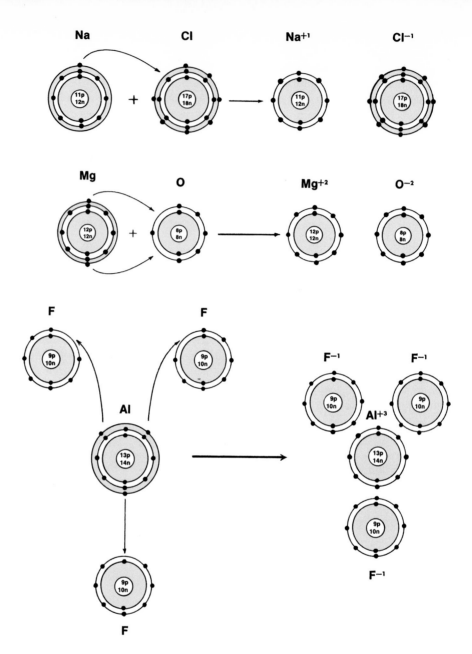

It is significant that all the ions produced in the three reactions above have eight electrons in their outer shells, once again showing the unusual stability of an outer shell of eight electrons. Sodium, in column I A, lost one electron; magnesium, in column II A, lost two electrons; and aluminum, in column III A, lost three electrons. Note that Na^{+1}, Mg^{+2}, and Al^{+3} all have eight electrons in their outer orbit. On the other hand, chlorine and fluorine, in column VII A, gained one electron each to give them the needed total of eight electrons. Oxygen, in column VI A, gained two electrons to complete its outer shell at eight electrons.

Each compound formed (NaCl, MgO, and AlF_3) consists of positively and negatively charged ions. The positive and negative ions are held together by mutual attraction since unlike charges attract each other. Thus, the three -1 fluoride ions are attracted to the single $+3$ aluminum ion and vice versa. This mutual attraction holds the AlF_3 molecule together as a single unit. This bonding between oppositely charged ions is known as ionic bonding.

The above three reactions could be more conveniently written in a shorthand form as follows:

$$Na + Cl \longrightarrow NaCl \quad \begin{pmatrix} \text{One Na atom reacts with} \\ \text{one Cl atom} \end{pmatrix}$$

$$Mg + O \longrightarrow MgO \quad \begin{pmatrix} \text{One Mg atom reacts} \\ \text{with one O atom} \end{pmatrix}$$

$$Al + 3F \longrightarrow AlF_3 \quad \begin{pmatrix} \text{One Al atom reacts} \\ \text{with three F atoms} \end{pmatrix}$$

NaCl is the formula for the compound sodium chloride (common table salt) and shows that the compound is made up of equal numbers of sodium and chloride ions. The formula for aluminum fluoride is AlF_3 and shows that the compound is composed of three fluoride ions bound to one aluminum ion. Although the three compounds formed consist of ions, each molecule of the compounds is electrically neutral as a unit.

$$Na^{+1}Cl^{-1} \qquad (\text{one } +1 \text{ and one } -1)$$
$$Mg^{+2}O^{-2} \qquad (\text{one } +2 \text{ and one } -2)$$
$$Al^{+3}F_3{}^{-1} \qquad (\text{one } +3 \text{ and three } -1)$$

MOLECULES AND EQUATIONS

Even these equations are not written as the chemist would write them since chlorine, oxygen, and fluorine should be shown as Cl_2, O_2, and F_2 respectively. On the other hand, atoms of sodium, magnesium, and aluminum exist as individual atoms.

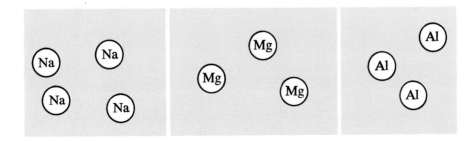

The three gases used in the equations exist as diatomic molecules i.e., two atoms to each molecule.

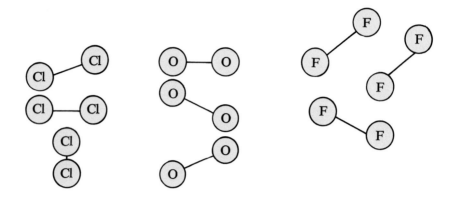

The reader should be very careful in distinguishing between atoms and molecules. A molecule is composed of two or more atoms. Thus, water,

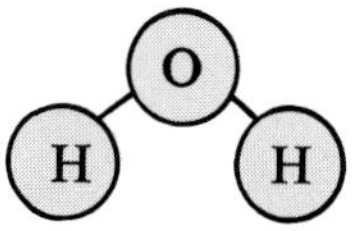

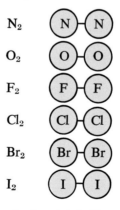 , consists of molecules containing one oxygen atom and two hydrogen atoms.

In general, then, atoms are the basic units of elements, and molecules are the basic units of compounds. The most obvious exceptions to this statement are those elements that react with themselves to form diatomic molecules that contain only atoms of one element. Examples of these elements are shown below:

N_2 (N)-(N)

O_2 (O)-(O)

F_2 (F)-(F)

Cl_2 (Cl)-(Cl)

Br_2 (Br)-(Br)

I_2 (I)-(I)

The numbers that are right-hand subscripts indicate the number of atoms present in one molecule. Thus, O_2 shows there are two atoms of oxygen in one molecule of oxygen, and H_2O shows that there are two hydrogen atoms and one oxygen atom in one molecule of water. If an atom in a molecule does not have a right-hand subscript, it is assumed to be a single atom and is not written as H_2O_1 but simply as H_2O. If one wishes to indicate more than one molecule of a substance, a left-hand coefficient is used. For example, $3H_2O$ shows that there are three molecules of water, and each molecule is made up of two hydrogen atoms and one oxygen atom.

Now, at last, we are able to write the three equations that we first started on page 259 as a chemist would record them.

$$2Na + Cl_2 \longrightarrow 2\,NaCl$$
$$2Mg + O_2 \longrightarrow 2\,MgO$$
$$2Al + 3F_2 \longrightarrow 2AlF_3$$

In the "balanced" equations shown above, the ratio of Na atoms reacting with Cl atoms is $1:1$; the ratio of Mg atoms reacting with O atoms is $1:1$; and the ratio of aluminum atoms reacting with fluorine atoms is $1:3$. These are the same ratios employed in the original equations on page 259. The equations are called balanced equations since the number of atoms of each element is the same on both sides of the equation arrow.

HOW TO "READ" A CHEMICAL EQUATION

Like the Periodic Chart, chemical equations have been designed so that a large amount of information can be obtained from them by someone skilled in "reading" them. Consider the reaction of carbon with oxygen in the presence of heat, i.e. the combustion of carbon.

$$C + O_2 \xrightarrow{\text{heat}} CO_2$$

First, in a qualitative sense the equation tells us that carbon burns in the presence of oxygen. The carbon may be in the form of coal, charcoal, graphite, lampblack, or, if one is both rich and foolish, diamond. Now let's consider the equation from a quantitative viewpoint. Although there are no coefficients shown in the equation, it is understood that the equation should be read as

$$1\,C + 1\,O_2 \xrightarrow{\text{heat}} 1\,CO_2$$

Now consider the different ways that the above equation can be "read."

1 C	+	1 O_2	$\longrightarrow$	1 CO_2
(1) one atom of carbon	plus	two atoms of oxygen	give	one atom of carbon and two atoms of oxygen combined as carbon dioxide
(2) one atom of carbon	plus	one molecule of oxygen	give	one molecule of carbon dioxide

atoms, molecules, and chemical change

(3) one atomic weight of carbon	plus	two atomic weights of oxygen	give	one atomic weight of carbon + two atomic weights of oxygen
(4) one atomic weight of carbon	plus	one molecular weight of oxygen	give	one molecular weight of carbon dioxide
(5) 12.011 grams of carbon	plus	31.998 grams of oxygen	give	44.009 grams of carbon dioxide
(6) 12.011 pounds of carbon	plus	31.998 pounds of oxygen	give	44.009 pounds of carbon dioxide

Once the equation is written out and balanced, the ratio of the weight of carbon to the weight of oxygen that will react to form carbon dioxide is determined by using the atomic weights of the atoms present in the carbon dioxide molecule. Thus, one atomic weight of carbon is 12.011, and two atomic weights of oxygen are $2 \times 15.999 = 31.998$. The molecular weight of carbon dioxide is one atomic weight of carbon plus two atomic weights of oxygen or $12.011 + 31.998 = 44.009$, or 12.011 weights of carbon will react with 31.998 weights of oxygen to form 44.009 weights of carbon dioxide.

ANOTHER WAY THAT ATOMS REACT

Not all atoms react by losing or gaining electrons due to differences in electronegativity. Atoms of one element may have almost the same attraction for electrons as atoms of a different element; that is, they have similar electronegativities. In such a case neither atom can capture the other's electron. What each can do, however, is to mutually share electrons.

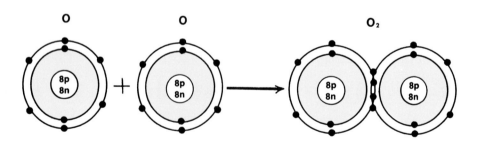

In the case of the O_2 molecule, both oxygen atoms are sharing the four electrons shown between them. The four electrons surround the nuclei of both oxygen atoms. This sharing of electrons produces a strong bond which holds the two oxygen atoms together. Bonds formed by electron sharing are known as covalent bonds. After the reaction occurs both oxygen atoms have eight electrons in their outer shell because of the electron sharing. Covalent bonds exist in many chemical compounds. Let's look at the electron structure of water.

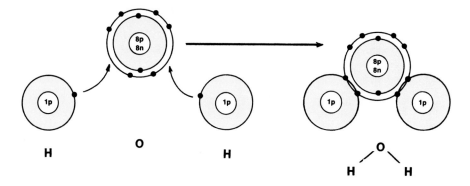

H   O   H

The sharing of a pair of electrons between two atoms, that is, a covalent bond, is represented by a dash. Thus, the water molecule can be shown as

$$H—O—H$$

and the oxygen molecule can be shown as

$$O = O$$

Of course, these and other molecules containing covalent bonds can be represented in a shortened form as O_2, H_2O, HCl, H_2, etc. The same is true for molecules containing ionic bonds, as we saw previously, as in NaCl, MgO, and AlF_3.

CHAPTER SUMMARY

When the elements are arranged in order of their atomic numbers, similar chemical and physical properties are exhibited by elements at regular intervals. This similarity of chemical and physical properties is the result of the similarity of the arrangement of the electrons of the atoms of these elements. Thus, the atoms of elements with only two electrons in their outermost shell all have similar properties.

Atoms with fewer than four electrons in their outermost shell tend to lose these electrons to become positively charged *ions*. Atoms with more than four electrons in their outermost shell tend to gain enough electrons to give a total of eight electrons in their outermost shell and become negatively charged ions. Atoms such as the noble gases with eight electrons in their outermost shell are relatively unreactive chemically because of the stability of the arrangement of eight electrons in the atom's outermost shell.

The atoms of some elements neither lose nor gain electrons but share their electrons with other atoms to form covalent chemical bonds.

An *ionic bond* results from the electrostatic attraction between positively charged ions. The ions are the result of the transfer of one or more electrons from one atom to another.

A *covalent bond* is produced when a pair of electrons is shared by two atoms.

atoms, molecules, and chemical change   **263**

QUESTIONS AND PROBLEMS

1. Which two elements should most resemble silver (element 47) in chemical and physical properties? (p. 254)

2. Why do all the elements of Group IA in the Periodic Chart have similar chemical properties? (p. 252)

3. Show the reaction between potassium (element 19) and bromine (element 35). Indicate any loss and gain of electrons by the atoms. (p. 258)

4. The atoms of which elements will have the same number of electrons in their outer shell as atoms of calcium. (p. 254)

5. What would be the formula of the compound produced when boron (B) reacts with fluorine (F)? (p. 258)

6. What charge would you expect to be associated with the ions of the elements Ba, S, Ga, and Cc? (p. 257)

7. Balance the following reactions: (p. 261)

    (a) $KClO_3 \xrightarrow{\text{heat}} KCl + O_2$

    (b) $H_2 + Cl_2 \longrightarrow HCl$

    (c) $H_2SO_4 + NaOH \longrightarrow Na_2SO_4 + H_2O$

basic
chemistry
12

THE ORIGINS OF CHEMISTRY
CHEMICAL AND PHYSICAL CHANGES
MIXTURES AND SOLUTIONS
THE STRUCTURE OF SOLIDS
IONIC CRYSTALS
ATOMIC CRYSTALS
MOLECULAR CRYSTALS
METALLIC CRYSTALS
COMBUSTION
OXIDATION AND REDUCTION

vocabulary

Density	Mixture
Oxidation	Reduction

THE ORIGINS OF CHEMISTRY

Man first practiced chemistry when he learned how to perform the rather high temperature oxidation of cellulose products—that is, when he learned to build a fire. Though he was not aware, of course, of the chemical reactions occurring, he nonetheless produced the necessary conditions for the reaction of oxygen with molecules of cellulose and other molecules present in wood. Later in time, in a section of the world far removed from man's beginnings, he accomplished a similar reaction using black "rocks" instead of just wood. Because man had seen fires started by lightning, he surely set out determinedly to learn how to start a fire from wood. Then he probably learned by sheer accident about the black "rocks" that burn. Imagine early man's astonishment when the black "rocks" on which he had built his fire burst into flame and burned with a heat more intense than that of the burning wood. Thus, man discovered coal as a source of heat and light. Quite often in science, one discovery leads to another. Such was the case with man's discovery of the making of fire which eventually and inevitably led to a number of other discoveries. One such tremendously important discovery was early man's next use of chemical reactions. These reactions were the partial thermal decomposition of various proteins, carbohydrates, fats, and so forth combined in living or recently-living organisms—or the cooking of meat

There is no way to know when man further learned to use heat to dehydrate certain clays to form pottery and bricks, but burned clay artifacts dating from 30,000 to 20,000 B.C. have been found. Later between 4000 and 3000 B.C. man learned that when certain blue stones (copper sulfate) were heated in a hot fire, the metal copper was produced. By 2500 B.C. man had learned to alloy copper with tin to produce bronze. This discovery ushered in the Bronze Age which had bronze fighting instruments far superior to any stone weapons that man had ever known. By 1600 B.C. man had learned to smelt iron ore, to burn the excess carbon out of the resulting soft iron to produce a fairly hard grade of iron, and to increase considerably the hardness of iron by plunging it into water after heating it to a dull red heat.

For thousands of years after the beginning of the iron age man continued to improve on the discoveries of the ancients and expanded the existing chemical technology. From about 500 to 1600 A.D. the development of chemistry was in the hands of the alchemists whose chief aims were to transmute the baser metals such as iron, copper, lead, tin, and zinc into gold and to find the elixer of eternal youth. Most of the work of the alchemists involved a trial and error technique mingled with considerable mysticism and secrecy. Consequently, very few advances in chemistry of a significant nature resulted.

A great number of charlatans flourished during the alchemy stage of the development of chemistry. One case in particular that occurred in England gives an insight into the ways in which an enterprising alchemist would use his secret knowledge of chemistry to play the rogue and swindle whole villages out of considerable sums of money. The alchemist was aware that when iron is placed into a solution of copper sulfate, the iron slowly goes into solution and the copper ions in solution plate-out on the iron to give a copper plated piece of iron. The alchemist would go into a village and promise the villagers that for a certain sum of money he would show them how to transmute their iron utensils into gold. He would then have them scrupulously clean their iron utensils and place them in a nearby stream that he knew contained copper sulfate. During the next day he would chant various incantations over the submerged utensils. By nightfall it was obvious to the villagers that their utensils were becoming gold-colored, and they assumed that the iron was being changed to gold. About this time the deceiving alchemist would collect the villagers' money and ride off into the night.

Since the seventeenth century, however, chemistry has made steady and even spectacular advances both experimentally and theoretically. Today, chemistry, like all the other sciences, is on a very firm basis of logic and experimentation.

CHEMICAL AND PHYSICAL CHANGES

What is the difference between the melting of wax and the burning of wax and between the dissolving of sugar in water and the heating of sugar until it chars? The answers lie in the difference between physical changes and chemical changes. Physical changes do not alter the chemical composition of a substance but they do alter one or more of its physical properties such as color, structure, density, transparency, and hardness. Physical changes would not, however, alter the temperature at which substances melt or boil. Examples of physical changes are the melting of wax, the dissolution of sugar, and the freezing of water.

All samples of a given substance will show identical physical properties under similar conditions, and, as a result, physical properties can be used to identify unknown substances. If one had samples of the metals cadmium, silver, and platinum, all of which are silver-colored, and wished to determine which sample was which metal, the physical property of density would be sufficient to determine the identity of all three.

Substance	Density (g/cm³)
cadmium	8.6
silver	10.5
platinum	21.5

Other examples of the use of physical properties to distinguish one substance from another are as follows:

Substances	Physical Property Used to Identify
sugar and salt	taste
copper and cadmium	color
oxygen and hydrogen sulfide	odor

Chemical changes result in the formation of one or more new substances which have new physical properties that differ from those of the original substance. Thus, when moist iron reacts with oxygen to produce rust (iron oxide), the rust formed is red-brown in color, has no structural strength, and has a density of 5.2 gm/cm³; whereas the original iron is gray in color, has considerable structural strength, and has a density of 7.9 gm/cm³.

In order to better understand chemical changes, consider the following reaction:

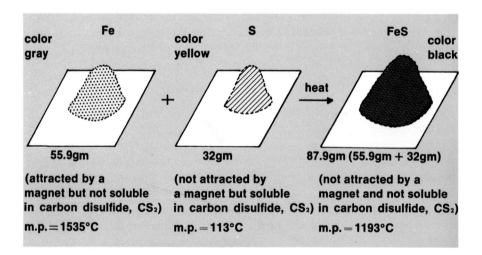

The new substance (FeS) formed from the chemical reaction of iron with sulfur has physical properties that are different from the physical properties of the reactants.

A comparison of the chemical and physical properties of the three substances, sulfur, sugar, and iron, is shown in Table 12.1.

MIXTURES AND SOLUTIONS

There are very few naturally occurring substances that are pure elements or compounds. Nature has produced small quantities of pure copper, silver, gold, platinum, and carbon in the form of diamonds, but most substances found in nature are mixtures. A mixture, which may be homogeneous or heterogeneous,

TABLE 12.1

Different substances have different properties.

Substance	Chemical Properties	Physical Properties
Sulfur	Burns in air. Reacts with metals when heated. No reaction with acid.	Yellow in color. Soluble in CS_2. Insoluble in water.
Sugar	Chars on heating. Reacts with acid.	White solid. Soluble in water. Tastes sweet.
Iron	Reacts with acid. Reacts with moist oxygen to form iron oxide.	Silver-colored solid. High strength. High melting point.

contains two or more compounds or elements. A heterogeneous mixture such as salt-pepper is a mixture of individual salt grains and individual pepper grains, whereas a homogeneous mixture such as salt-water is a mixture at the molecular level—salt molecules intimately mixed with water molecules. Some examples of heterogeneous mixtures are salt-pepper, sugar-flour, concrete, and dirt. Homogeneous mixtures are given a special name—solutions. Some examples of solutions or homogeneous mixtures are shown in Table 12.2.

TABLE 12.2

Some examples of types of solutions.

Example	Type of Solution
sugar-water	solid in liquid
air	gas in gas
carbonated drink	gas in liquid
alcohol-water	liquid in liquid

How does a mixture differ from a compound? The various elements in a compound are chemically bonded to each other in definite proportions by weight. This is the *law of definite proportions.* Thus, in the compound H_2O all molecules of water contain two hydrogen and one oxygen atoms; or stated another way $2 \times 1.008g$ or $2.016g$ of hydrogen always combine with $15.999g$ of oxygen. On the other hand, a mixture can consist of elements or compounds in any weight ratio. For instance, one can mix salt and sugar in any proportions.

Solutions are actually mixtures at the molecular level; that is, the particles of each component of the mixture are molecules rather than large or small particles. When salt is dissolved in water, the salt is called the solute and the water is called the solvent. In general, the solvent is understood to be the medium that the solute is dissolved in and usually it is not too difficult to tell the role both

basic chemistry

substances serve. Thus, when a small amount of alcohol is dissolved in a lot of water, the alcohol is the solute and the water is the solvent. Similarly, if a small amount of water is dissolved in a lot of alcohol, the water is the solute and the alcohol is the solvent. But if one has a fifty-fifty mixture of alcohol and water, which is the solute and which is the solvent? You can name them either way you want.

THE STRUCTURE OF SOLIDS

Just as most of the elements that exist in the universe are solids, so are most of the compounds that nature has produced or that man has synthesized. With the exception of water, man sees relatively few liquids, and the ones that he sees are mostly solutions or suspensions of water or oil. He hardly ever sees gases since most gases are not colored and can therefore not be seen. Even though we are completely immersed in the gaseous solution called air, we are not particularly aware of the air.

Many solids such as wood, rocks, or iron are not particularly attractive in appearance. Other solids such as a snowflake or ice crystal, gold, diamonds, or colored glass are extremely attractive. Everyone is familiar with the general properties of solids—solids resist a change in shape and are relatively firm and compact. This resistance to deformation varies considerably with different solids. Some solids, such as glass, are extremely brittle because they are amorphous and have no definite structural order. Therefore, they break under the slightest deformation. Other solids, such as gold, can be beaten into foil that is only one millionth of an inch thick and can be drawn into wire so thin that one mile of the gold wire weighs only one half a gram!

IONIC CRYSTALS

Why should solids resist a change in shape and be quite firm and compact? These characteristics of a solid are the direct result of the attractive forces between the ions or molecules in a solid and their three-dimensional crystal arrangement. Figure 12.1 shows the regular geometric arrangement of Na^+ and Cl^- in solid sodium chloride that constitutes the ionic crystal lattice of the solid.

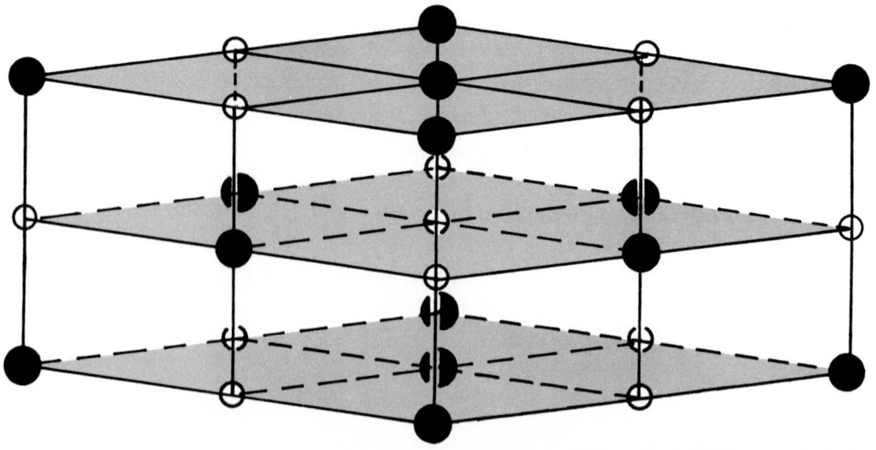

Fig. 12.1 **The ionic crystal lattice of sodium chloride. The black balls represent Cl⁻ and the white balls represent Na⁺.**

Since Figure 12.1 shows only a portion of the sodium chloride lattice, we should realize that the Na^+ and Cl^- would be continued in all three directions. With this understanding, we notice that each sodium ion has six neighboring chloride ions and each chloride ion has six neighboring sodium ions. In short, the sodium chloride crystal consists of Na^+ and Cl^- arranged in such a manner as to produce maximum mutual attraction between the positive and negative ions. This maximum mutual attraction results in a compact, rigid solid that resists deformation.

ATOMIC CRYSTALS

A diamond is an example of a solid consisting of atoms rather than ions. Since atoms are electrically neutral, what then holds the atoms in place? Each atom of carbon in the diamond crystal shares a pair of electrons (covalent bond) with four other carbon atoms; see Figure 12.2. This covalent bonding of every carbon atom in the diamond crystal with four other carbon atoms results in a very hard solid.

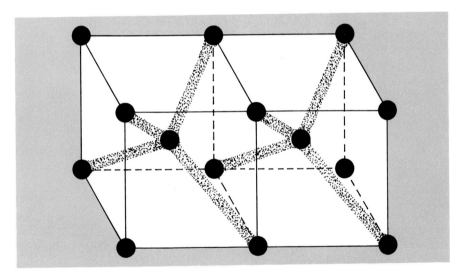

Fig. 12.2 The atomic crystal lattice of diamond.

MOLECULAR CRYSTALS

Molecular crystals are solids constructed of molecules held together by relatively weak forces. These weak forces result in a solid easier to break than ionic or atomic solids. As an example of a molecular crystal, consider the solid *ice*. Ice is composed of water molecules held together in a rigid three dimensional pattern by relatively weak forces. Therefore ice can be easily broken.

The electron pair bond between the hydrogen atoms and the oxygen atom of the water molecule is not evenly distributed between the hydrogen and the oxygen atoms. Because the electronegativity of oxygen is much greater than the

basic chemistry　　　　　　　　　　　　　　　　　　　　　　**273**

electronegativity of hydrogen, the electron pair is pulled closer to the oxygen atom as shown below:

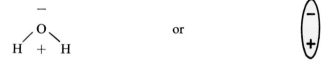

This unequal charge distribution results in a high electron density around the oxygen atom and a low electron density around the hydrogen atoms. This unequal charge distribution produces what is called a polar molecule and can be represented as follows:

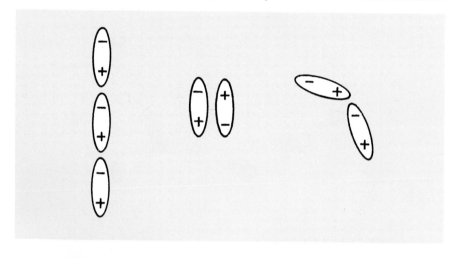

or

The water molecules in ice then would align themselves in a fashion so that the high electron density (oxygen end) of one molecule would be near the low electron density (hydrogen end) of an adjacent molecule as shown below:

Fig. 12.3 **Water crystals (snow-flakes). Most snowflakes have a hexagonal shape.**

(Courtesy of Bailey Library, University of Vermont)

basic chemistry

Some other compounds that form molecular crystals are sugar, waxes, and dry ice (solid CO_2).

METALLIC CRYSTALS

In a way, the most unusual arrangement of crystal lattices occurs in metals. Since atoms of metals generally have rather low electronegativity, the outer electrons of metal atoms are bound to the nucleus by rather weak forces. Indeed, the outer electrons of the atoms of a metal can move within the metal rather freely, so that it would be incorrect to say that a particular electron belongs to a specific nucleus. One way to look at the metal crystal is to consider the positive metal ions in a geometric pattern. These ions are completely surrounded by a cloud of freely moving electrons. It is this freedom of movement of electrons in a metal that makes metals such excellent conductors of electricity.

COMBUSTION

As pointed out at the beginning of this chapter, combustion (fire) was the first chemical reaction man performed. At the same time it was one of man's most momentous accomplishments.

Fire gave man a means of protection against wild animals, of warming himself during periods of cold weather, of cooking his food, and of eventually smelting metal ores to produce the metals essential to a technological civilization.

Combustion is no less important to man today than in the distant past, and for many of the same reasons. Heat from the combustion process is still used to heat most of our homes, to run our cars, to smelt metal ores, and to generate electricity.

Early man did not understand the chemical processes going on when substances burned. The Greeks were the first to attempt an explanation of combustion that did not assume the mystical workings of a god. The Greeks explained that all objects had both earth and fire in them and that during the combustion process the fire was released and the earth, as ashes, remained. It was not until thousands of years later, during the end of the eighteenth century, that the chemical explanation of combustion was known.

Essentially, combustion is the rather rapid combination of oxygen from the air with various materials such as wood or coal. For example, the natural gas that is piped into many homes is essentially methane (CH_4), and it burns according to the following equation.

$$CH_4 + 2O_2 \xrightarrow{\text{flame}} CO_2 + 2H_2O + \text{heat}$$

Wood burns by the same type of reaction, but the complexity of compounds present in wood makes it impossible to show the combustion of wood with a single equation.

OXIDATION AND REDUCTION

Actually, combustion is a specific example of the chemical process of oxidation. One example of combustion with which we are all familiar is the burning of magnesium ribbon in a photographic flash bulb. The flash bulb contains a mass of fine magnesium ribbon in an oxygen atmosphere. When the camera is operated, an electric current is sent through the magnesium ribbon. The passage of the current through the ribbon heats up the ribbon until it begins to burn in the oxygen. The magnesium burns with such an intense flame that for a short period of time a brilliant light is produced. This reaction of oxygen with an element or compound is known as oxidation. In the case of the combustion of magnesium in oxygen, the magnesium is said to be oxidized to magnesium oxide.

We find, however, that combustion may occur in the absence of oxygen if another reactive gas is present. Magnesium ribbon also burns quite easily in the presence of chlorine gas and in doing so produces a very hot, intense flame.

$$2\ Fe + 3Cl_2 \xrightarrow{\text{flame}} 2\ FeCl_3 + heat$$

Since some material will burn in gases other than air (oxygen), as well as chlorine, it is better if the term *oxidation* is defined to include more reactions than combustion in air.

To see if there is an underlying principle that can be used to define more generally the term *oxidation,* let's look more carefully at the equations for the burning of magnesium in oxygen and the burning of magnesium in chlorine.

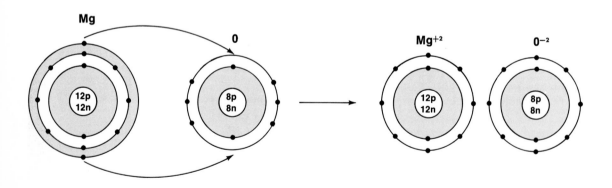

In the above reaction, magnesium loses electrons (two), and oxygen gains electrons (two). When magnesium reacts with chlorine, magnesium loses

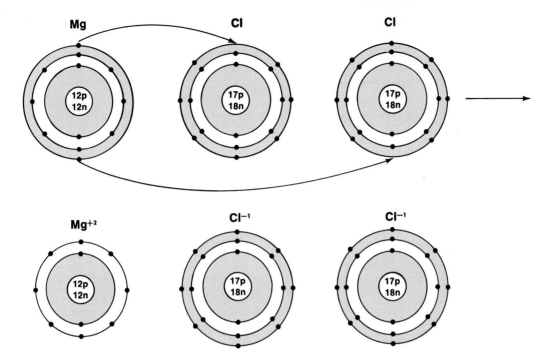

electrons (two), and each chlorine atom gains an electron. In both reactions magnesium is oxidized and loses electrons. The important change that magnesium undergoes in both oxidations is the loss of electrons. Consequently, oxidation is broadly defined as the loss of electrons. The gain of electrons is called *reduction*. In a more restricted sense, the loss of oxygen by a compound is also called reduction. If carbon is heated to a very high temperature with magnesium oxide, the magnesium oxide is reduced to metallic magnesium.

$$MgO + C \xrightarrow{\quad 2000°C \quad} Mg + CO$$

The magnesium oxide loses oxygen and, therefore, is reduced. The magnesium ion of magnesium oxide also gains electrons (reduction) as is shown below.

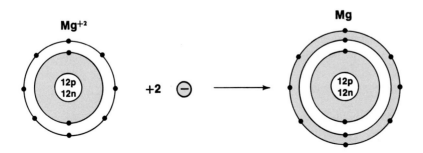

When reduction occurs, electrons are gained by atoms or ions. In order for electrons to be gained, they must first be lost by the process of oxidation. Likewise, when electrons are lost in an oxidation process, the lost electrons are always taken up (gained) by other atoms or ions present in the reaction. It can be said, then, that oxidation is always accompanied by reduction, and reduction is always accompanied by oxidation. If in a chemical reaction one substance is oxidized, another substance must be reduced and vice versa.

CHAPTER SUMMARY

The different elements present in a compound are chemically bound to each other in definite proportions by weight. This is known as the *Law of Definite Proportions*.

A *mixture* is a combination of two or more substances which are not chemically united and which exist in no fixed proportion to each other.

A *solution* is a mixture on the molecular level.

The atoms or molecules of a *solid* are held together by various forces so as to give the solid a definite shape and to resist deformation.

An *ionic crystal* is a solid resulting from the three-dimensional arrangement of ions. This arrangement gives maximum mutual attraction of the positive and negative ions in the crystal.

An *atomic crystal* is a solid in which the atoms are covalently bound to each other in a definite three-dimensional pattern.

Molecular crystals are solids constructed of molecules held together by dipolar forces.

Metallic crystals are made up of positive metal ions arranged in a fixed geometrical pattern and immersed in a group of freely moving electrons.

The chemical combination of oxygen with another element is known as *oxidation*. Oxidation is defined in more general terms as the loss of electrons by an element or compound.

The rapid oxidation of a substance is known as *combustion*.

When an oxygen-containing compound chemically loses its oxygen, the process is called *reduction*. A more general definition of reduction is the gain of electrons.

QUESTIONS AND PROBLEMS

1. What were the goals of the alchemists? (p. 268)
2. What is the difference between a physical change and a chemical change? (p. 269)
3. How could one separate a mixture of powdered iron and powdered sulfur? (p. 270)
4. What is the difference between a mixture and a solution? (p. 271)
5. Why do solids resist deformation? (p. 272)

6. How does the unequal charge distribution in water molecules affect the resistance of ice to deformation? (p. 274)

7. What is the principle behind quenching flames by a foam-type fire extinguisher?

8. Explain the oxidation that occurs when methane burns in air. (p. 275)

9. Explain the oxidation that occurs when iron reacts with sulfur. (p. 276) Write equations showing the processes of oxidation and reduction for the above reaction. (p. 276)

10. Is the chemistry of respiration in humans an example of oxidation? Explain.

11. Explain the difference in an element, a compound, and a mixture. (p. 271)

(Courtesy of F. Byrne)

water
and
ionization
13

WATER: MAN'S MOST IMPORTANT CHEMICAL
THE ACTION OF WATER ON IONS AND MOLECULES
NON-IONIC MOLECULES THAT PRODUCE IONS
ACIDS
BASES
NEUTRALIZATION AND SALTS

vocabulary

Acid | Base
Ion | Ionization
Neutralization

In the preceding chapter we discussed the formation of ions by the loss and gain of electrons. In addition, we saw how the ions of an ionic compound were arranged in the solid state. Now, we will be concerned with how these same ions behave when they are dissolved in water. Do they maintain the rigid three-dimensional pattern they have in the solid state, or do they freely move about among the water molecules? Do the water molecules affect the attraction between the ions? These are some of the questions we will answer in this chapter.

WATER: MAN'S MOST IMPORTANT CHEMICAL

The chemical compound that exists in overwhelmingly greater volume than any other compound on the face of our planet is water. There are four hundred thousand million billion (4×10^{20}) gallons of water in the ocean basins of Earth and twelve thousand million billion (12×10^{18}) gallons of frozen water in the polar caps and ice sheets of our globe. This ubiquitous substance covers about 71% of Earth's surface in the form of water and approximately 10% in the form of huge ice sheets covering vast stretches of land and water. Water is present in gaseous solution in our atmosphere and physically trapped and chemically combined within the rocks of Earth's crust and mantle. Water falls from the skies as soft, delicate snowflakes or as hail in the form of hard fragments of ice or as pelting rain. Of course, most of the water that falls upon the land inexorably finds its way back to the great ocean basins of our planet.

Most naturally occurring substances known to man exist in varying concentrations in solution in our oceans, which are approximately a 3.5% solution of dissolved materials. These dissolved substances are present mostly in the form of ions. The ions that constitute a major portion of the dissolved matter are Na^{+1}, Mg^{+2}, Ca^{+2}, K^+, Cl^-, SO_4^{-2}, and HCO_3^-. If there is such a thing as a universal solvent, surely it must be water.

Man is quite dependent on water for his very existence. He must drink water daily in order to live. He must use huge quantities to maintain the vast industrial structure of his society and he must have water available for purposes of cleanliness and sanitation. It is this very dependence on water by man that makes the pollution of rivers, lakes, and oceans so dangerous in terms of man's future on this planet. Not only does our society's continued contamination of its bodies of water affect the beauty of our globe; pollution also destroys, in large quantities, the foods that we must take from the sea in order to feed even partially our present population. Much of the food which we are presently harvesting from the seas is contaminated by the pollution that surrounds it. Indeed, our society must immediately change its attitudes and redirect its resources if man is to preserve in a usable state the most important and precious chemical compound that we have. Our Earth is the only "water" planet in the solar system, and surely we must protect it if man as the social creature we know today is to survive.

THE ACTION OF WATER ON IONS AND MOLECULES

As we have seen in the preceding chapter, water molecules are polar. This polarity of water molecules arises from the differences in electronegativity of oxygen and hydrogen. This electronegative difference causes the build-up of electron density around the oxygen atom with a resulting low electron density around the hydrogen atoms. Even though the water molecule is electrically neutral overall, there is an unequal distribution of the electrons within the molecule. The above situation is usually referred to as unequal charge distribution, and the molecule is said to be polar. It was further pointed out in the preceding chapter that the water molecules align themselves so that the high charge density of the oxygen end of one water molecule is next to the low electron density of the hydrogen atoms of an adjacent water molecule as shown below:

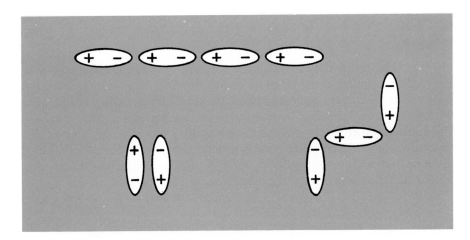

If water molecules are grouped together because of this polarity, certainly they should group around an ion that is in water solution. Let's consider what happens when a grain of salt is dissolved in water. In the solid crystalline state, Na^+ and Cl^- are held together by mutual attraction. When the solid NaCl is placed in water, the positive sodium ions on the outside of the solid attract the high electron density end of the solvent water molecules, and the negative chloride ions attract the low electron density end of the solvent water molecules. The result is that the Na^+ and Cl^- are surrounded by the appropriate ends of water molecules. This arrangement of water molecules weakens the attraction of Na^+ for Cl^- with the result that eventually the Na^+ and Cl^- bonds break.

water and ionization   **283**

The water molecules continue to align themselves around the ions (solvate the ions) until an envelope of water molecules completely encases the separate ions as shown below.

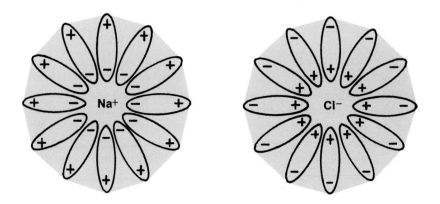

It turns out, then, that negative and positive ions in solution are not held rigidly together by ionic bonds as they are in the solid state, but are completely free to move about at random among the water molecules. Also, the ions do not exist as individual ions but carry with them a complete envelope of water molecules.

The behavior of molecules that are not ionic in nature when dissolved in water is fairly similar to the behavior of ions in water solution. Thus, though sugar molecules are not composed of ions, the sugar molecules, like water molecules, have unequal distribution of electrons and are, therefore, polar in nature. Consequently, when sugar molecules are dissolved in water, water molecules align themselves in appropriate fashion to encase the sugar molecule in a water envelope. Indeed, it is this ability of water to align itself around ions or polar molecules that accounts for the solubility of various substances in water.

NON-IONIC MOLECULES THAT PRODUCE IONS

The vinegar that one finds on the kitchen shelf at home is approximately a five percent solution of acetic acid in water. This vinegar will conduct an electric current. If, however, one uses 100% pure acetic acid with no water present, the pure acetic acid will not conduct an electric current. We know that pure water does not conduct an electric current whereas salt water does. We might assume then that since salt water contains sodium and chloride ions, it is the presence of ions in the salt water that is responsible for the conduction of an electric current. Since acetic acid molecules themselves do not consist of ions, how then do water molecules produce ions from the non-ionic acetic acid molecules?

water and ionization

The formula for acetic acid is

$$H-\underset{\underset{H}{\overset{\overset{H}{|}}{|}}}{C}-\overset{\overset{O}{||}}{C}-O-H$$

The hydrogen-oxygen bond of acetic acid, like the hydrogen-oxygen bond of water, is polar and, therefore, water molecules align themselves around the acetic acid hydrogen-oxygen bond as shown below:

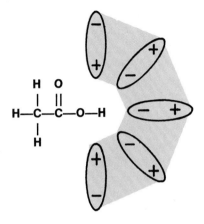

This arrangement of water molecules around the covalent hydrogen-oxygen bond weakens the bond to such an extent that the bond ruptures to produce a positively charged hydrogen ion and a negatively charged oxygen ion as shown below:

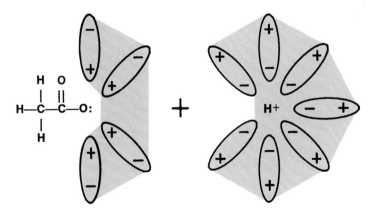

Acetic acid undergoes solvation by water molecules to cause the covalent acetic acid structure to form ions. This process of ion formation from molecules is called *ionization*. It is now easy to understand why pure acetic acid does not conduct an electric current whereas acetic acid dissolved in water does.

water and ionization   **285**

ACIDS

Acetic acid has a sour, tart taste and changes the color of the dye litmus from blue to pink. In fact, vinegar is sometimes added when cooking purple cabbage to change the cabbage color from purple to pink. The purple color of the cabbage is caused by a dye similar to litmus. The sour taste of a substance, coupled with its ability to change the color of litmus from blue to pink, characterizes substances called acids. Some acids are strong and some are weak. Sulfuric acid (H_2SO_4) is a strong acid and attacks skin, clothing, metal, and most other substances. On the other hand, carbonic acid (H_2CO_3), made by dissolving carbon dioxide in water, is a very weak acid and is so harmless that it is used in many beverages. In fact, the term *carbonated beverage* comes from the fact that the beverage contains carbon dioxide and, therefore, carbonic acid. Hydrochloric acid (HCl), present in our stomachs, is a strong acid, but our stomachs contain only a dilute solution of HCl. It is interesting that whereas the hydrochloric acid in our stomach attacks and partially decomposes the food that we eat, it does not attack our stomach walls, a very fortunate situation. We are conscious of the acid content of certain citrus fruits, such as the grapefruit and the lemon, because of their sour taste. We are not aware, however, of the acid content of the orange or the tangerine. Yet all four of the above-mentioned citrus fruits contain about the same concentration of acid (citric acid), but the orange and the tangerine have enough natural sugar present to mask the sour taste of the citric acid so that it is not noticeable.

Why are acids sour tasting, and why do they turn blue litmus to pink? The reason is that in water solution acids ionize to produce hydrogen ions (H^+). It is the hydrogen ions that are responsible for the acidic properties of acids.

Actually, the hydrogen ions produced by acids do not remain by themselves in water solution. The hydrogen ions react with a neutral water molecule to form what is called the hydronium ion as shown below.

$$H^+ + H_2O \longrightarrow H_3O^+$$

The reaction of a hydrogen ion with water is not very unusual considering that the hydrogen ion is actually a lone proton. The +1 charge of the proton is concentrated in an incredibly small volume and, therefore, has a very high charge concentration in its small volume. It is natural that this bare proton with its very high charge density be attracted to the high electron density of the orbital electrons of the oxygen atom of water. To be more precise in our equations showing the reaction of non-ionic acids with water, we should show the formation of hydronium ions rather than hydrogen ions.

$$HCl + H_2O \longrightarrow H_3O^+ + Cl^-$$
$$H_2SO_4 + 2H_2O \longrightarrow 2H_3O^+ + SO_4^{-2}$$

BASES

Bases are compounds that have a bitter taste and a soapy feel and that change the color of litmus from pink to blue. Whereas most acids are non-ionic liquid compounds, most bases are ionic solid compounds. As evidence of the ionic nature of bases, a pure base that has been melted will conduct an electric current. Some bases, such as sodium hydroxide (NaOH), are strong bases whereas others, such as magnesium hydroxide (milk of magnesia), are relatively weak bases. Strong bases used in various commercial products to unstop the drains of sinks and commodes consist mostly of the base sodium hydroxide. Many liquid preparations used to remove wax from floors contain the weak base ammonium hydroxide. The depilatory agents used to remove unwanted body hair are fairly strong solutions of a base.

When bases are dissolved in water, they all yield one or more hydroxide ions as shown below:

$$NaOH \longrightarrow Na^+ + {}^-O\text{–}H$$
$$Ca(OH)_2 \longrightarrow Ca^{+2} + 2 \ {}^-O\text{–}H$$

However, some bases such as ammonia (NH_3) are not ionic and do not contain hydroxide ions in the pure state. When ammonia is dissolved in water, however, it chemically reacts with the water molecules to produce hydroxide ions.

$$NH_3 + H_2O \longrightarrow NH_4^+ + {}^-O\text{–}H$$

NEUTRALIZATION AND SALTS

One characteristic of acids and bases is that they react with each other, and if equal amounts of acid and base are added together, the resulting solution is neither acidic nor basic, but neutral. The reaction of an acid with a base is known as *neutralization*. Thus, sodium hydroxide is neutralized by hydrochloric acid to produce sodium chloride and water.

$$NaOH + HCl \longrightarrow NaCl + H_2O$$

This type of reaction is quite general as shown below:

$$Ca(OH)_2 + H_2SO_4 \longrightarrow CaSO_4 + 2H_2O$$
$$Mg(OH)_2 + 2HCl \longrightarrow MgCl_2 + 2H_2O$$

Although the common name of NaCl is "salt," the term *salt* refers to a whole family of compounds. For our purposes here, we can consider a salt as the ionic substance produced when an acid and a base react. Thus, calcium sulfate

($CaSO_4$), formed from the reaction of sulfuric acid (H_2SO_4) with calcium hydroxide ($Ca(OH)_2$), and magnesium chloride ($MgCl_2$), formed from the reaction of magnesium hydroxide ($Mg(OH)_2$) with hydrochloric acid, are both salts. Salts are characterized as ionic solids with extremely high melting points. Some examples of salts, other than table salt, are baking soda ($NaHCO_3$) and epsom salts ($MgSO_4$).

CHAPTER SUMMARY

Water is a polar molecule and, therefore, dissolves other molecules that have unequal charge distributions such as ionic and covalent polar compounds. In the dissolution of a compound in water, water molecules align themselves so that the low electron density of the water molecule is near the high electron density of the dissolving polar molecule or ion and vice versa. Consequently, the solution of a substance in water results in the substance's being completely encased in water molecules.

An *acid* produces hydrogen ions (H^+) in water solution. Once the hydrogen ion is present, it reacts with a water molecule to form a hydronium ion (H_3O^+).

Bases are compounds that produce a hydroxyl ion (HO^-) in water solution.

When an acid reacts with a base to produce a salt and water, the process is termed *neutralization.*

Salts are ionic crystalline solids consisting of positive metallic ions and negative nonmetallic ions. Salts can be produced by the reaction of an acid with a base.

When a compound has a bond that is sufficiently polar, water molecules align themselves around the polar bond, and the alignment of the water molecules causes the bond to rupture and to produce positive and negative ions in solution; that is, *ionization* occurs.

QUESTIONS AND PROBLEMS

1. Explain what happens when HCl dissolves in water. (p. 286)
2. Why does pure liquid HCl not conduct an electric current but HCl dissolved in water does conduct an electric current? (p. 284)
3. Explain the difference between an acid and a base. (p. 286)
4. How could one prepare a solution of protons? (p. 286)
5. How would you define a salt? (p. 287)
6. Why are curdled milk, grapefruit juice, and vinegar all sour tasting? (p. 286)

organic
chemistry
14

THE CENTRAL ROLE OF CARBON
HYDROCARBONS
ALCOHOLS AND ETHERS
ORGANIC ACIDS
ESTERS, FATS, AND OILS
ALDEHYDES AND KETONES
AMINES, AMINOACIDS AND PROTEINS
CARBOHYDRATES
DNA – THE FUNDAMENTAL MOLECULE OF LIFE

vocabulary

Aromatic Compounds	Enzymes
Hydrocarbons	Isomers
Organic Compound	Polymer

There must exist over a million compounds containing the element carbon. There are probably more individual compounds of carbon in the world than all other compounds combined. Most of the chemical compounds that make up our bodies, and the substances of all living creatures, are compounds containing carbon.

Early scientists used the term *organic compounds* to describe compounds closely related to living processes. Since, for instance, urea was found only in the urine of animals, and sugar was found only in living plants, and vinegar and ethyl alcohol were produced only by the fermentation of various parts of living plants, it was natural for man to assume that these organic compounds were different in some mysterious way from inorganic compounds from which non-living matter is constructed. It was further believed that it was impossible for man ever to synthesize organic compounds because only a god-like vital force could produce the compounds associated with a living organism. In 1829, however, the German chemist Friedrich Wohler found that the inorganic compound ammonium cyanate could be converted by heat into the organic compound urea. Since that time a vast number of compounds present in living organisms have been duplicated in the laboratory by man. Today the term *organic compound* is used to indicate any compound containing carbon (other than a few simple compounds such as carbon dioxide) rather than just compounds found in living cells.

THE CENTRAL ROLE OF CARBON

Why should the element carbon be so favored that it can form such a vast array of different compounds whereas other elements form comparatively few compounds? It is because carbon, like only a few other elements, can combine with itself to form chains consisting of carbon atoms. In fact, diamond consists solely of carbon atoms bonded to each other. Silicon can also combine with itself to form chains but not with the versatility exhibited by carbon.

Carbon has four electrons in its outer shell. Each of these electrons combines with an electron from another atom to form a total of four electron pair covalent bonds. There are many elements that carbon combines with, such as hydrogen, oxygen, sulfur, chlorine, and iodine, but carbon also very readily combines with itself. An example of a compound containing a carbon chain is shown below. Only the electrons actually involved in covalent bonding are shown for each atom.

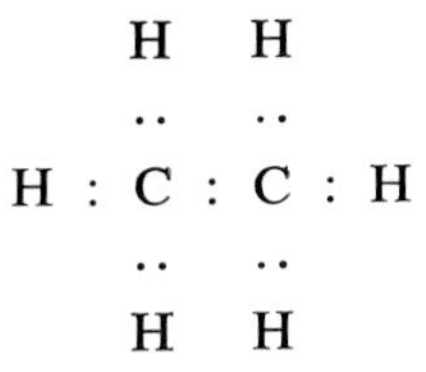

```
        H   H
        ..  ..
   H :  C : C : H
        ..  ..
        H   H
```

ethane

Generally, the covalent bonds are represented by dashes as shown below:

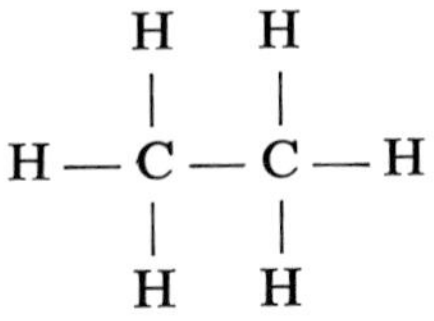

```
        H   H
        |   |
   H —  C — C — H
        |   |
        H   H
```

ethane

The above formula is sometimes written as $CH_3 — CH_3$ and occasionally condensed into an empirical formula C_2H_6.

HYDROCARBONS

The least complex group of organic compounds contains only hydrogen and carbon and is, therefore, called a hydrocarbon. Without this class of organic compounds our mechanized society could not exist as we know it since included in the hydrocarbons are the various gasolines and oils essential as fuels and lubricants for our internal combustion engines. From an economic point of view, finding petroleum, a complex mixture of hydrocarbons, has captured the popular imagination second only to finding gold as a way to quick and great wealth.

The simplest hydrocarbon, methane, contains a single carbon atom and four hydrogen atoms.

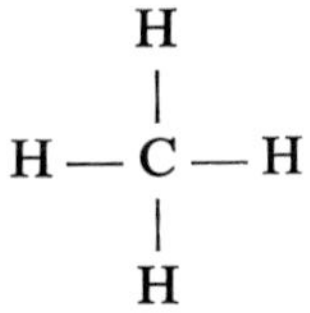

```
        H
        |
   H —  C — H
        |
        H
```

Methane is the main constituent of natural gas, which is used to heat homes and supply energy for industries. Methane is constantly being released into the atmosphere in small quantities by the anaerobic (without oxygen) decay of organic matter from the bottoms of swamps. Bubbles of methane gas, often called *marsh gas,* can be seen rising to the surface of swamps and marshes. Also, the deadly gas called *fire damp* that causes explosions in coal mines is methane. Although there are only trace amounts of methane in our atmosphere today, methane was one of the major constituents of Earth's primitive atmosphere before life came to our planet.

"Bottled" gas, used to heat homes, is a mixture of propane and butane.

butane   propane

Gasoline, the petroleum fraction which boils between 40° and 200°C, is a mixture of hydrocarbons which have from five to ten carbons per molecule. Kerosene, used in tractors and jet engines, is a mixture of hydrocarbons having from twelve to eighteen carbon atoms per molecule. Other products obtained from petroleum are lubricating oils, petroleum jelly (Vaseline), petroleum greases, paraffin waxes, and asphalt.

The hydrocarbon formulas given so far in this chapter show only a single covalent bond between any two carbon atoms. Hydrocarbons having this particular feature are referred to as saturated hydrocarbons. Saturated hydrocarbons burn very readily in air and under proper conditions react with chlorine, but, in general, they are extremely inert to most reagents. The reaction of the saturated hydrocarbon methane with chlorine in the presence of ultraviolet light substitutes some of the hydrogen in the methane molecule for chlorine to produce among other things $CHCl_3$ (the general anaesthetic, chloroform) and CCl_4 (carbon tetrachloride—used in fire extinguishers and dry cleaning).

When chemists were first investigating the chemistry of hydrocarbons, they became aware of an initially confusing situation. They found that two different compounds existed, both of which had the empirical formula of butane. One butane boiled at −0.6°C, and the other butane boiled at −10°C. It was soon learned that the two butanes differed in the way in which their carbon chains were constructed, as shown below:

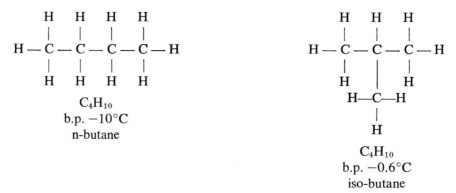

Different compounds that have the same empirical formula are called isomers. Thus, n-butane and iso-butane are isomers of each other. Whereas butane (C_4H_{10}) has two isomers, pentane (C_5H_{12}) has three isomers.

$$CH_3 - CH_2 - CH_2 - CH_2 - CH_3$$
n-pentane

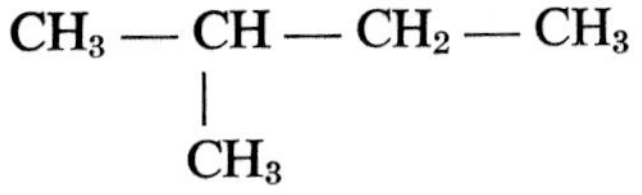

$$CH_3 - CH - CH_2 - CH_3$$
$$|$$
$$CH_3$$
iso-pentane

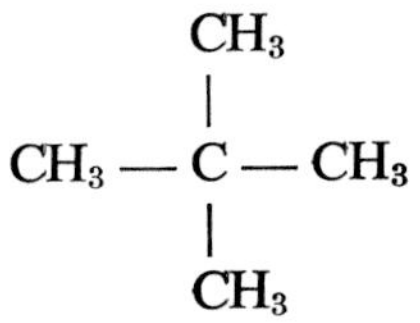

$$CH_3$$
$$|$$
$$CH_3 - C - CH_3$$
$$|$$
$$CH_3$$
neo-pentane

A hydrocarbon with twenty carbons in its chain has approximately one third of a million isomers, and a hydrocarbon with thirty carbons in its chain has an unbelievable four million isomers! Of course, only a very few of the isomers mentioned above have ever been isolated.

Hydrocarbons that have more than one bond between two carbon atoms are called unsaturated hydrocarbons, the simplest members of which are shown below:

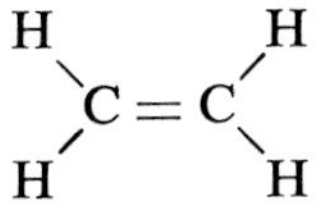

H   H
\   /
C = C
/   \
H   H
ethylene
(used to produce polyethylene)

$$H - C \equiv C - H$$
acetylene
(used in acetylene torches)

$$CH_2 = CH - CH = CH_2$$
butadiene
(used in the synthesis of synthetic rubber)

Unsaturated hydrocarbons, in contrast to saturated hydrocarbons, are very reactive organic molecules. This reactivity is the result of the extra bond(s) between carbon atoms in the molecule. These "double" and "triple" bonds present in unsaturated hydrocarbons add other molecules across the bonds as shown below:

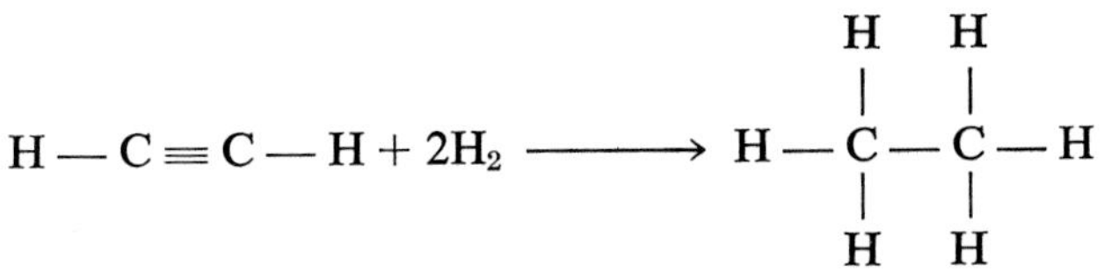

$$H - C \equiv C - H + 2H_2 \longrightarrow$$

H   H
| |
H - C - C - H
| |
H   H
ethane

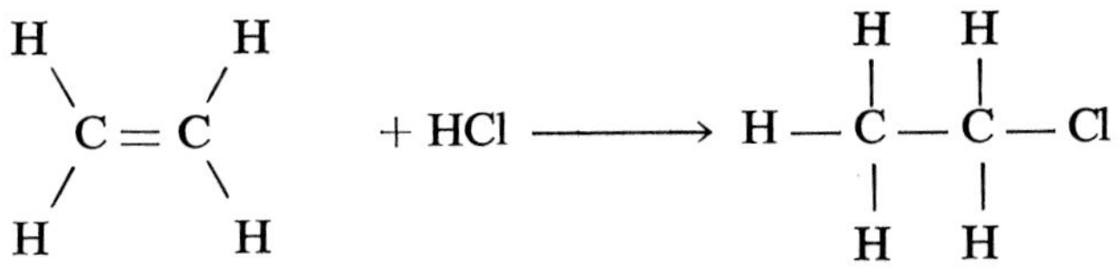

H   H
\   /
C = C   + HCl $\longrightarrow$
/   \
H   H

H   H
| |
H - C - C - Cl
| |
H   H
ethyl chloride

Unsaturated compounds can also react with themselves to produce long chains made up of many units of the starting material. The starting material is called the *monomer,* and the long chain product is called the *polymer.* The process by which a monomer is converted into a polymer is known as *polymerization.* Thus, ethylene polymerizes to produce the well-known polymer polyethylene (poly=many) which is widely used as a transparent packaging material. Two molecules of ethylene at first react as shown below:

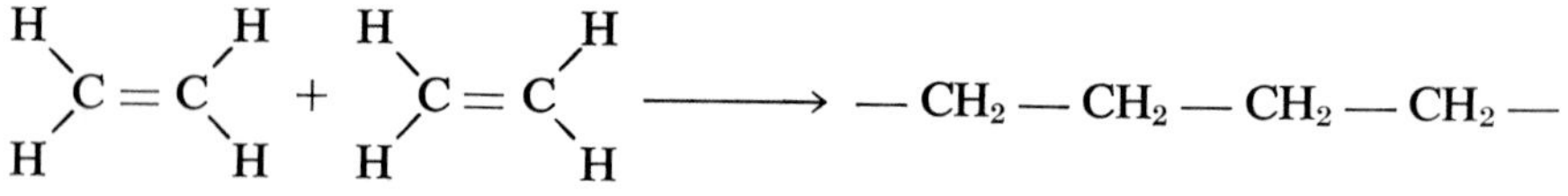

$$
\underset{H}{\overset{H}{>}}C=C\underset{H}{\overset{H}{<}} + \underset{H}{\overset{H}{>}}C=C\underset{H}{\overset{H}{<}} \longrightarrow -CH_2-CH_2-CH_2-CH_2-
$$

Other ethylene molecules react in turn to build up a saturated hydrocarbon consisting of a long chain.

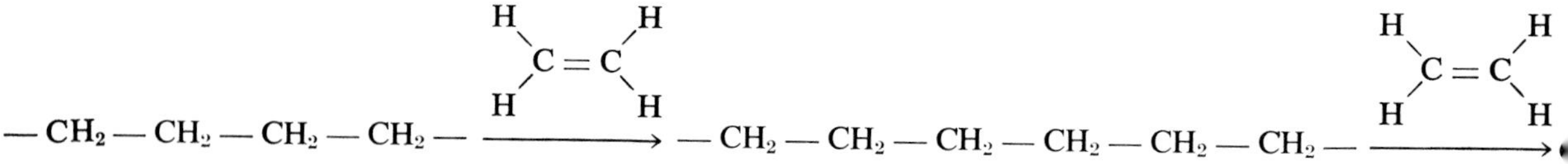

$$
-CH_2-CH_2-CH_2-CH_2- \quad \underset{H}{\overset{H}{>}}C=C\underset{H}{\overset{H}{<}} \longrightarrow -CH_2-CH_2-CH_2-CH_2-CH_2-CH_2- \quad \underset{H}{\overset{H}{>}}C=C\underset{H}{\overset{H}{<}} \longrightarrow
$$

In a fashion similar to the formation of polyethylene, vinyl chloride polymerizes to polyvinyl chloride, which is used to make phonograph records and waterproof wearing apparel.

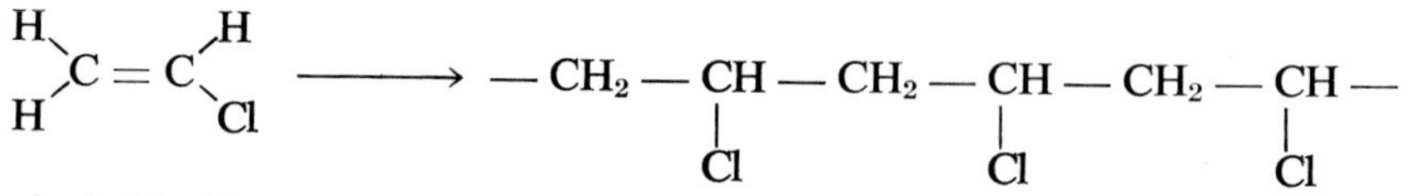

$$
\underset{H}{\overset{H}{>}}C=C\underset{Cl}{\overset{H}{<}} \longrightarrow -CH_2-CH-CH_2-CH-CH_2-CH-
$$

vinyl chloride   polyvinyl chloride

(with Cl substituents below each CH)

A special type of unsaturated hydrocarbon that has unique chemical properties is benzene.

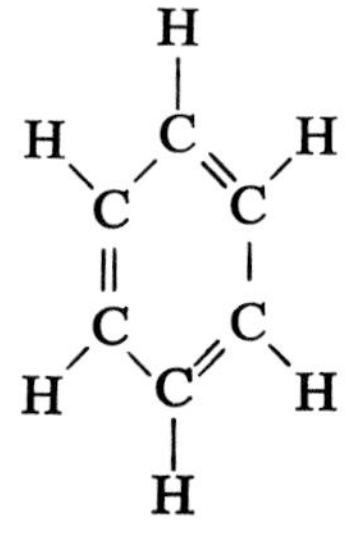

benzene

Benzene compounds like it are known as *aromatic hydrocarbons* and do not add other molecules to their double bonds, nor do they polymerize with themselves. Benzene's unusual properties are the result of its double bonds alternating with single bonds in a cyclic structure. This particular arrangement allows some of the electrons in the double bonds of benzene to circulate in a circle around the benzene molecule with the result that the electrons are not available to react in the fashion of normal unsaturated hydrocarbons. Instead, benzene undergoes substitution reactions similar to the reactions of saturated hydrocarbons in which the hydrogens of benzene are substituted for other atoms or groups of atoms. For instance, the aromatic compound toluene reacts with nitric acid to form trinitrotoluene (TNT).

$$
\text{(toluene)} + 3HNO_3 \longrightarrow \text{(trinitrotoluene)} + 3H_2O
$$

trinitrotoluene
(TNT)

ALCOHOLS AND ETHERS

Organic compounds that have a hydroxyl group ($- O - H$) in place of one of the hydrogen atoms of a hydrocarbon are called alcohols. Methyl alcohol (wood alcohol)

$$
H - \overset{\displaystyle H}{\underset{\displaystyle H}{\overset{|}{\underset{|}{C}}}} - O - H
$$

is produced by heating wood in the absence of air to yield charcoal, methyl alcohol, and other chemicals. Large quantities of methyl alcohol are also made by reacting carbon monoxide with hydrogen. Methyl alcohol is used mainly as an antifreeze and as a solvent. If consumed, methyl alcohol can cause blindness or death.

Perhaps the most widely known alcohol is ethyl alcohol,

$$
H - \overset{\displaystyle H}{\underset{\displaystyle H}{\overset{|}{\underset{|}{C}}}} - \overset{\displaystyle H}{\underset{\displaystyle H}{\overset{|}{\underset{|}{C}}}} - O - H
$$

the alcohol present in all alcoholic beverages. The quantity of ethyl alcohol used annually in alcoholic beverages, however, is only a very small fraction of the total quantity of ethyl alcohol used for industrial purposes in this country. Although a large percentage of the ethyl alcohol produced in this country is made synthetically from ethylene, the most well-known method of production is fermentation of various starches and sugars obtained from corn, potatoes, rye, or molasses.

Isopropyl alcohol,

$$
\begin{array}{ccc}
\text{H} & \text{H} & \\
| & | & \\
\text{H}-\text{C} \!\!-\!\!\!-\!\!\!- \text{C}-\text{O}-\text{H} \\
| & | & \\
\text{H} & \text{H}-\text{C}-\text{H} \\
& | & \\
& \text{H} &
\end{array}
$$

also known as rubbing alcohol, is used extensively as a solvent. Although methyl alcohol and ethyl alcohol do produce intoxication if consumed, isopropyl alcohol, as well as higher molecular weight alcohols, is not intoxicating.

Ethylene glycol,

$$
\begin{array}{cc}
\text{H} & \text{H} \\
| & | \\
\text{H}-\text{O}-\text{C}-\text{C}-\text{O}-\text{H} \\
| & | \\
\text{H} & \text{H}
\end{array}
$$

a molecule containing two hydroxyl groups, is the main ingredient in "permanent antifreezes." Glycerol,

$$
\begin{array}{ccc}
\text{H} & \text{H} & \text{H} \\
| & | & | \\
\text{H}-\text{O}-\text{C}-\text{C}-\text{C}-\text{O}-\text{H} \\
| & | & | \\
& \text{O} & \\
\text{H} & \text{H} & \text{H}
\end{array}
$$

which contains three hydroxyl groups, finds widespread use in tobacco products and cosmetics because of its moisture retaining property. Glycerol reacts with nitric acid to produce nitroglycerine from which dynamite is made.

$$
\begin{array}{ccc}
\text{H} & & \text{H} \\
| & & | \\
\text{H}-\text{C}-\text{O}-\text{H} & & \text{H}-\text{C}-\text{O}-\text{NO}_2 \\
| & & | \\
\text{H}-\text{C}-\text{O}-\text{H} \quad +\,3\text{HNO}_3 \longrightarrow & \text{H}-\text{C}-\text{O}-\text{NO}_2 \quad +\,3\text{H}_2\text{O} \\
| & & | \\
\text{H}-\text{C}-\text{O}-\text{H} & & \text{H}-\text{C}-\text{O}-\text{NO}_2 \\
| & & | \\
\text{H} & & \text{H} \\
\text{glycerol} & & \text{nitroglycerine}
\end{array}
$$

organic chemistry

Nitroglycerine is too easily detonated by shock to be used safely as an explosive, but Alfred Nobel, a Swedish industrialist who established the Nobel prizes, found that mixing nitroglycerine with inert substances such as sawdust produces a product (dynamite) that allows it to be transported and handled safely. The patents that Nobel obtained on the manufacture of dynamite from nitroglycerine resulted in great wealth for him. Part of this wealth, in accordance with Nobel's wishes, is given each year to men who have made outstanding contributions to science, the arts, and humanities.

A very unusual type of alcohol in that it has two — O — H groups attached to the same carbon atom is chloral hydrate, a water solution which is commonly known as "knock-out drops."

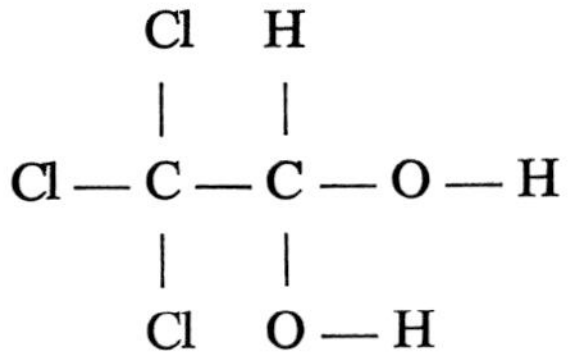

$$
\begin{array}{c}
\underset{\displaystyle|}{Cl}\;\;\underset{\displaystyle|}{H} \\
Cl - C - C - O - H \\
\underset{\displaystyle}{Cl}\;\;O - H
\end{array}
$$

chloral hydrate

In ether compounds, two carbon atoms are joined to a single oxygen atom. Ethyl ether, usually just called *ether,* has been used as a general anesthetic in hospitals since 1846.

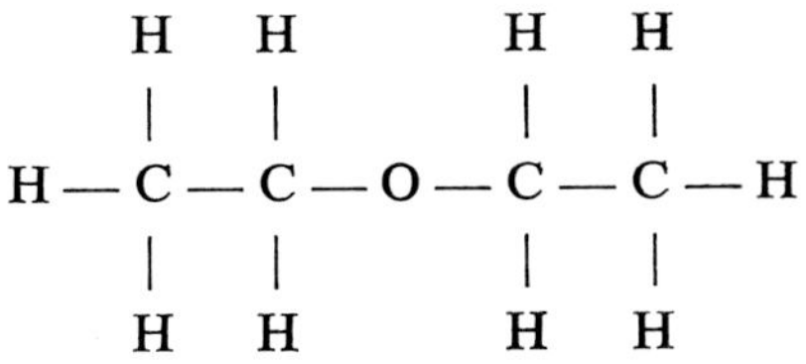

$$
\begin{array}{c}
H\;\;\;HH\;\;\;H \\
|\;\;\;\;||\;\;\;\;| \\
H - C - C - O - C - C - H \\
|\;\;\;\;||\;\;\;\;| \\
H\;\;\;HH\;\;\;H
\end{array}
$$

ethyl ether ("ether")

ORGANIC ACIDS

Molecules of organic acids all contain the carboxyl group $\left(- \overset{\displaystyle O}{\overset{\displaystyle \|}{C}} - O - H\right)$. Like an inorganic acid, an organic acid (a carboxylic acid) is a covalent compound that ionizes in water to produce a hydronium ion. A carboxylic acid does not give up its proton as readily as an inorganic acid and is, therefore, a weaker acid than hydrochloric or sulfuric acid.

The simplest possible carboxylic acid is formic acid,

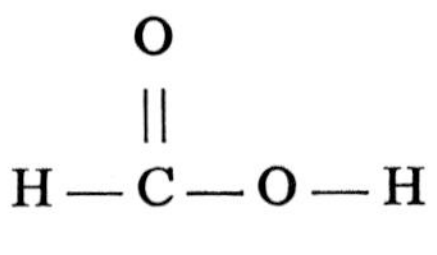

$$
H - \overset{\displaystyle O}{\overset{\displaystyle \|}{C}} - O - H
$$

which was first isolated from the distillation of red ants and got its name from the Latin meaning *ant*. In fact, the pain produced by the bite of a red ant is the result of the ant's giving a hypodermic injection of formic acid to its victim.

Acetic acid,

```
      H    O
      |    ||
  H — C — C — O — H
      |
      H
```

got its name from the Latin meaning *vinegar,* since vinegar is a five percent solution of acetic acid in water. Actually, acetic acid is one of the few organic acids that we purposely put on our foods.

Butyric acid has a very unpleasant odor, and the objectionable odors of perspiration, rancid butter, and certain strong cheeses are due to the presence of butyric acid.

```
      H    H    H    O
      |    |    |    ||
  H — C — C — C — C — O — H
      |    |    |
      H    H    H
```
butyric acid

Capric acid is named from the Latin meaning *goat* since it is this acid in the skin secretions of goats that gives goats their unpleasant smell.

```
      H    H    H    H    H    H    H    H    H    O
      |    |    |    |    |    |    |    |    |    ||
  H — C — C — C — C — C — C — C — C — C — C — OH
      |    |    |    |    |    |    |    |    |
      H    H    H    H    H    H    H    H    H
```
capric acid

It might be interesting to blue cheese and Roquefort cheese lovers that the rank smell of these cheeses is due mainly to the presence of the "goat acid," capric acid.

Citric acid occurs in many plants and animals but is most prominent in certain citrus fruits (from which citric acid gets its name) where the acid tartness is so easily detected. Lemon juice is a seven percent solution of citric acid. Because of citric acid's tartness and pleasant flavor, it is widely used in soft drinks and foods.

Neutralization of long-chain organic acids with bases forms salts that find various uses in our everyday lives. Thus, sodium and potassium salts of long-chain carboxylic acids are the main ingredients of bath soaps; and the calcium salt of propionic acid, called calcium propionate, is added to bakery products as a fungicide to retard spoilage.

$$
\begin{array}{ccccc}
\text{H} & \text{H} & \text{O} & & \text{O} \quad \text{H} \quad \text{H} \\
| & | & || \\
\text{H} - \text{C} - \text{C} - \text{C} - \text{O} \quad \text{Ca} \quad \text{O} - \text{C} - \text{C} - \text{C} - \text{H} \\
| & | & & & | \quad | \\
\text{H} & \text{H} & & & \text{H} \quad \text{H}
\end{array}
$$

+2

calcium propionate

ESTERS, FATS, AND OILS

A carboxylic acid reacts with an alcohol to produce a compound called an ester.

$$
\begin{array}{cccccc}
\text{H} & \text{O} & & \text{H} & \text{H} \\
| & || & & | & | \\
\text{H} - \text{C} - \text{C} - \text{O} - \text{H} + \text{H} - \text{O} - \text{C} - \text{C} - \text{H} & \longrightarrow \\
| & & & | & | \\
\text{H} & & & \text{H} & \text{H}
\end{array}
$$

$$
\begin{array}{cccccc}
\text{H} & \text{O} & & \text{H} & \text{H} \\
| & || & & | & | \\
\text{H} - \text{C} - \text{C} - \text{O} - \text{C} - \text{C} - \text{H} + \text{H}_2\text{O} \\
| & & & | & | \\
\text{H} & & & \text{H} & \text{H}
\end{array}
$$

ethyl acetate (an ester)

An ester then is a compound in which the proton of a carboxylic acid has been replaced with a group containing one or more carbon atoms. Esters are widespread in nature and are mainly responsible for the pleasant fragrances of flowers and the delightful aromas and savory flavors of fruits. The aromatic bouquet of brandies and wines is also produced by esters. Although flowers and fruits generally contain a complex mixture of esters, quite often a single ester present in predominant amounts is mainly responsible for their fragrances. Thus, there are four esters that contribute to the aroma and taste of the pineapple, but it is one ester, ethyl acetate,

$$
\begin{array}{cccccc}
\text{H} & \text{O} & & \text{H} & \text{H} \\
| & || & & | & | \\
\text{H} - \text{C} - \text{C} - \text{O} - \text{C} - \text{C} - \text{H} \\
| & & & | & | \\
\text{H} & & & \text{H} & \text{H}
\end{array}
$$

ethyl acetate

organic chemistry

present in overwhelming quantity (97%), that dictates the predominant odor and flavor characteristics of pineapples. The characteristic smell and flavor of bananas are due to the two esters isoamyl acetate and butyl acetate.

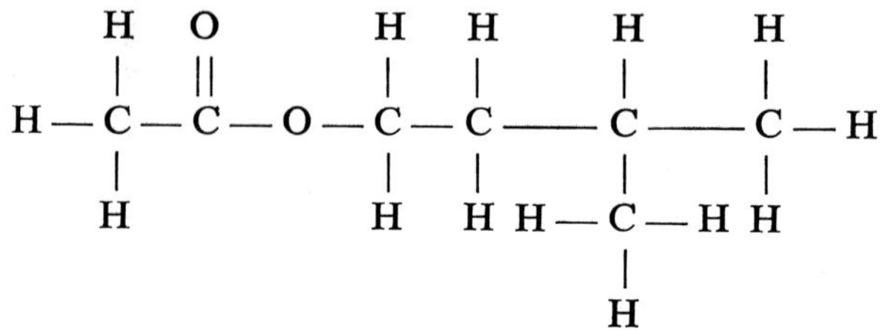

$$
\begin{array}{cccccccc}
& H & O & & H & H & & H & & H \\
& | & || & & | & | & & | & & | \\
H- & C- & C- & O- & C- & C & \underline{\hspace{1em}} & C & \underline{\hspace{1em}} & C-H \\
& | & & & | & | & & | & & | \\
& H & & & H & H & & H-C-H & H \\
& & & & & & & | & \\
& & & & & & & H &
\end{array}
$$

isoamyl acetate

$$
\begin{array}{cccccccc}
& H & O & & H & H & H & H \\
& | & || & & | & | & | & | \\
H- & C- & C- & O- & C- & C- & C- & C-H \\
& | & & & | & | & | & | \\
& H & & & H & H & H & H
\end{array}
$$

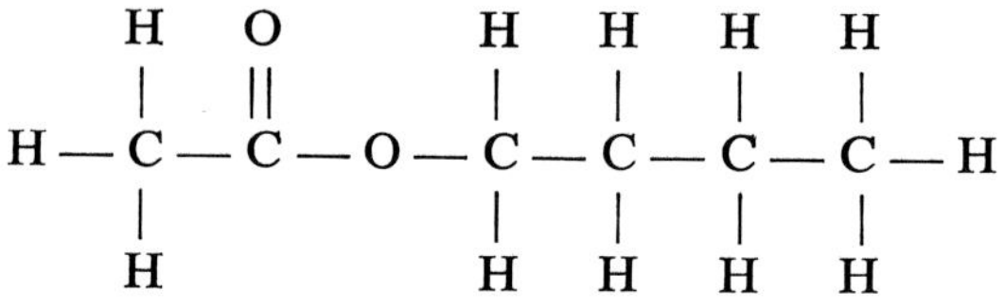

butyl acetate

Pears contain the ester amyl acetate.

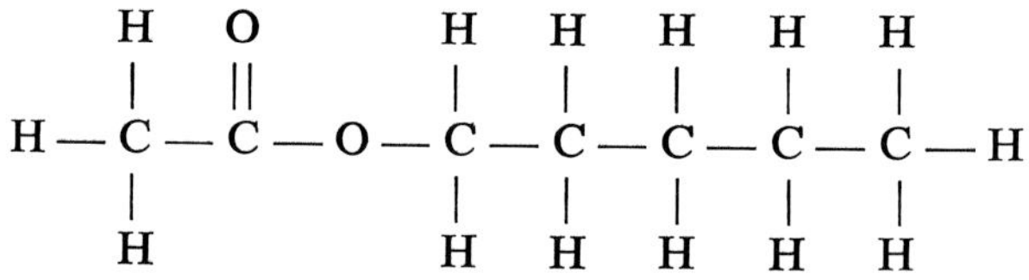

$$
\begin{array}{ccccccccc}
& H & O & & H & H & H & H & H \\
& | & || & & | & | & | & | & | \\
H- & C- & C- & O- & C- & C- & C- & C- & C-H \\
& | & & & | & | & | & | & | \\
& H & & & H & H & H & H & H
\end{array}
$$

amyl acetate

In view of the considerable difference in the odor and flavor of pears and bananas, it is amazing how similar in structure the esters are that are responsible for the odor and flavor of the two fruits. Even the delicate fragrance of roses is due to a bewildering mixture of many esters although the main component is amyl undecanoate.

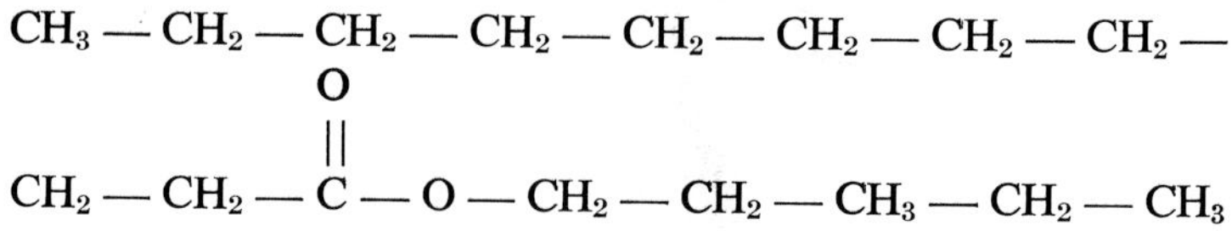

$$CH_3 - CH_2 - CH_2 - CH_2 - CH_2 - CH_2 - CH_2 - CH_2 -$$
$$\overset{\displaystyle O}{\underset{\displaystyle ||}{}}$$
$$CH_2 - CH_2 - C - O - CH_2 - CH_2 - CH_3 - CH_2 - CH_3$$

amyl undecanoate

Since esters may be snythesized very readily in the laboratory, they are used in various mixtures in the preparation of perfumes and artificial food flavors. One reason for synthesizing certain perfumes and flavors is that they are often much less expensive than are the natural products.

Fats and oils are special types of esters formed from the reaction of carboxylic acids with an alcohol (glycerol) containing three hydroxyl groups. Shown below is the reaction between three molecules of a carboxylic acid and one molecule of glycerol. The "R" in the carboxylic acid molecules represents a long carbon chain.

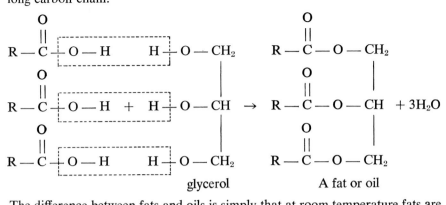

glycerol   A fat or oil

The difference between fats and oils is simply that at room temperature fats are solids and oils are liquids. Fats and oils may be of animal origin (such as beef tallow or lard) or of vegetable origin (such as peanut oil or soy bean oil). Vegetable oils contain a greater number of unsaturated hydrocarbon chains than do animal fats or oils. A certain amount of unsaturated vegetable oils (polyunsaturated oils) is essential to one's diet since a dietary deficiency of unsaturated fats or oils results in impairment of growth and in development of skin lesions. Furthermore, the consumption of unsaturated oils produces a lowering of cholesterol content in the blood. Since a high cholesterol content in the blood results in hardening of the arteries (known as arteriosclerosis), the fat or oil necessary in a healthful diet should be the unsaturated type. On the other hand, one should not include in his diet excessive quantities of fats and oils since they have over twice the calorie content of all other foods. Consequently, we should have a relatively low fat diet.

Many of the waxes found in nature are high molecular weight esters. Thus, beeswax is mostly myricyl palmitate, one molecule of which contains a total of forty-seven carbon atoms.

$$C_{15}H_{31} - \overset{\overset{\displaystyle O}{\|}}{C} - O - C_{31}H_{63}$$

myricyl palmitate

A benzene compound containing two carboxylic acid groups reacts with ethylene glycol to produce a polyester that is sold under the name *Dacron* when it is made into textile fibers and is sold under the name *Mylar* when it is made into thin transparent sheets used for packaging.

ALDEHYDES AND KETONES

If the hydroxyl group of a carboxylic acid is replaced by hydrogen, the resulting compound is called an *aldehyde,* and when the hydroxyl group is replaced by a carbon group, the resulting compound is called a *ketone.*

acetaldehyde
(an aldehyde)

acetone
(a ketone)

As can be seen from their formulas, aldehydes and ketones are characterized by the presence of

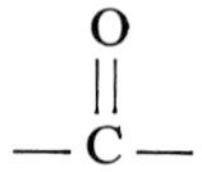

$$\begin{matrix} & O \\ & \| \\ - & C & - \end{matrix}$$

which is known as the *carbonyl group*. Aldehydes have at least one hydrogen atom attached to the carbonyl group whereas ketones have only carbon atoms attached to the carbonyl group. Aldehydes and ketones in general have rather pleasant odors, and those aldehydes having approximately ten carbon atoms in their molecular chain are used in perfume formulations. Ketones are widely used as solvents. One of the ketones, methyl amyl ketone, is partially responsible, along with carboxylic acids, for the distinctive odor of blue cheese. The characteristic flavor of butter is due to the presence of the two ketones shown below:

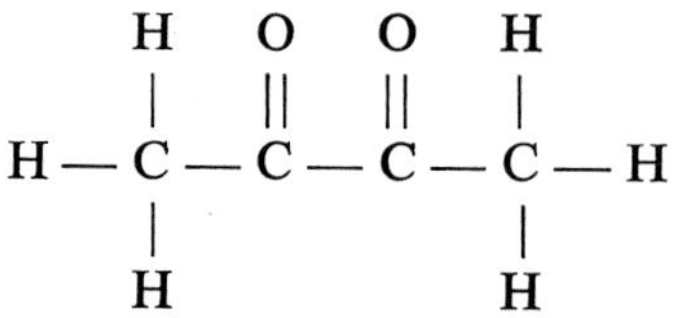

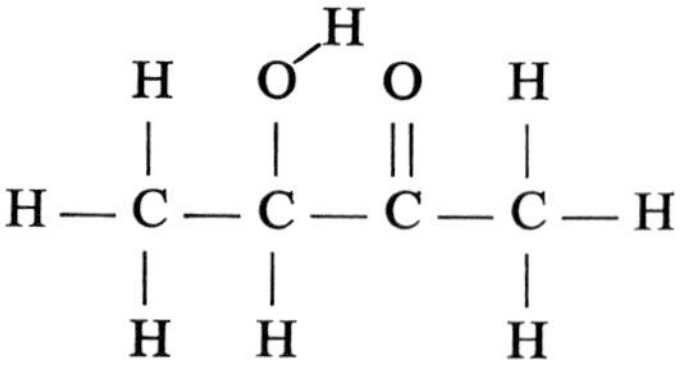

AMINES, AMINO ACIDS, AND PROTEINS

Amines are compounds that have anywhere from one to three carbon atoms attached to nitrogen. Amines as such are not used to a great extent by man, because of their objectionable odors. The decay of flesh, for example, is accompanied by the liberation of various amines. Thus, part of the unpleasant odor of decaying fish is due to trimethylamine

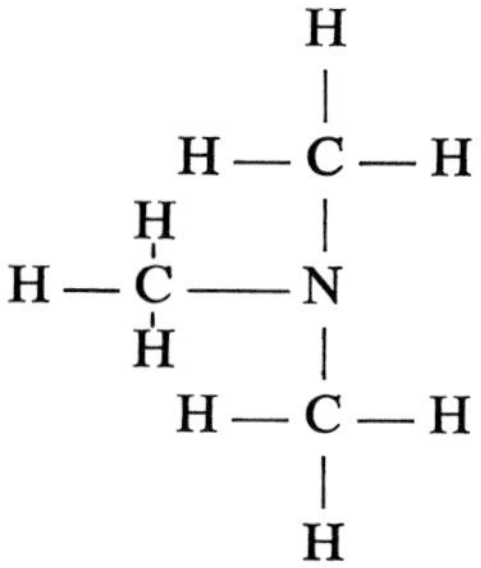

and much of the disagreeable smell of other types of putrefying flesh is due to the presence of two amines (cadaverine and putrescine), each of which contains two amino groups ($-NH_2$). The names of these two amines obviously are intended to reflect their offensive odors.

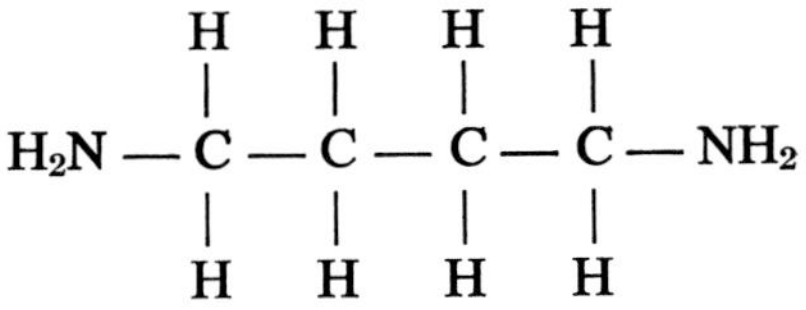

$$H_2N - \underset{\underset{H}{|}}{\overset{\overset{H}{|}}{C}} - \underset{\underset{H}{|}}{\overset{\overset{H}{|}}{C}} - \underset{\underset{H}{|}}{\overset{\overset{H}{|}}{C}} - \underset{\underset{H}{|}}{\overset{\overset{H}{|}}{C}} - NH_2$$

putrescine

$$H_2N - \underset{\underset{H}{|}}{\overset{\overset{H}{|}}{C}} - \underset{\underset{H}{|}}{\overset{\overset{H}{|}}{C}} - \underset{\underset{H}{|}}{\overset{\overset{H}{|}}{C}} - \underset{\underset{H}{|}}{\overset{\overset{H}{|}}{C}} - \underset{\underset{H}{|}}{\overset{\overset{H}{|}}{C}} - NH_2$$

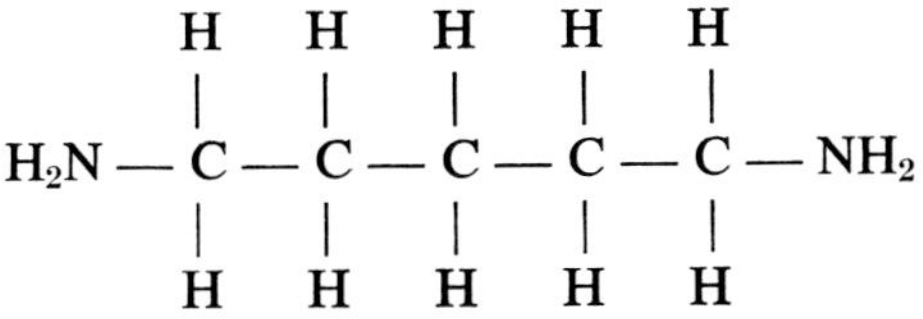

cadaverine

Amines can react with carboxylic acid to form compounds called *amides*.

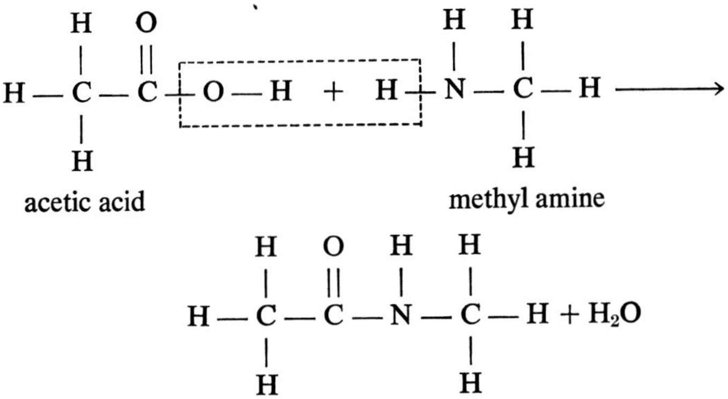

$$H - \underset{\underset{H}{|}}{\overset{\overset{H}{|}}{C}} - \overset{\overset{O}{\|}}{C} + O - H \quad + \quad H + \underset{\underset{H}{|}}{\overset{\overset{H}{|}}{N}} - \underset{\underset{H}{|}}{\overset{\overset{H}{|}}{C}} - H \longrightarrow$$

acetic acid   methyl amine

$$H - \underset{\underset{H}{|}}{\overset{\overset{H}{|}}{C}} - \overset{\overset{O}{\|}}{C} - \overset{\overset{H}{|}}{N} - \underset{\underset{H}{|}}{\overset{\overset{H}{|}}{C}} - H + H_2O$$

an amide
(N-methyl acetamide)

Nylon is a polymer containing many amide units and has the general formula shown below:

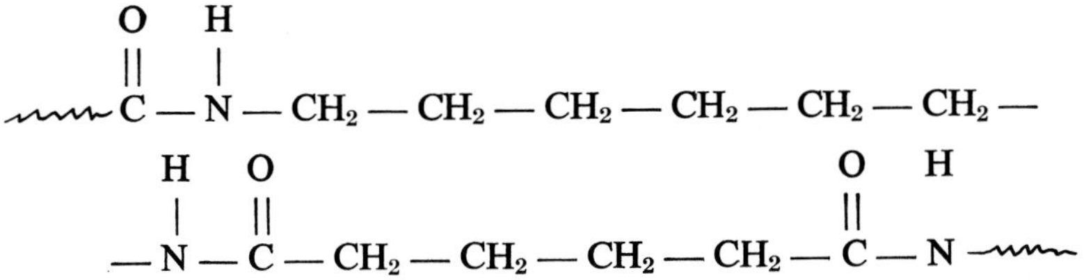

$$\sim\sim\sim \overset{\overset{O}{\|}}{C} - \overset{\overset{H}{|}}{N} - CH_2 - CH_2 - CH_2 - CH_2 - CH_2 - CH_2 -$$

$$- \overset{\overset{H}{|}}{N} - \overset{\overset{O}{\|}}{C} - CH_2 - CH_2 - CH_2 - CH_2 - \overset{\overset{O}{\|}}{C} - \overset{\overset{H}{|}}{N} \sim\sim\sim$$

Amino acids are compounds that contain both an amino group and a carboxylic acid group. Although the total number of amino acids that can be synthesized in the laboratory is virtually unlimited, there are only approximately a hundred that occur in nature. Of these hundred amino acids, about twenty to twenty-five occur in overwhelming abundance. The formula of amino acids can in general be represented as

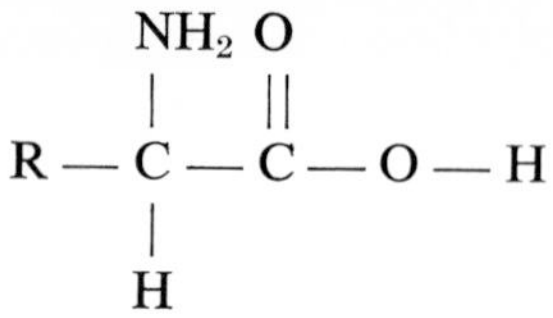

$$\begin{array}{c}\quad\ \ \overset{\displaystyle NH_2}{|}\ \ \overset{\displaystyle O}{\parallel}\\ R-C-C-O-H\\ \overset{|}{H}\end{array}$$

where R represents a large variety of structures. If the R-group of an amino acid is a hydrogen atom, the resulting molecule is the simplest possible amino acid, glycine.

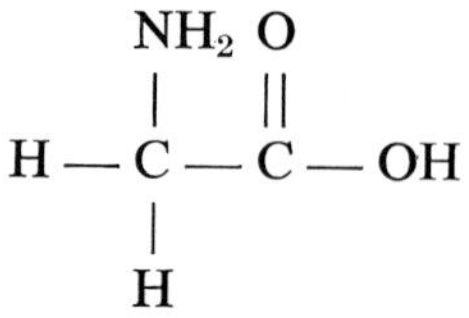

$$\begin{array}{c}\quad\ \ \overset{\displaystyle NH_2}{|}\ \ \overset{\displaystyle O}{\parallel}\\ H-C-C-OH\\ \overset{|}{H}\end{array}$$

glycerine

If the R-group is $H-O-\overset{\displaystyle O}{\overset{\parallel}{C}}-CH_2-CH_2-$, the resulting molecule is glutamic acid.

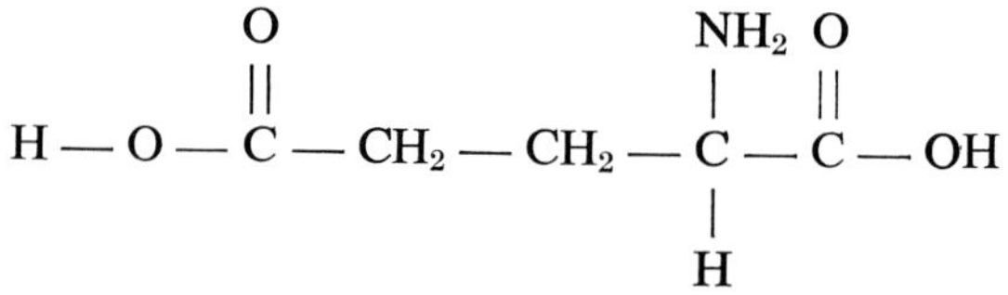

$$H-O-\overset{\displaystyle O}{\overset{\parallel}{C}}-CH_2-CH_2-\overset{\displaystyle NH_2}{\overset{|}{C}}-\overset{\displaystyle O}{\overset{\parallel}{C}}-OH$$
$$\underset{H}{|}$$

glutamic acid

The single sodium salt of glutamic acid is called monosodium glutamate which is used extensively in foods to increase the sensitivity of the tongue's taste buds with the result that flavors naturally present in foods are considerably enhanced. If R is $CH_3-S-CH_2-CH_2-$, the amino acid methionine results.

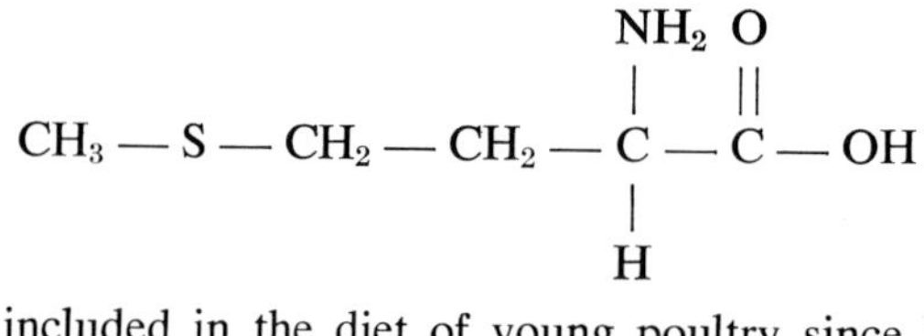

$$CH_3-S-CH_2-CH_2-\overset{\displaystyle NH_2}{\overset{|}{C}}-\overset{\displaystyle O}{\overset{\parallel}{C}}-OH$$
$$\underset{H}{|}$$

Methionine is included in the diet of young poultry since small amounts of methionine considerably increase their growth rate.

All living things require the presence of amino acids in their diet to carry out various biochemical functions necessary to life. Plants can synthesize all the amino acids they require from inorganic substances present in the air and soil. Animals, however, can synthesize approximately only half of the amino acids that they require and must obtain the remaining amino acids that they need by eating plants or by eating animals that have consumed plants.

In essentially the same fashion that molecules of ethylene react to form polyethylene and molecules of amines and carboxylic acids react to form the polymer Nylon, amino acids can react to form large polymeric molecules known as *proteins*. Proteins have the general formula

$$
\begin{array}{ccccccccc}
\text{H} & \text{H} & \text{O} & \text{H} & \text{H} & \text{O} & \text{H} & \text{H} & \text{O} \\
| & | & || & | & | & || & | & | & || \\
\sim\!\!\sim\!\!\text{N} & -\text{C} & -\text{C} & -\text{N} & -\text{C} & -\text{C} & -\text{C} & -\text{N} & -\text{C}\sim\!\!\sim \\
& | & & & | & & | & & \\
& \text{R} & & & \text{R} & & \text{R} & &
\end{array}
$$

Since proteins are of such fundamental importance to life forms (the word protein is from the Greek, *proteios,* first in importance) and since proteins are made up of amino acids, amino acids are frequently called the "building blocks" of life. Proteins may contain as few as fifty-one amino acids as in the hormone insulin (a deficiency of which causes diabetes) or as many as millions of amino acids, strung like beads on a string, as in virus proteins.

Although all parts of living organisms contain varying amounts of proteins, some parts contain much higher concentrations than others. Thus, the flesh, skin, hair, finger nails, toe nails, horns, and hoofs of animals and the seeds of plants have high protein concentrations. Proteins make up approximately three-fourths of the dry weight of animals and are the structural components of animal bodies. Also, all biochemical reactions that occur in living cells are controlled by special proteins called *enzymes*. (See page 309.)

There are literally thousands upon thousands of different protein molecules that exist in nature. How is it possible that such a great variety of separate protein molecules can be produced from only twenty-five different amino acids? In order to answer this question, we must first answer the question "Since proteins are made of only twenty to twenty-five amino acids, how do the proteins differ from one another?" The difference in various proteins is a result of the difference in the sequence in which the various amino acids are arranged. Thus, if we let letters of the alphabet represent different amino acids in a protein molecule, one protein may have an amino acid sequence of ACAGGHACY . . . whereas another protein may have a sequence of ACCHAAYGG. . . . It is because of this difference in possible sequences of the amino acids in protein molecules that there can be such a great variety of protein molecules built up from the same twenty to twenty-five amino acids. If only five different amino acids existed from which to make protein molecules, they could be arranged in various sequences to produce 120 different molecules. Double the number of different amino acids available and 3,628,800 different molecules could be constructed. Triple the number (15) of starting amino acids and over a million million different molecules could be prepared! With twenty to twenty-five amino acids available in nature, we are not surprised then at the large diversity of protein molecules found in different life forms.

CARBOHYDRATES

What do wood, cotton, starch, sugar, and the exoskeletons of insects and crabs all have in common? They all are made up to a large extent of a class of organic compounds called carbohydrates. Carbohydrates are molecules containing many hydroxyl groups (alcohol groups) plus usually an aldehyde or a keto group. The simplest carbohydrates all exhibit varying degrees of sweetness, and for this reason carbohydrates in general are referred to as saccharides from the Latin meaning sugar. By the way, the artificial sweetener, saccharin, is not a carbohydrate but a compound isolated from coal-tar. Saccharin is 400 times sweeter than cane sugar but has a bitter taste in high concentrations. This section of organic chemistry will deal briefly with monosaccharides, disaccharides (carbohydrate molecules containing two monosaccharide units), and polysaccharides (carbohydrate molecules containing many monosaccharide units).

The most important monosaccharide is glucose which either as free glucose or combined glucose in disaccharides and polysaccharides is the most abundant organic compound on earth. Glucose is the monomer in the natural polymers starch and cellulose and is also present in sugar, milk, honey, syrups, flower nectars, and many fruits. The structure of glucose

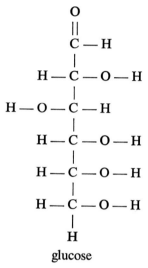

glucose

is typical of many of the monosaccharide structures. Note that glucose is a polyalcohol with an aldehyde group whereas fructose

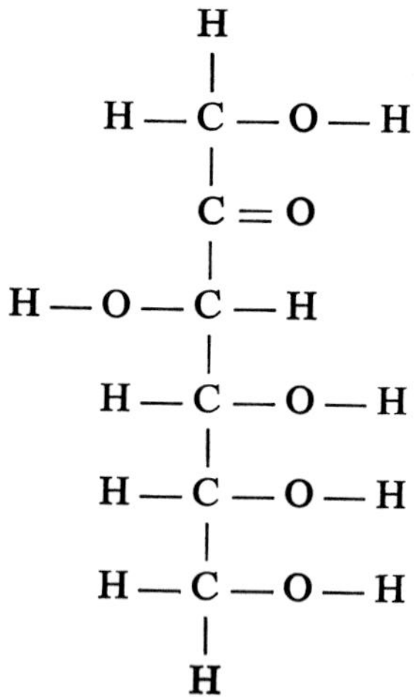

fructose

is a polyalcohol with a keto group. Fructose is found along with glucose in honey and fruits. Fructose is the sweetest of all carbohydrates, being sweeter even than cane sugar.

Ribose and deoxyribose

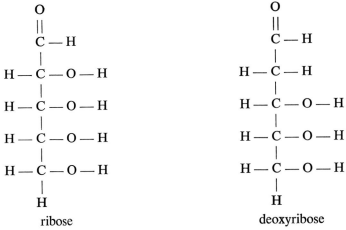

ribose   deoxyribose

are present in various organic catalysts, called enzymes, and deoxyribose is a vital constituent of deoxyribonucleic acid (DNA), the double helix structure that carries the genetic code for the chromosomes of cells.

Lactose and sucrose are the two most important disaccharides. Lactose, sometimes called *milk sugar,* is present in the milk of mammals. Cow milk is about 5% lactose and human milk is approximately 7% lactose.

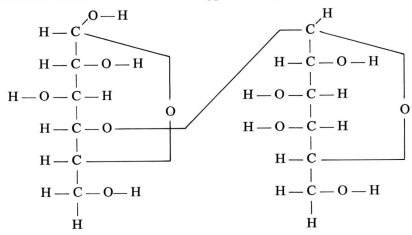

lactose (milk sugar)

Since sucrose is extracted from the juices of sugar cane and the sugar beet, sucrose is also called cane sugar and beet sugar. Most people, however, omit the part of the name that indicates the source of sucrose production and simply call sucrose, *sugar.* Sucrose was probably the first organic compound to be isolated in a pure state by man. The first recorded purification of sucrose occurred in India about 300 A.D.

organic chemistry   **309**

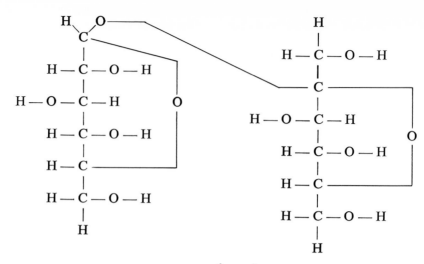

sucrose (sugar)

The polysaccharides of greatest significance and distribution in nature are starch, glycogen, and cellulose. All three of these polysaccharides are natural polymers composed of the monomer glucose. The difference between starch and cellulose is essentially the difference in which the glucose units are chemically bound together. Starch and glycogen have the same chemical bonds but differ in the size and shape of their molecules.

Starch is a reserve food supply that plants store in various parts of their structures such as roots, seeds, and fruit. Some seeds, in fact, are almost 70% starch. The main commercial production of starch is from corn, potatoes, and

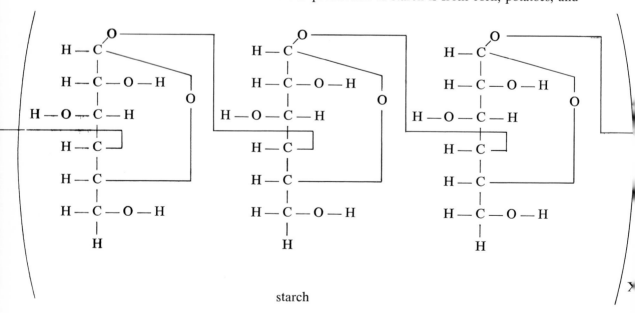

starch

organic chemistry

rice. Approximately four billion pounds of starch are produced annually in the United States; starch from corn accounts for 75% of this production. In the starch structure shown on previous page the *x* indicates that the glucose units, three of which are shown here, are repeated many times. The number of glucose units present in starch varies from two hundred to a million.

Glycogen, like starch, is made up of glucose units, but in general the number of units per glycogen molecule is greater by a factor of ten than for starch. Glycogen, sometimes called *animal starch,* serves a function for animals similar to the function starch serves for plants. Glycogen acts as a reserve supply of glucose units in various animal tissues. When the level of glucose in blood drops below the critical value of 0.1%, glycogen from the liver is broken down into glucose. The glucose then enters the blood stream to maintain the necessary blood glucose level. Glycogen stored in muscle tissue is used as a source of energy for muscle contractions.

The fibrous parts of plants are made up of cellulose; consequently dry wood is approximately 50% cellulose, and cotton is 98% cellulose. It is the cellulose part of wood that is utilized in the production of paper. Whereas starch is quite easily digested by the enzyme systems of animals, cellulose cannot be directly decomposed by animal enzymes into glucose in the manner that starch is decomposed. This inability of the enzyme systems is a direct result of the difference in which the glucose units are held together in cellulose as compared to starch. Thus, in the structure of cellulose shown below, note that the manner in which the glucose units are bound to each other differs from the method of bonding in starch.

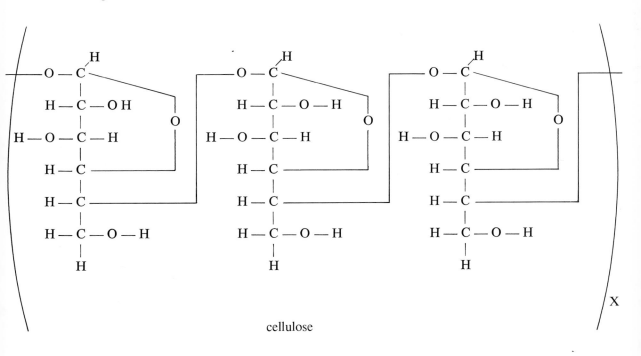

cellulose

Although cows and other ruminants eat grass, and therefore cellulose, as their source of food, they do not have the proper enzymes in their digestive system to utilize cellulose. Instead, there are specific bacterial cultures in the digestive tract that break down cellulose into glucose units which the cow can then use as a food source.

The treatment of pure cellulose with nitric acid produces cellulose known as *gun cotton* which is used as smokeless powder. *Rayon* and *cellophane* are cellulose that has been chemically processed and then converted back to cellulose.

DNA—THE FUNDAMENTAL MOLECULE OF LIFE

Inheritance is controlled by genes that reside in the chromosomes of the nucleus of all cells. Each gene carries a single characteristic, such as "free bleeding" (hemophilia) or color blindness. It is now known that genes are simply DNA (deoxyribonucleic acid) molecules or parts of DNA molecules. It has been mentioned earlier in this chapter that each organic chemical reaction occurring in cells is controlled by an enzyme (organic catalyst). Each particular enzyme present accounts for one particular hereditary trait, and each enzyme is the result of an individual gene being present in the cell's chromosomes. Each individual gene is nothing more or less than a DNA molecule or a portion of a DNA molecule.

The DNA molecule is a very large polymer made up of three different essential units: (1) a carbohydrate sugar, deoxyribose, (2) an inorganic acid, phosphoric acid and (3) an organic nitrogen base of which four different types are used in the DNA molecule (see Table 14.1). The DNA molecule consists of the units shown in Table 14.1 arranged, for example, in the following manner:

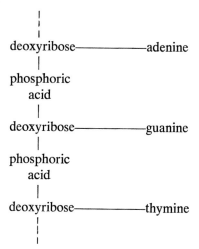

Actually, however, DNA exists as two separate molecules held together by hydrogen bonds. The two molecules are twisted together to form a double helix.

TABLE 14.1

The Chemical Components of DNA.

Type of Component	Structure
Sugar	$$\begin{array}{c} O \\ \parallel \\ C-H \\ \mid \\ H-C-H \\ \mid \\ H-C-O-H \\ \mid \\ H-C-O-H \\ \mid \\ H-C-O-H \\ \mid \\ H \end{array}$$ deoxyribose
Inorganic Acid	H_3PO_4 phosphoric acid
Organic Nitrogen Bases	adenine · guanine · cytosine · thymine

The two molecules are arranged together as shown below where - - - H - - - represents hydrogen bonds that hold the organic nitrogen bases together.

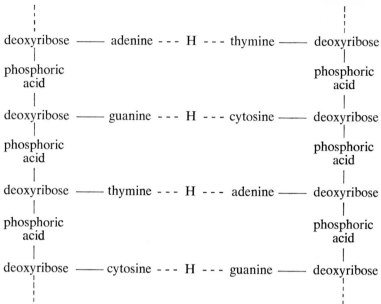

The genetic code is the result of the sequence in which the organic nitrogen bases are arranged in a DNA molecule. That is, DNA molecules contain a "blueprint" by which new proteins are synthesized. Thus, in the synthesis of a protein molecule, each amino acid present in the molecule is predetermined by a single code unit of three organic nitrogen bases. For instance, the following sequence of three organic nitrogen bases, constituting a single code unit,

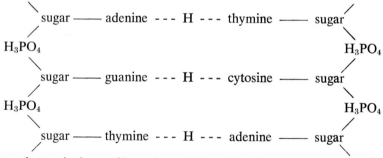

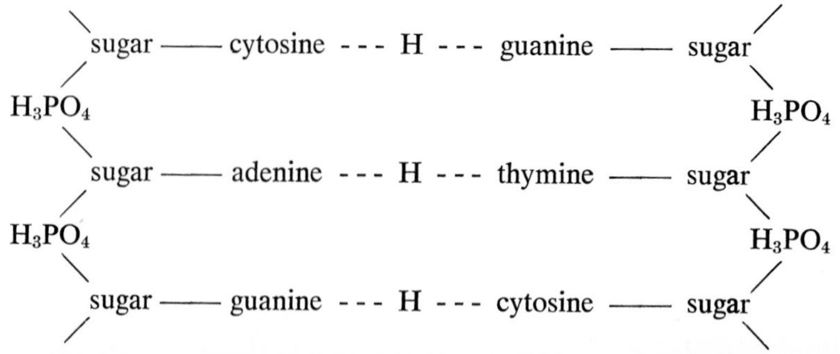

dictates that a single specific amino acid is placed in a certain position in the protein molecule being synthesized. On the other hand, the following single code unit dictates that a different specific amino acid be placed next to the amino acid controlled by the adenine-guanine-thymine code unit. By varying the sequence of the single code units and by varying the organic nitrogen bases within each code unit, the specific amino acids present in a protein molecule as well as the particular position of each amino acid in the molecule are precisely controlled. Bear in mind that there are only four different organic nitrogen bases in DNA and that a set of only three of the four in a definite sequence determines the code for a single amino acid. There are sixty-four ways in which the four bases can be grouped into units of three bases. Since there are only twenty to twenty-five amino acids used in synthesizing proteins, this "triplet" code system is more than adequate for the synthesis of all known proteins. Consequently, if a protein molecule containing 100 amino acids arranged in a definite sequence is synthesized, the synthesis is controlled by a portion of a DNA molecule consisting of 100 single code units or 300 individual organic nitrogen bases. The arrangements of the triplet units of bases in each single code unit constitute a coded message that is "read" by the cell. The cell in turn translates the code into various proteins. Some proteins produced are incorporated into the structural portions of the cell. Other proteins that are produced are enzymes that control the many processes of cell operation. These operations include the repair of damaged proteins and the production of energy by the breakdown of carbohydrate sugars.

DNA molecules are able to duplicate themselves exactly so that when a cell divides into two daughter cells each daughter cell is genetically identical to the parent cell. The parent double helix DNA molecule accomplishes this replication by separating its constituent DNA strands. Each strand, as it unwinds, synthesizes the other strand so that by the time the parent double helix is completely unwound two new and exact copies of the parent DNA molecule are produced. Then when the cell divides, each daughter cell has an exact copy of the DNA molecule of the parent cell. In this manner the hereditary genetic code is passed on to all new cells.

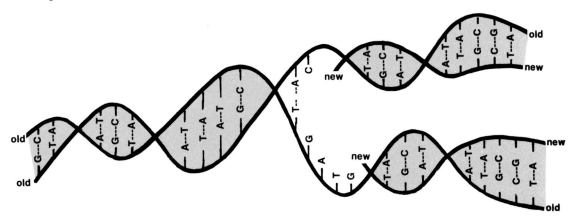

Fig. 14.1 A parent DNA molecule duplicates itself to produce two identical daughter DNA molecules.

CHAPTER SUMMARY

Organic chemistry is the chemistry of carbon compounds. The fact that carbon can form over a million different compounds is the result of the ability of carbon atoms to react with other carbon atoms to form carbon chains and rings of all shapes and sizes. Organic compounds consist mainly of carbon atoms covalently bound to other carbon atoms and various other elements. Carbon forms single, double, and triple covalent bonds.

Isomers are compounds that have the same empirical formulas but different structural formulas and, consequently, different chemical and physical properties.

The process whereby many small molecules are reacted together to produce a very long chain, called a polymer, is called polymerization. Chemists produce many different polymers in the laboratory such as polyethylene, rayon, and Dacron. Nature also produces many polymers such as proteins, starch, and cellulose.

All organic compounds, with the exception of hydrocarbons, have functional groups attached to a carbon chain or ring. Some of the more common functional groups are listed below:

$— O — H$	alcohol
$— O —$	ether
$\overset{\displaystyle O}{\overset{\|}{— C}} — O — H$	carboxylic acid
$\overset{\displaystyle O}{\overset{\|}{— C}} — O — \overset{\|}{\underset{\|}{C}} —$	ester
$\overset{\displaystyle O}{\overset{\|}{— C}} — H$	aldehyde
$\overset{\displaystyle O}{\overset{\|}{— C}} —$	ketone
$— NH_2$	amine

Deoxyribonucleic acid (DNA) is a large natural polymer made up of a carbohydrate sugar (deoxyribose), an acid (phosphoric acid), and an organic nitrogen base (one of four different nucleic acids). The DNA molecules carry the genetic code of life that is passed on from one generation to the next.

QUESTIONS AND PROBLEMS

1. Why are there so many more compounds containing carbon than compounds containing other elements? (p. 292)

2. In what way do saturated hydrocarbons differ from unsaturated hydrocarbons? (p. 295)

3. How do aromatic hydrocarbons differ from unsaturated hydrocarbons? (p. 296)

4. Which class of organic compounds would you expect to have chemical properties similar to water? (p. 297)

5. What do alcohols, sugars, starch, and cellulose all have in common? (p. 308)

6. What do esters, fats, oils, and some waxes all have in common? (p. 302)

7. In what way does one type of protein differ from another type of protein? (p. 307)

8. What organic compounds are present in deoxyribonucleic acid, DNA? (p. 313)

stars
15

OUR CLOSEST STAR – THE SUN
WHAT MAKES THE STARS SHINE
TYPES OF STARS
PULSARS
BIRTH, EVOLUTION, AND DEATH OF STARS
BLACK HOLES

vocabulary

Carbon Cycle	Eclipsing Binary
Proton – Proton Reaction	Pulsars
Supernovae	White Dwarf

When man climbed the evolutionary ladder to change from a mere sentient being to a creature with the glimmerings of thought and reason, he must have pondered about the mysterious uncountable points of light that were sprinkled over the vaulted night sky. Just as a light inevitably attracts a moth, man's curiosity has been attracted throughout the ages by those distant specks of brightness that man calls stars. For a major fraction of man's total existence, it seemed that he would never know the answers to the many questions asked about the stars. It was undoubtedly then a great event in the lives of scientists when it was first realized that contrary to previous understanding there was a star relatively near Earth that could be observed and studied. How marvelous must have been the realization that our sun was in fact a star and that the countless numbers of stars that are trillions of miles from our planet are much like our star—the sun. The understanding that the distant impersonal stars are in reality suns made all mankind feel less alone and isolated in this endless universe, for man now knows that a certain percentage of the other suns must also have a following of planets on which life might exist.

OUR CLOSEST STAR—THE SUN

Man's telescopic photographs of the sun reveal the restless turmoil that exists on the sun's "surface." As we will learn in the chapter "Our Solar Family," gigantic explosions and upheavals are incessantly occurring on the sun's surface to produce the magnificent flaming loops and arches of star material that sometimes are hurtled a quarter of a million miles into space. Huge black blemishes, sun spots, occasionally stain the sun's surface causing greater than normal ejections of charged particles. Radio waves, infrared rays, light rays, and x-rays are unremittingly discharged in all directions. Since our sun is a star, then it is reasonable to assume that other stars undergo the same general processes as those mentioned above for our sun.

WHAT MAKES THE STARS SHINE

We will learn in the chapter "The Birth of Earth" that once the core temperature of a star with the mass of our sun reaches approximately eighteen million degrees Fahrenheit, the hydrogen nuclei present are transmuted into nuclei of the element helium by a nuclear fusion process called the *proton-proton reaction*. The Proton-Proton Reaction:

$$^{1}_{1}\text{H} + ^{1}_{1}\text{H} \longrightarrow ^{2}_{1}\text{D} + e^{+} + \text{energy}$$

$$^{2}_{1}\text{D} + ^{1}_{1}\text{H} \longrightarrow ^{3}_{2}\text{He} + \text{energy}$$

$$^{3}_{2}\text{He} + ^{3}_{2}\text{He} \longrightarrow ^{4}_{2}\text{He} + ^{1}_{1}\text{H} + ^{1}_{1}\text{H} + \text{energy}$$

During the proton-proton reaction the conversion of mass into energy furnishes the enormous quantities of energy needed to maintain the immense outpouring of energy that is characteristic of stars.

For stars more massive than our sun the core temperatures are correspondingly higher. When the core temperature of a star reaches the vicinity of twenty-seven million degrees Fahrenheit, another fusion reaction called the *carbon cycle* becomes a major source of energy for the stars. The Carbon Cycle:

$$^{12}_{6}\text{C} + ^{1}_{1}\text{H} \longrightarrow ^{13}_{7}\text{N} + \text{energy}$$

$$^{13}_{7}\text{N} \longrightarrow ^{13}_{6}\text{C} + e^{+} + \text{energy}$$

$$^{13}_{6}\text{C} + ^{1}_{1}\text{H} \longrightarrow ^{14}_{7}\text{N} + \text{energy}$$

$$^{14}_{7}\text{N} + ^{1}_{1}\text{H} \longrightarrow ^{15}_{8}\text{O} + \text{energy}$$

$$^{15}_{8}\text{O} \longrightarrow ^{15}_{7}\text{N} + e^{+} + \text{energy}$$

$$^{15}_{7}\text{N} + ^{1}_{1}\text{H} \longrightarrow ^{12}_{6}\text{C} + ^{4}_{2}\text{He} + \text{energy}$$

The overall result of the carbon cycle is the conversion of four hydrogen nuclei into one helium nucleus ($4\ ^{1}_{1}\text{H} \longrightarrow ^{4}_{2}\text{He} + \text{energy}$), and in this sense the carbon cycle accomplishes the same result as the proton-proton reaction. In the carbon cycle, the carbon-12 isotope ($^{12}_{6}\text{C}$) acts as a catalyst to bring about the fusion of the four hydrogen nuclei into a helium nucleus. A catalyst is involved in a reaction but is eventually regenerated without having undergone a permanent change.

During the "old age" of a star a considerable percentage of the star's hydrogen has been converted into helium, and the core temperature of the star will have reached 200 million degrees Fahrenheit. At this temperature a nuclear fusion reaction known as the *triple-alpha process* becomes an important source of energy for a star. The triple-alpha process amounts to the conversion of three helium nuclei into one carbon nucleus.

$$3\ ^{4}_{2}\text{He} \longrightarrow ^{12}_{6}\text{C} + \text{energy}$$

Since a helium nucleus is known as an alpha particle and since three of these alpha particles are involved in the reactions, the process is called the triple-alpha process.

One should realize that while the triple-alpha process is occurring, the carbon cycle and the proton-proton reaction are also still actively producing energy for the star. Other nuclear reactions are also carried on at very high temperatures. It is easy, therefore, to understand how all the elements known in the universe are produced from the original hydrogen that was probably the primeval material present at the birth of our universe.

TYPES OF STARS

Stars are classified according to their photosphere ("surface") temperatures or according to color, which is a direct consequence of their photosphere temperatures. The temperatures of the photospheres of stars range from above 25,000°C (45,032°F) to as low as 3000°C (5432°F). Those stars with photosphere temperatures above 10,000°C (18,032°F) are bluish-white in color whereas a temperature of about 10,000°C results in a white star. A star appears yellow-orange with a photospheric temperature of about 5,000°C (9032°F), and at temperatures 3,500°C (6332°F) stars are red. Our sun, which has a photosphere temperature of about 10,300°F, is yellow.

There is an almost unbelievable variation in the sizes and densities of stars. Some *red supergiant* stars are so huge that if they were placed where our sun is, the star's surface would extend beyond the orbit of Mars. On the other hand some *white dwarf* stars are no larger than half the size of Earth, and *neutron stars* have a diameter of no more than a few miles! A neutron star consists solely of neutrons. The protons and electrons initially present in the star's matter were subjected to such fantastically high pressures that they coalesced to form neutrons. The density of red supergiant stars is so low that they can be considered a respectable vacuum whereas a basketball-size portion of a white dwarf may weigh twenty-five thousand tons, and the same volume of a neutron star would weigh ten thousand million tons!

Present-day information indicates that 38% of all the stars we see are not single stars but are multiple star systems. Most of these star systems contain only two stars and are called *binary systems,* but a very small percentage contain three or more stars per system. The stars of binary star systems mutually revolve around a common center of gravity. The first binary system to be discovered was the star Mizar in the middle of the handle of the Big Dipper (Ursa Major, the Great Bear). In 1650 an astronomer trained his telescope on Mizar and noted

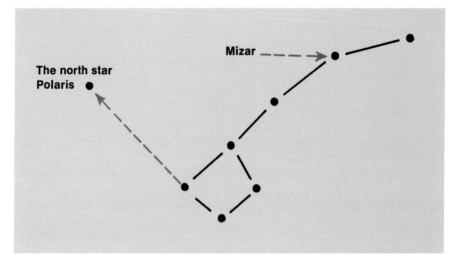

Fig. 15.1 **The binary system Mizar in the handle of the Big Dipper.**

that what had appeared to the naked eye as a single star was in reality two stars revolving around a common barycenter. During the next 200 years it was found that each of the two stars (Mizar A and Mizar B) was itself a binary system. Thus, the star we see in the middle of the handle of the Big Dipper is actually four stars all mutually bound together by gravitation. Polaris, the north star, is actually a triple star system.

Occasionally Earth lies in the plane formed by the two revolving members of a binary system. As a result, with each revolution one star passes in front of the other star with the consequence that each star eclipses the other star. Such a system is called an *eclipsing binary* system. This rhythmic and periodic eclipsing results in periodic fluctuation in the light intensity coming from the binary system. When one star eclipses its companion, the total light we receive is from the front star only, but when the two stars are not in an eclipse formation, we receive the light from both stars with the result as shown in Figure 15.2.

The light from some stars shows periodic variation in intensity even when the stars are not binary systems. Stars that are not binary systems but show periodic fluctuations in light intensity are called *pulsating variables*. The outer

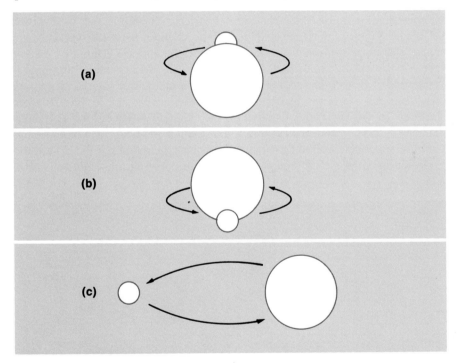

Fig. 15.2   An eclipsing binary system. In (a) the smaller star is behind the larger star, and in (b) the smaller star is in front of the larger star. In (a) and (b) we receive approximately the light intensity given off by the larger star. In (c) we receive the light from both stars. The mutual revolution of the two stars results in a periodic fluctuation of light intensity given off by the binary system.

stars   **323**

surface of pulsating variable stars is continually expanding and contracting. During these pulsations, the light intensity of the star increases and decreases in rhythm with the star's pulsations. Most pulsating variables are red and yellow giant and supergiant stars. Their periods of pulsations range from twelve hours to two years. Polaris, for example, is a pulsating variable with a period of about four days between times of maximum light intensity. So far 20,000 variable stars have been found in our galaxy.

The most spectacular stars that inhabit our universe are the *novae* and *supernovae* stars. Nova comes from the Latin meaning *new* and the term is applied to stars whose brightness suddenly increases by a factor of thousands and occasionally becomes as bright as 100 million suns. A star becomes a nova or a supernova star when it undergoes, with titanic violence, an explosion in which its outer layers expand at fantastic speeds. The nova reaches maximum brightness in a matter of days. The star then begins to slowly fade back to its original brightness. About twenty-five novae occur in our galaxy each year, but most are visible only through telescopes.

The brightest supernova known in recorded history was the "guest" star recorded by Chinese and Japanese astronomers in 1054 A.D. that appeared in the constellation Taurus. At the height of the "guest" star's brilliance it outshown the planet Venus and was visible during the daytime for three weeks. Some supernovae have been known to increase in brightness over a hundred million times. A supernova explosion throws out so much stellar material that an expanding shell of material surrounds the star. The expanding material may travel away from the star at speeds of up to 1000 miles per second. In the 920 years since the Taurus supernova burst forth, the expanding material has covered a diameter of twenty trillion miles and is today called the Crab Nebula. Since the Crab Nebula is about 5000 light years from Earth, man first saw the light of the explosion in 1054 A.D., although the explosion occurred approximately 4000 B.C.

Fig. 15.3 **The Crab nebula in the constellation Taurus is the remnant of a supernova explosion that occurred in 1054.**

(Photograph Courtesy of the Hale Observatories)

PULSARS

There is much more to the story of the Crab Nebula than has been taken up so far. Careful observation of the Crab Nebula shows that there is a star in the center of the nebula that acts as a giant celestial beacon whose light flashes on and off thirty times a second. More careful scrutiny reveals that the star in the center of the Crab Nebula is sending out pulses from the x-ray, infrared, and radio wave regions of the electromagnetic spectrum in the same time interval. The central star of the Crab Nebula is not the only pulsating beacon in our galaxy. There are many such stars. Stars that emit electromagnetic radiation in definite pulses are called *pulsars*. Pulsars are rapidly rotating and highly magnetic neutron stars.

BIRTH, EVOLUTION, AND DEATH OF STARS

We will learn in the chapter "The Birth of Earth" that stars are formed by the gravitational condensation of dust and gas in interstellar space. During the birth of a star, gravitational potential energy is converted into kinetic energy, and part of the kinetic energy is converted into heat and light. This increase in heat results in an increase in pressure within the star. Eventually the internal pressure of the star is high enough to stop the continued gravitational collapse of the star's material. An equilibrium is then established between the internal pressure of the star, which tends to expand the star, and the inward gravitational pull of the star's matter, which tends to contract the star.

If the initial mass of material that gravitationally collapses to form a star is less than 0.07 the mass of our sun, then the internal temperature of the star when it reaches equilibrium is not high enough to ignite nuclear fusion reactions. If the internal temperature of a star is not high enough for nuclear fusion reactions to occur, the star shines for only a few hundred million years by the conversion of gravitational energy into heat and light. The star eventually shrinks to a white dwarf. On the other hand if a star starts its life with a mass five times that of our sun, it takes only between fifty to sixty million years for it to become a red giant. This comparatively short lifetime as a "normal" star is due to the high core temperatures which result in the star's using up its nuclear fuel at an excessive rate. A star with a mass approximately the mass of our sun has a more moderate rate of nuclear fuel consumption. Such a star has a "normal" life span of about ten billion years before any catastrophic changes start taking place. Our star, the sun, has used up approximately only half of its "normal" lifetime and, there-fore, it has approximately five billion more years before it will undergo changes that will result in its becoming a red giant.

After approximately ten billion years of fusing hydrogen into helium, the central core of a star similar to our sun becomes mostly helium. The outer por-tions of the star expand over a period of about a billion years until the star's diameter becomes fifty times greater than the diameter of our sun. The star is then a red giant. During this expansion process, the helium core contracts with a resultant increase in temperature until eventually a temperature of about 100 million degrees Celsius is reached. At this fantastically high temperature three helium nuclei are fused together to produce one carbon nucleus. This reaction proceeds very rapidly, and soon the star has a carbon core. By continuation of nuclear fusion processes similar to those responsible for converting a hydrogen core into a helium core and a helium core into a carbon core, all the elements known in the universe have probably been formed in these stellar furnaces.

The star is now ready to relinquish its role as a red giant star, and it begins a process of ridding itself of a portion of its mass. In general, stars of the sun's mass and less eject part of their mass into interstellar space by various mechanisms. Perhaps the most common method used by stars to lose mass is the ejection of part of their outer layers to produce what is called a *planetary nebula*. The word *planetary* here has nothing to do with planets but is simply part of the descriptive term used to identify stars with expanding shells of material. The star of a planetary nebula ejects mass in all directions to produce a shell of material surrounding the star. The shell moves away from the central star. When we look at a planetary nebula, we see what appears to be a giant smoke ring surrounding the star, but in reality it is a spherical shell of expanding matter. The most spectacular way a star has of expelling mass is by a nova explosion. The ejection of large quantities of mass by a red giant is followed by contraction of the remaining portion of the star until eventually a star that originally had a diameter near that of our sun (864,000 miles) shrinks by gravitational contraction until its diameter is no greater than between one-half and four times the diameter of Earth. The resulting diminutive star is called a *white dwarf*.

If a star is considerably more massive than the sun, it may undergo a cataclysmic explosion and throw off enormous quantities of matter to produce a *supernova* similar to the 1054 A.D. supernova that produced the Crab Nebula. The remaining mass of the star would gravitationally collapse to form a neutron star.

Most stars, however, become white dwarfs after they lose part of their mass. When a star has reached the white dwarf stage, it is in its "old age." However, the old age of a star turns out to be the longest period of its life. It takes millions of years for the star to be born from the star dust of interstellar space. But it takes billions of years for the average star to consume its hydrogen fuel, and it takes a billion years for the star to go through a "transition period." During a star's transition period it becomes a red giant and ejects some of its matter back into interstellar space. This ejected matter eventually gives birth to new stars. But it takes not billions but trillions of years for a white dwarf to cool sufficiently that it no longer gives off heat and light. When this stage is reached in a star's life, it is a *black dwarf*. This burned out cinder, in order to once again experience "life," must await the rebirth of the universe. Since the universe may be only ten to twelve billion years old and since it takes trillions of years to produce a black dwarf, there are probably no black dwarfs in the universe as yet. Toward the end of the present cycle of expansion and contraction of our universe, black dwarfs may be rather common celestial objects.

BLACK HOLES

As pointed out earlier, a contracting star continually increases its density. This increase in density is accompanied by an increase in expansion forces. Eventually continued gravitational contraction is halted to produce a white dwarf. In cases of extreme gravitational contraction, due to an originally large mass, a neutron star is formed. If the original mass of the contracting star is great enough, however, the forces within the star that usually stop further gravitational contraction are completely overwhelmed, and uninhibited gravitational collapse results. When this ultimate stellar event occurs, there is produced what is called a *black hole*. Since the diameter of the resulting mass of the star is for all practical purposes negligible, the density of the matter in a black hole is so unimaginably large that it defies comprehension.

The gravity field surrounding a black hole is so great that the escape velocity from a black hole exceeds the speed of light! Thus, no matter or even electromagnetic radiation can escape from a black hole if it is a non-rotating body. If, on the other hand, it is a rotating black hole, there is a region surrounding the black hole in which matter, though profoundly affected by the mass of the black hole, can escape into space. If a piece of matter enters the special area surrounding the black hole, the matter splits into two parts, one of which is captured by the black hole and the other is ejected into space with tremendous acceleration. The energy needed to accelerate the escaping particle is supplied from the rotational energy of the black hole.

CHAPTER SUMMARY

Our sun and other stars produce energy by the very high temperature fusion of four hydrogen nuclei into one helium nucleus, $4H \longrightarrow 1 He$. The difference in the mass of four hydrogen nuclei and one helium nucleus shows up as energy. At very high temperatures hydrogen is converted into helium by the *carbon cycle* and at even higher temperatures these helium nuclei are fused into a carbon nucleus by the *triple alpha process*.

Some yellow and red supergiant stars have diameters that exceed 300 million miles whereas a star such as our sun has a diameter of only about three quarters of a million miles. *White dwarfs,* no larger than approximately half the size of Earth, are still huge in comparison to *neutron stars* that have a diameter of only a few miles. The ultimate minimum-size star is the so-called *"black hole,"* which has for all practical purposes a negligible diameter and a density high enough to produce an escape velocity exceeding the speed of light.

Approximately one third of all the stars we can see from Earth are *binary star systems*, and a very small percentage are systems of three or more stars.

Some stars fluctuate in brightness. There are three main reasons for light intensity fluctuations of stars. A star's photosphere may expand and contract with a resulting pulsation of its light output. Other pulsating stars are the result of *eclipsing binary systems* where the pulsations result from the eclipsing of one star by another as they mutually revolve around a common gravity center. Some stars, such as the star in the center of the Crab Nebula, are highly magnetic neutron stars that rotate very rapidly with the result that very short bursts of light and radio energy are emitted by the star. Such stars are called *pulsars*.

A star is formed out of the "star dust" of space when a large cloud of dust and gas slowly coalesces due to the mutual gravitational attraction of the dust and gas particles. During the contraction, gravitational potential energy is converted into heat energy in sufficient quantities that the star begins to shine. When the internal temperature of the new star reaches many millions of degrees, the fusion of hydrogen atoms into helium atoms begins with subsequent conversion of mass into energy. The average star then has a relatively long life of about ten billion years after which it expands to become a *red giant*. Larger stars, during this period of expansion may expand with explosive violence to form a *nova* or a *supernova*. After the explosion of a supernova, contraction of the supernova remnant produces a *neutron star* or a *black hole*. The average star, however, after it expands to the red giant stage, slowly contracts to become a *white dwarf* star. The white dwarf slowly cools over many billions of years to produce a burned-out cinder of a star called a black dwarf.

QUESTIONS AND PROBLEMS

1. What would our solar system be like if we did not have the sun? (p. 320)
2. Since the sun is so very hot, why hasn't the hydrogen of the sun escaped into outer space?
3. How do stars produce their energy? (p. 327)
4. What determines whether a star eventually becomes a white dwarf, a neutron star, or a black hole? (p. 326)
5. Why do old stars have a greater percentage of helium than young stars? (p. 325)
6. What is the difference in an eclipsing binary system and a pulsating variable star? (p. 323)
7. Discuss the old age of our sun. (p. 326)

our restless universe

16

THE MILKY WAY
OTHER "ISLAND UNIVERSES"
ASTRONOMIC DISTANCES
COSMOLOGY AND THE REDSHIFTS
QUASARS

vocabulary

Galaxy	Globular Cluster
Milky Way	Quasars
Redshift	Spiral Galaxy

One of the great beauties of the universe that man has always been able to see without the aid of telescopes is the archway of soft gentle light that meanders across the vault of our night sky from horizon to horizon—The *Milky Way*. This "lighted passage to heaven" dominates the moonless night sky. When one looks at the Milky Way through a telescope, what appears to the naked eye as a soft diffused glow is revealed to be an incredible multitude of individual stars and patches of nebulae. Obviously, the Milky Way is actually an aggregation of billions of stars.

(Photograph Courtesy of the Hale Observatories)

Fig. 16.1 **The Milky Way Galaxy as seen from Earth.**

THE MILKY WAY

Our sun is one of 100 billion stars or more that make up our home galaxy which is called the Milky Way. Our Milky Way Galaxy is a lens shaped galaxy some 100,000 light years in diameter and 10–15 thousand light years thick at its nucleus or center. Figure 16.2 shows what our galaxy would look like if viewed from "above" (or "below") and from the side.

When at night we look up and see the luminous path of light called the Milky Way, we are looking edge on through our galaxy as shown in Figure 16.3. When we look ninety degrees from the Milky Way, we are looking in a direction "above" or "below" our galaxy.

Our galaxy rotates around its nucleus or center, and the flattened shape of our galaxy is the result of this rotation. Our sun revolves around the galactic

(Photograph Courtesy of the Hale Observatories)

(a)

(Photograph Courtesy of the Hale Observatories)

(b)

Fig. 16.2 (a) The spiral galaxy, Messier 74, which is similar in shape to our Milky Way Galaxy. (b) An edge-on view of the spiral galaxy in Coma Berenices. Our Milky Way Galaxy would have a similar appearance if viewed edge on.

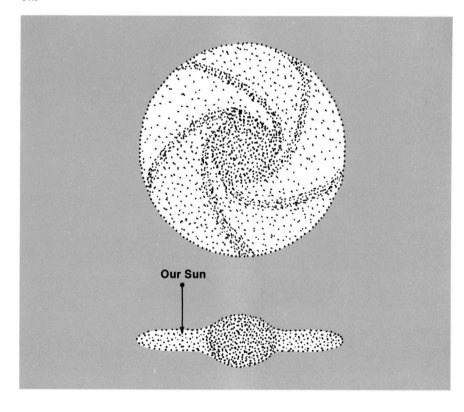

Our Sun

Fig. 16.3 A schematic representation of our Milky Way Galaxy showing the general position of our sun.

center at a speed of a half million miles per hour, and even at this fantastic speed it takes our sun 200 million years to complete one revolution! That the galactic "year" is 200 million Earth years is an indication of how immense our galaxy is. In fact, our galaxy has had time to make only fifty complete revolutions since the galaxy's birth about 10 billion years ago. Our sun is about two thirds of the way out from the galactic center (about 30,000 light years) in one of the galaxy's spiral arms. The spiral arms contain much cosmic dust and gas and relatively "young" and bright stars, whereas the nucleus of our galaxy contains older, and therefore redder, stars.

Our flattened galactic disk is embedded in a spherical veil made up of beautiful, symmetrical clusters of stars called *globular clusters* as shown in Figure 16.4. The globular clusters surrounding the spiral-armed flattened disk of our galaxy, as well as a "mist" of individual stars surrounding the nucleus of the disk, make up the *corona* of our galaxy. The corona extends 50,000 light years "above" and 50,000 light years "below" the disk of our galaxy.

Fig. 16.4 A picture of a globular star cluster made with the 200 inch telescope at Mount Palomar.

(Photograph Courtesy of the Hale Observatories)

OTHER "ISLAND UNIVERSES"

When one stops and thinks that our Milky Way Galaxy is so immense that light traveling at 186,000 miles per second takes 100,000 years to go from one end of our galaxy to the other, it is difficult to comprehend that any structure could be larger and more impressive. Actually, our galaxy is like a speck of dust

our restless universe

floating in a sea of empty space, and there are tens of billions of other galaxies much like ours scattered throughout the limits of our cosmos. The lonely isolation of galaxies in the void of space has resulted in galaxies being described as "island universes."

What do the other galaxies in our universe look like? Would they appear like the Milky Way Galaxy to an intergalactic traveler? Is our galaxy of average size? There are a number of different types of galaxies based on differences in galactic shapes. Perhaps we should first consider the type of galaxies to which our Milky Way Galaxy belongs—*spiral galaxies*. In a spiral galaxy the considerable rotation of the galaxy causes flattening of a major portion of the galaxy into a disk or lens shape. In the nucleus of the galaxy, stars were formed from the cosmic dust of the original nebula at an early stage and, therefore, the nucleus today contains old stars in general. Because of the rotation of the galaxy, the cosmic dust and gas outside the nucleus have sufficient motion to slow the formation of stars by gravitational accretion. Consequently, the spiral arms of such a galaxy contain younger stars and even stars presently in the process of formation. The stars in the globular clusters surrounding a spiral galaxy also include old stars that probably formed about the same time that the stars of the galactic nucleus were formed. The nuclei of some spiral galaxies are elongated into a "bar" shape, and these spiral galaxies are consequently referred to as *"barred" spiral galaxies*. An example of a spiral galaxy is shown in Figure 16.5. Perhaps

Fig. 16.5   A barred spiral galaxy.

(Photograph Courtesy of the Hale Observatories)

the most famous and most spectacular spiral galaxy other than our own Milky Way is the *Great Andromeda Galaxy*. The number of stars in spiral galaxies ranges from one billion to over 200 billion.

Fig. 16.6 The Andromeda Galaxy.

(Photograph Courtesy of the Hale Observatories)

The most common type of galaxy in the universe is the *elliptical* galaxy. The shape of elliptical galaxies varies from spherical to near lens shaped. The difference in the shapes of elliptical galaxies is the direct result of the difference in the speed of rotation of these galaxies. If a galaxy does not rotate to any appreciable extent, its shape is essentially spherical, but if the galaxy has a reasonable rotation, then it is flattened out and approaches the shape of the faster rotating spiral galaxies. Elliptical galaxies consist almost entirely of old stars whose ages approximate the ages of stars in the nuclei of spiral galaxies and in globular clusters. The largest galaxies in the universe are elliptical galaxies.

Not all galaxies have symmetrical shapes, and a third class of galaxies called *irregular galaxies* are amorphous in appearance. Irregular galaxies are relatively rare as compared to elliptical and spiral galaxies. The most renowned examples of irregular galaxies are the so-called *Large Magellanic Cloud* and *Small Magellanic Cloud* of the southern sky. These galaxies are our nearest neighboring galaxies and were named after the explorer Ferdinand Magellan since he was the first European to see and describe these southern sky objects as a result of his circumnavigation of our planet.

ASTRONOMIC DISTANCES

There is no way for the unaided eye to detect the difference in distances of various celestial objects. The sun and the moon appear to be about the same size to an observer on Earth, yet the sun is 389 times farther from Earth than is the moon. Some stars are brighter than other stars, a fact which might make us believe that the brighter stars are closer, but this assumption is not necessarily true. Indeed, a very bright star may be much farther away than a faint star, but if the bright star is basically very bright, it still outshines the faint closer star and appears brighter to us.

In order to get some feeling for the distances from us of various objects in our universe, let's consider the length of time it takes light to reach us from different bodies within the universe. To start with, let's remember that light travels at a speed of 186,000 miles/sec. or 669,600,000 miles/hr. At this incomprehensible speed, light takes only a little more than 0.1 of a second to travel around Earth. Light from the moon, which is 239,000 miles away, reaches us in a mere 1⅓ seconds. The sun sends us its life-giving warmth and light from a distance of 93,000,000 miles in about 8⅓ minutes. Pluto, our lonely sentinel at the outpost of our solar system, sends us the sun's reflected light in 5⅓ hours over a distance of 3,583,000,000 miles.

As an example of the solitary isolation of stars even within our galaxy, the star nearest Earth, Proxima Centauri of the Alpha Centauri group, is twenty-five thousand billion miles from the planet Earth. Light from that lonely beacon in space must travel for 4⅓ years before it reaches us. If Proxima Centauri were to be annihilated in a great cataclysmic explosion, it would be 4⅓ years later before we knew that the star no longer existed. Light from the closest galaxy to our own Milky Way, the Large Magellanic Cloud, must travel for 163,000 years before reaching our planet. Thus the light we see today from the Large Magellanic Cloud must have left that galaxy about the time that Neanderthal Man and Cro-Magnon Man were first struggling into existence!

The lovely and spectacular Andromeda Galaxy, shining like a spiral of diamonds set in the darkness of space, is so distant from us that light from Andromeda must travel over two million years before it reaches Earth. When we look at the Andromeda Galaxy today, we are seeing light that left Andromeda tens of thousands of years before the beginning of the great ice age in which immense sheets of ice covered most of the northern portion of the United States.

How far into the depths of space can man look and still find galaxies? This question cannot be answered at this time. As man has built larger and larger telescopes, he has been able to see farther and farther toward the "edges of the Universe," and each new telescope reveals even more galaxies. With the 200 inch telescope at the Hale Observatory on Mount Palomar in the U. S. (a 236 inch telescope is being constructed at the Crimea Observatory in the U.S.S.R.) man is able to photograph galaxies so remote that the light we see from these galaxies left those galaxies when our solar system was just forming out of the "star dust" of space.

Fig. 16.7 **A cluster of galaxies in the constellation Hercules.**

(Photograph Courtesy of the Hale Observatories)

The farther out our telescopes allow us to probe, the farther back into the dim past we can see. It could be said that astronomy is the scientific study of the past history of our universe.

What is the universe like "today"? Man does not know! Since the light from our nearest neighboring galaxy left that galaxy about 160,000 years ago and the light from one of the farthest known galaxies left about six billion years ago, the universe that we are "seeing" today is actually the universe as it existed hundreds of thousands or billions of years ago. We can only assume that the present universe is doing approximately the same sort of things as the incredibly ancient universe that we "see" today.

COSMOLOGY AND THE REDSHIFTS

In looking night after night and year after year at the "unchanging" night sky, one gets the impression of a static universe in which all the stars and galaxies have been carefully and permanently placed. It would appear that celestial objects in the heavens will remain in their designated places, forever unchanging, until the end of time. Nothing could be further from the truth. True, casual observations of the universe do leave one with the impression of an unchanging cosmos, but more sophisticated detection methods reveal that the objects in our galaxy as well as in all the galaxies in our universe are moving at

our restless universe

very high speeds. The planets, comets, and asteroids of our solar system are racing around the sun. At the same time, our sun, as well as all the other stars in our neighborhood, are whirling around the galactic center at an average speed of half a million miles per hour. Simultaneously, all the galaxies in the universe are mutually rushing away from each other at staggering velocities. It is only the enormous distances separating celestial objects that are responsible for the impression of no movement or a snail-pace movement of objects in the universe.

When the light from certain stars in various galaxies is analyzed by a spectrograph, it is found that the spectral lines representing certain gaseous elements present in the stars are not in the same position in which one finds them when light from these elements are passed through a spectrograph in the laboratory. Instead, the spectral lines present in the light of galaxies are shifted toward longer wavelengths (and lower frequencies): that is, shifted toward the red end of the spectrum. This *redshift* is due to the Doppler effect on the light sent to us from these distant, receding galaxies.

As was pointed out at the end of Chapter 9, if a source of light is moving toward an observer or if the observer is moving toward a source of light, the observer sees the wave front of the light "bunched" closer together than normal and spectral lines are shifted toward shorter wavelengths (higher frequencies) and, therefore, toward the blue end of the spectrum. If a source of light is traveling away from an observer or if an observer is traveling away from a source of light, the observer sees wave fronts of the light "spread" apart more than normal, and spectral lines are shifted toward longer wavelengths (lower frequencies), that is, shifted toward the red end of the spectrum.

With the exception of only a few galaxies close to the Milky Way which are members of what is called the "local group," all galaxies show a shift of spectral lines toward the red end of the spectrum. This red shift can only mean that all the galaxies that we can see (except a few members of our local group) are receding from our Milky Way Galaxy. Interestingly, there is a direct relationship between the recessional speed of a galaxy and its distance from us. Thus, the more distant a galaxy is, the faster it is receding from us. This distance–velocity relationship is known as the *law of redshifts*. One of the most distant galaxies that has been checked for a redshift shows a recessional speed of 270 million miles per hour (41% of the speed of light!). In order for the galaxy to show this high recessional speed, the law of redshifts indicates that the galaxy must be over five billion light years away!

It is rather disconcerting to realize that all the galaxies in the universe are rushing away from us. Does this mean that we are at the center of the universe and that all other galaxies are expanding away from the center? It is as if other galaxies have shrewdly decided that it is foolish to be too close to the Milky Way Galaxy which harbors such a destructive humanoid species as Homo sapiens. Actually, an observer on any other galaxy in the universe would have the illusion that all the other galaxies in the universe were rushing away from his galaxy. How can this illusion be? A very simple analogy can explain this apparent anomaly. Consider a balloon that has dots painted on it. As the balloon is blown up, the distance between the dots increases. An ant on any one of the dots would observe that all the other dots were receding from his dot as the balloon continues to increase in size (see Fig. 16.8).

Fig. 16.8
Separation of dots on a balloon during expansion. An ant on any dot would see all other dots recede from him during the expansion.

We are forced then to conclude that all the galaxies in the universe are involved in a general expansion from some central site in our universe. Calculations indicate that the expansion started about ten to twelve billion years ago. We can assume then that some ten to twelve billion years ago (time "zero" for our universe) all the matter in the universe was concentrated into a colossal primeval "egg." This original "egg" must have undergone an explosion, the likes of which man cannot even conceive, with the result that matter was sent hurtling outwards at unimaginable speeds to nowhere and everywhere. During the continued expansion of this matter, galaxies slowly formed and the distances between these galaxies continued to increase. This theory of the birth of the universe is commonly known as the *"big bang theory."* At present there is no way to know how long this universal expansion will continue, but most scientists today believe that the expansion is slowing down. Ultimately, the expansion will stop, and the universe will begin to contract due to mutual gravitational attraction between the galaxies. Then all the galaxies of the universe will move back toward the place of their birth at ever-increasing velocities until eventually they once again fuse together to form a new "egg." We have, therefore, a picture of a limited universe with a definite maximum diameter. If after the formation of the new egg, the egg again explodes to begin a new expansion and contraction cycle, we have the concept of a never-ending cyclic universe. Is the cycle we are presently in the first cycle for the universe or have there been untold numbers of cycles before the cycle we find ourselves in? Perhaps even from a philosophical point of view it makes no difference whether this is the first cycle or not, because the start of the present cycle was the "beginning" of the universe as far as we will ever know.

How large is our present universe? Since the distance and speed of various galaxies are known, one can back-calculate and determine how long these galaxies must have taken at their present speeds to travel from the "initial egg" to the positions that they presently occupy. Such a calculation gives a value of about ten to twelve billion years. With this age of the universe we can calculate the maximum distance that light could have traveled in a straight line from the center of the explosion ten-twelve billion years ago to the present time. This distance, then, would be the maximum theoretical radius of the universe. Since light travels 5.88×10^{12} miles per year then $(5.88 \times 10^{12} \frac{miles}{year}) (12 \times 10^9$ years$) = 7 \times 10^{22}$ miles or seventy billion trillion miles. Since matter cannot travel at the speed of light, the actual radius of the present universe must be something less than the above value.

QUASARS

In 1960, two faint star-like objects were found to emit enormous quantities of radio waves. An analysis of these two star-like objects revealed that they had characteristics quite different from those of regular stars. Consequently, they were named quasi-stellar radio sources, which was soon shortened to *quasars*. More recent studies indicate that there are millions of quasars in the universe. Quasars have been found to exhibit extraordinarily large redshifts. The redshifts of quasars greatly exceed the redshifts of galaxies so far investigated. Amazingly, one quasar has been found that is receding from us at the fantastic rate of 550 million miles per hour, which is about 82% of the speed of light. As astonishing as is the recessional speed of quasars, there is yet another property that these celestial puzzles have that arouses awe even in dispassionate astronomers. This property is the astonishing quantities of energy that quasars eject into space. Although quasars in general are only about as large as our solar system, they pour out energy at the rate of ten thousand billion suns or more energy than 100 Milky Way galaxies! So far astronomers have been unable to unravel the riddle of the paradoxical quasars.

Not all quasars emit large quantities of radio energy. In fact, only one quasar in 300 is an extraordinary radio emitter. In spite of the rarity of quasars that are very active in the emission of radio waves, the name quasars, implying a radio source, has been retained.

CHAPTER SUMMARY

Our galaxy, the *Milky Way,* is a lens-shaped collection of a hundred billion stars isolated in the almost endless space of our universe by vast distances from the other galaxies. Our sun is approximately two thirds out from the center of our galaxy. At night we can see a portion of our galaxy by looking at the Milky Way that meanders from horizon to horizon across the sky.

The flattened lens-shaped galaxies, such as the Milky Way, are called *spiral galaxies. Elliptical galaxies* vary in shape from spherical to near lens shaped and are the most common type of galaxy in the universe. Other galaxies have no particular symmetrical shape and are referred to *as irregular galaxies.*

The most widely accepted theory of the history of the universe is the *"big bang" theory.* About twelve billion years ago all the matter in the universe was collected into a gigantic single mass which exploded with titanic violence. All the matter in the universe is still expanding outward from that initial explosion. Light from other galaxies shows a *Doppler redshift* that indicates that the universe is still expanding today. It is assumed that eventually the expansion will slow to a stop and then the universe will begin to contract. The contraction will continue until once again all the matter will be compressed into a new single mass. From this mass a new cycle of explosion, expansion, and contraction will occur.

QUESTIONS AND PROBLEMS

1. How many times larger is the diameter of our galaxy than the diameter of our solar system? (p. 332)
2. What would the night sky of a planet of a star in a globular cluster look like? (p. 334)
3. Based on your understanding of the formation of our solar system, discuss the formation of the Milky Way Galaxy. (p. 396)
4. How would the Milky Way appear in the night sky if we were near the center of our galaxy? (p. 333)
5. What would the night sky look like if we lived on a planet of a star on the outer rim of our galaxy? (p. 333)
6. Why can it be said that astronomy is the scientific study of the past history of our universe? (p. 338)

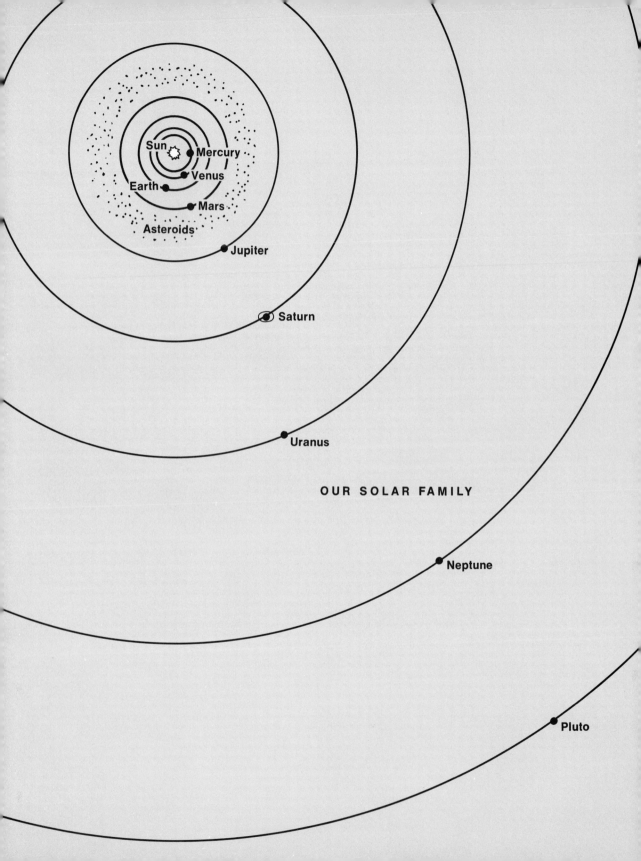

Sun

Mercury

Venus

Earth

Mars

Asteroids

Jupiter

Saturn

Uranus

OUR SOLAR FAMILY

Neptune

Pluto

our solar family

17

THE SUN
THE PLANETS
MERCURY
VENUS
EARTH
MARS
JUPITER
SATURN'S RINGS AND ROCHE'S LIMIT
URANUS
NEPTUNE
PLUTO
THE ASTEROIDS
COMETS
METEORS

vocabulary

Asteroid	Comet
Inferior Planet	Perihelion
Retrograde Rotation	Sunspots

Our solar system consists of a central star, the sun, its nine planets, and their thirty-two satellites or moons, tens of thousands of asteroids, innumerable comets, and myriads of meteors. Consider, however, what an intragalactic visitor would observe as he approached our solar system. Until he was quite close to our system, he would notice only our sun. Even if he came closer, our sun would still reign supreme against the black void of space. By comparison with the sun, the sun's planets, comets, and asteroids would appear as mere specks, since the sun has 99.86% of all the mass in our solar system! On closer inspection, Jupiter, Saturn, and perhaps Uranus would appear reasonably impressive, but the rest of the planets, including our Earth would look rather inconsequential and be difficult to locate. Unless some of the comets happened to be close to the sun, they would be almost impossible to detect since at great distances from the sun they have no tail and reflect barely any light from the sun. The great variation in the sizes of the sun's planets would be startling to our space visitor. He would observe that the largest planet, Jupiter, is so massive (larger than the mass of all the other planets combined) that one of its moons is larger than the planet Mercury.

After the visitor had had time to study in detail and to catalog the various components of our solar system, he would notice that as he looks down on our solar system from the vicinity of the North Star, all the planets revolve around their star in a counterclockwise direction and, with only one exception, spin on their axes in a counterclockwise direction. With a few exceptions, the thirty-two satellites or moons of the planets also revolve around their mother planets in a counterclockwise manner. The visitor would further notice that the planets circle the sun in approximately the same plane whereas, on the other hand, the comets orbit the sun at all possible angles to the plane of the planets' orbits.

Further examination of the planets' orbits would reveal what Kepler was the first to recognize in 1607; that is, the planets' orbits are slightly elliptical. This ellipticity of planetary orbits is called Kepler's first law of planetary motion. This law means that the distance between a planet and the sun varies depending on the position of the planet in its orbit. When the planet is nearest the sun, it is said to be at *perihelion,* and when it is farthest from the sun, it is at *aphelion.*

Kepler's second law of planetary motion states that the closer a planet is to the sun, the faster it moves in its orbit; and the farther it is from the sun, the slower it moves in its orbit. Consequently, a planet speeds up as it approaches perihelion and slows down as it approaches aphelion. It is obvious then that the closest planet to the sun, Mercury, should move the fastest (30 miles per second) in its orbit.

Kepler's third law states that there is a relationship between the time it takes for a planet to make one complete revolution (one period) around the sun and the average distance that the planet is from the sun.

In more precise terms Kepler's three laws can be stated as:

1. Planets move about the sun in paths that are elipses.
2. As a planet revolves around the sun, a straight line joining the sun and the planet would sweep out equal areas in space in equal intervals of time.
3. The square of the period of revolution of a planet is directly proportional to the cube of the planet's average distance from the sun.

Now we will proceed to look at the various members of the solar system in more detail. The information that man has accumulated about the solar system has been collected from visual, telescopic, and radio observations. Man has sent out fly-by space probes to photograph and collect information about Venus, Mars, and the moon and has also landed space probes on all three. Finally, man has landed on several occasions on the moon and carried out limited surface explorations.

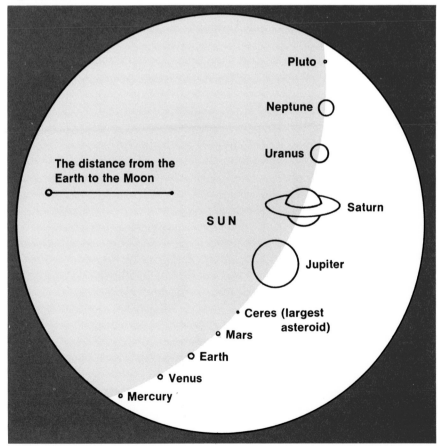

Fig. 17.1 **The relative sizes of the planets as compared to the size of the sun.**

THE SUN

The realm of the solar system is ruled by an immense and resplendent monarch—the sun. The sun is a gigantic globe of incandescent gas, whose "surface" temperature of 5,700°C (10,292°F) gradually increases to between fifteen million and twenty million °C in the interior. The diameter of the sun is 864,000 miles, which is about four (3.6) times the distance from Earth to the moon! Spectroscopic analysis of the sun's light reveals that the sun is made up of 87% hydrogen and between 8 to 12% helium. All the other ninety naturally occurring elements then constitute only 1 to 5% of the sun's mass.

That portion of the sun which we normally see is called the *photosphere* and has a temperature of about 5700°C. Surrounding the photosphere and extending outward for about one or two thousand miles is a region called the *chromosphere,* which can be seen as a red glow for an instant before and an instant after the totality of a solar eclipse. The element helium was discovered in the chromosphere of the sun long before it was known to exist on Earth. In fact, helium was named to show that it was a sun element since its name comes from the Greek word *helio* meaning sun and the Latin suffix *ium* meaning element. The shining, ivory-white halo that surrounds the totally eclipsed sun is called the *corona* (see Fig. 17.3). The corona surrounds the chromosphere and extends outwards for millions of miles.

The smooth, calm appearance of the sun's surface, the photosphere, belies the violent turbulence that actually exists, as shown in Figure 17.4.

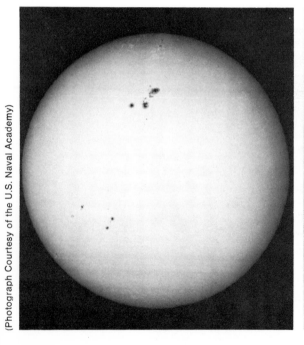

Fig. 17.2 **The photosphere of the sun. Many sun spots are visible.**

Fig. 17.3 **The corona of the sun visible during a total solar eclipse.**

our solar family

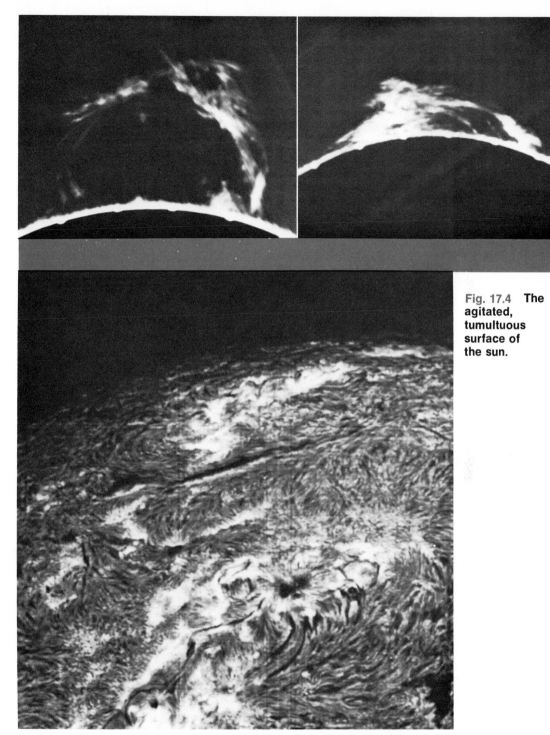

Fig. 17.4 **The agitated, tumultuous surface of the sun.**

(Photograph Courtesy of the Hale Observatories)

The sun continuously undergoes stormy and fiery upheavals that result in magnificent arches of fire that reach heights of a quarter of a million miles. On the photosphere these can be seen as *sunspots*. Sunspots have been observed since antiquity but were not seriously studied until Galileo trained his newly built telescope on them. They appear as great dark blotches on the sun's surface, and at any given time only a few (or none) may be present, or they may occur in large numbers appearing singly and in sizeable groups. Some sunspots are extremely large, attaining a diameter of about 100,000 miles, a size which is large enough to hold twenty-five Earths! The sunspots look dark only by comparison with the brighter photosphere which is 800°C (1472°F) hotter than the sunspots. Actually, since the sunspots are approximately 4,870°C (8798°F), they would appear very bright if they did not have the brighter photosphere as a background. In a similar fashion the lighted interior of a house appears dark if viewed in the daytime from outside but is brightly lit up if viewed at night from the outside. The presence of sunspots is responsible for the sun's emission of large quantities of radio waves that result in considerable static interference with radio transmissions here on Earth.

There are more sunspots visible on the sun in some years than in others. Actually there is a rhythm to the increase and decrease of the number of sunspots. Sunspot activity reaches a maximum approximately every eleven years. Sunspots allow one to determine the length of time it takes for the sun to complete one rotation on its axis since one needs only to notice when a sunspot disappears around the edge of the sun and how long it takes the sun to rotate the spot back around and to the edge again. Sunspots indicate that it takes the sun a little less than one of our months to make one rotation. Unlike the Earth, however, not all parts of the sun rotate at the same speed, and the value of less than one month is an average value.

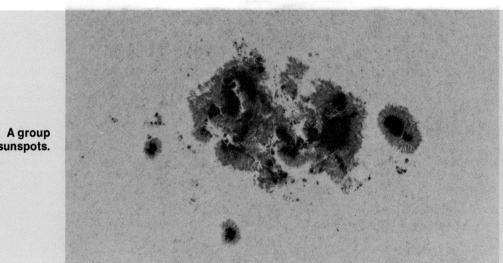

Fig. 17.5 A group of large sunspots.

(Photograph Courtesy of the Hale Observatories)

our solar family

Besides the light that the sun produces, it also emits infrared and ultraviolet radiation, radio waves, x-radiation, and charged particles. A large percentage of the radio waves directed toward Earth is reflected back into space by the ionized gases in our ionosphere. Ultraviolet radiation is mostly absorbed by the ozone layer in the upper regions of our atmosphere, a fact which, of course, is very fortuitous since if all the ultraviolet that reaches the ozone layer also reached the surface of our planet, Earth might become quite sterile. Light and heat from the sun reach Earth's surface in large quantities. To illustrate just how much heat the sun is constantly emitting, imagine a ball of ice the size of our planet. Then if all the heat which the sun emits were directed on the ice, the ice would completely melt in only twenty-three minutes!

The sun continuously emits large quantities of charged particles, mostly electrons and protons, to produce what is called the "solar wind." It is this solar wind, along with light pressure, that forces the tails of comets continuously away from the sun. During periods of high sunspot activity and solar flares, the number of particles emitted by the sun is greatly increased and, consequently, an intense rain of particles is received here on Earth. Some of the particles are trapped by the Earth's magnetic field to produce the *Van Allen* radiation *belts* that surround Earth (see page 373). The remaining particles bombard the atoms and molecules in the upper reaches of our atmosphere to produce the aurora at Earth's polar region. The auroras (the northern and southern lights) are caused by the high speed electrons and protons striking atoms and molecules of the atmosphere so that electrons in the atoms and molecules jump to higher energy levels. When the electrons drop back to their lower levels, they emit electromagnetic radiation in the visible range to produce the aurora in a process similar to the production of light in a neon sign. The aurora occurs between 50 and 150 miles up.

Fig. 17.6 The aurora borealis (Northern lights).

(Photograph Courtesy of NOAA)

THE PLANETS

The four planets closest to the sun, called the *"inner" planets* or *"earth-type" planets,* are all relatively small in size and have relatively high densities, that is from 4 to 5.5 g/cm³ (water has a density of 1 g/cm³). These "inner" planets (Mercury, Venus, Earth, and Mars) have either no atmosphere, such as Mercury, or oxidizing atmospheres somewhat similar to Earth's. An oxidizing atmosphere is one that in general contains varying amounts of oxygen, carbon dioxide, and nitrogen whereas reducing atmospheres, such as those of the "outer" planets, contain varying amounts of hydrogen, helium, methane, and ammonia. Because of the closeness of the "inner" planets to the sun, they are all either fairly warm or reasonably hot bodies. Mercury and Venus have daytime temperatures of about 340°C and 480°C (644°F and 896°F) respectively, and Earth and Mars have daytime temperatures of about 21°C and 4°C (70°F and 39°F) respectively. Of course, the above temperatures are average temperatures of the planets' sunlit sides. The nightside temperatures of Mercury and Mars can become quite low. Only Earth and Mars have temperatures that allow water to exist as a liquid, and in the case of Mars this temperature exists only during the daytime. The two planets Mercury and Venus, whose orbits lie between those of the Earth and the sun, are called *inferior* planets, and the remaining planets, Mars, Jupiter, Saturn, Uranus, Neptune, and Pluto, whose orbits are farther from the sun than Earth's orbit, are called *superior* planets.

Jupiter, Saturn, Uranus, Neptune, and Pluto are called the *"outer"* planets, although the planet Pluto is not like the other "outer" planets but is instead similar to the "inner" planets. The other "outer" planets are quite alike in chemical composition, size (in that they are quite large compared to the "inner" planets), temperature, and density (from 0.7 to 2 g/cm³). The planets Jupiter, Saturn, Uranus, and Neptune all have chemical compositions similar to that of the sun, other stars, and nebulae. Because of their high escape velocities (a result of their large masses) and low temperatures (because of their great distances from the sun), they have not lost much of their atmospheres and, consequently, still have approximately the same chemical makeup that they had at their birth when the solar system condensed out of a dark nebula four and a half billion years ago. On the other hand, the "inner" planets have lost their atmospheres of hydrogen, helium, methane, and ammonia.

Only three planets in our solar system do not have moons or satellites—Mercury, Venus, and Pluto.

MERCURY

The planet Mercury is the closest planet to the sun. Since its orbit, like the orbits of all planets, is an ellipse, its distance from the sun varies from about twenty-eight million miles at perihelion to about forty-three million miles at aphelion. Mercury was named after the swift, wing-footed Roman god who served as the messenger of the gods. This is an appropriate name since Mercury, being the closest planet to the sun, moves more swiftly than any of the other planets. Since Mercury's orbit is so close to the sun, we can see Mercury for only a short period of time after sunset and before sunrise. In fact, there are only about two weeks of each year during which Mercury is easily seen since most of the time it is lost in the solar glare. As a result, relatively few people have ever seen Mercury. Mercury is the smallest planet in the solar system with a diameter of only 3030 miles. The Earth's moon has a diameter of two-thirds that of Mercury, and Jupiter's largest moon is actually larger than the planet Mercury since it has a diameter of 3162 miles.

For a long time it was thought that Mercury kept the same face to the sun in the manner that our moon keeps the same face toward Earth. In order for Mercury to keep one side perpetually facing the sun, its period of rotation on its axis would have to equal its period of revolution around the sun. Mercury's period of revolution has been known for centuries to be about eighty-eight days, and in 1965 it was shown by radar observations that Mercury rotates on its axis in fifty-nine days. It cannot, therefore, keep the same face toward the sun. The sunlit side of Mercury has a temperature of 340°C (644°F), which is hot enough to melt metals such as zinc, tin, and lead. At the same time that Mercury's sunlit side has a temperature of 340°C, its nightside has a temperature of −120°C (−248°F). This is a difference of 460°C between the two sides.

Mercury does not have a satellite or an atmosphere.

VENUS

With the exception of the moon, Venus is the most brilliant and beautiful object in the night sky. At its brightest it can even be seen during the day if one knows where to look for it, and at night it can cast shadows. Venus' shining splendor has occasionally resulted in reports that the star of Bethlehem has reappeared. Surely Venus has been fittingly named after the Roman goddess of beauty. In Greek mythology, Venus would correspond to the goddess of love and beauty, Aphrodite.

Venus is often called Earth's twin since its diameter is 7600 miles as compared to Earth's diameter of 7917 miles. It takes Venus 225 Earth days to complete one revolution around the sun, but 243 Earth days to rotate once on its axis. A Venusian day then is longer than a Venusian year! Of the nine planets, only Venus rotates on its axis in a clockwise direction; that is, it rotates opposite to the direction of rotation of the other planets. A planet that rotates clockwise is said to undergo *retrograde* rotation. A visitor on Venus would see the sun rise in the west and set in the east. The splendid brilliance of Venus is due to its magnificent shroud of white clouds that eternally obscure its surface and to its closeness to Earth. At closest approach, twenty-five million miles, Venus is closer to Earth than any other planet. Since Venus is an inferior planet, it, like Mercury, goes through phases similar to the phases shown by our moon.

Venus is brighter when it is only a crescent than when it is more fully lit because it is closer to the Earth at crescent phase.

The veil of beautiful white clouds that encircle and conceal the surface of Venus is mainly responsible for the extraordinarily high surface temperature of this planet. Venus' surface temperature of 480°C (896°F) is the direct result of the high carbon dioxide content (90%) of its atmosphere. The high concentration of carbon dioxide produces a "hothouse effect" by trapping the infrared radiation of the sun in the same manner that greenhouses on Earth trap the sun's warmth. Venus' atmosphere also contains water vapor (1%), oxygen (about 1%), and nitrogen (7%). Because of the high reflectivity of the Venusian clouds, light from the sunlit side of Venus is probably reflected into all portions of the nightside so that instead of having dark moonless nights, Venus enjoys a perpetual twilight.

EARTH

Earth and its lone satellite are examined in detail in Chapter 18 and are mentioned here merely to stress their position in the solar system.

MARS

More popular science articles, newspaper stories, and science fiction books have been written about the "mysterious red planet" than about any other planet in the solar system. The mere mention of the planet Mars conjures up in most people's minds visions of an ancient race of strange and very intelligent people living near vast networks of water canals on a reddish Sahara-desert type of planet whose lonely, barren plains are swept by storms of swirling red dust.

The reddish appearance of Mars suggested blood to the ancient Babylonians, and they called Mars the Star of Death. Its present name is no less a suggestion of bloodshed and death. Mars is named after the Roman god of war, and Mars' two moons, Phobos and Deimos, are named after the two sons of Ares (the god of war in Greek mythology), who personified terror and fear. Terror and fear are just the type of retinue one would expect of a god of war. Phobos races along in its orbit so fast (it takes Phobos only about seven and a half hours to complete one revolution around Mars) that it rises in the west and sets in the east twice in one night. On the other hand, Deimos takes considerably longer than a Martian day to make one complete revolution and, consequently, rises in the east and remains in the sky two nights before setting in the west. Phobos is oblong in shape and may be the only non-spherical satellite in the solar system. Mars' moons are the smallest satellites in the solar system. Phobos is fifteen miles long and thirteen miles wide, and Deimos is about eight miles in diameter.

Since Mars rotates on its axis in about twenty-four and a half hours, the Martian day and night are about the same length as ours. The Martian year (687 days), however, is almost twice as long as an Earth year, a fact which means that each of the four Martian seasons is six months long. Since the planet's diameter is only 4,200 miles, Mars is only slightly more than half the size of Earth. This smaller mass results in a smaller surface gravity. Thus a man who weighs 180 pounds on Earth would weigh only sixty-eight pounds on Mars.

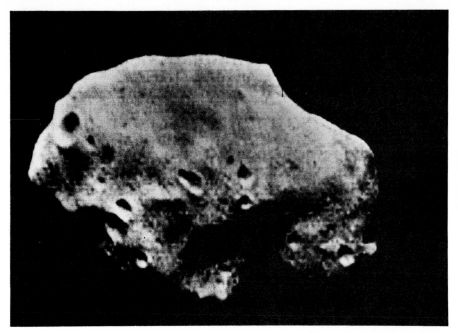

Fig. 17.7 **Mars' innermost satellite, Phobos which is about sixteen miles long and thirteen miles wide.**

(Photograph Courtesy of NASA)

The most prominent feature of Mars is its polar caps. During the Martian winter, the winter polar cap may reach half way to the Martian equator, but during the summer the polar cap may completely disappear. Since they shrink rapidly, the Martian polar caps must be very thin, perhaps only a few feet thick in most places. The winter Martian polar cap is probably similar to the winter snows on Earth that cover an area that extends almost half way to our equator but quickly melt off during our warm spring. Evidence suggests that the polar caps of Mars may not be composed entirely of ice but may contain appreciable quantities of "dry ice" (solid carbon dioxide).

It has been known for many years that Mars has an atmosphere because of the large dust storms that have been observed to blow across the Martian deserts, but it was not until the Mariner IV space probe fled past Mars in 1965 that it was realized just how tenuous and thin its atmosphere really is. Mariner IV reported that the atmosphere is only about 1% as dense as Earth's atmosphere and consists mostly of carbon dioxide with smaller quantities of water vapor and minute quantities of oxygen. Because of the thinness of the planet's atmosphere, the temperature changes which the Martian surface experiences are rather extreme. Thus, during the summer at the equator, surface temperatures range from 30°C (86°F) during the day to about −73°C (−100°F) at night.

The question of whether there is life on Mars has intrigued scientists and laymen alike for the last two centuries. As an example of the extent to which some people go in their speculation concerning intelligent life on Mars, there has even been the absurd speculation that since Mars' two moons are so small, these

our solar family   **355**

moons may be huge artificial satellites launched into orbit by a very advanced technological civilization. But by Earth standards the conditions that exist on Mars today are not very conducive to life forms as we know them. The amount of oxygen in the atmosphere is very small, the temperature variation between day and night is very large, and the amount of water available to possible plants and animals is quite minimal. These harsh conditions almost certainly rule out animal life of any size whatsoever, but do not eliminate the possibility of vegetative life on Mars. Mars surely must have gone through the same stages of formation and early chemical evolution in its atmosphere that Earth underwent, with the result that the precursor organic molecules of life were probably formed on Mars during the same time period that saw their formation on Earth. It very well may be that life did arise on Mars and has been able to maintain itself but on a very low evolutionary level. During the Martian spring when the polar caps are retreating, large areas that during the winter have been brown in color turn a bluish-green. This change in color follows the receding polar caps. Some scientists have interpreted this change in color of Mars' surface as the spring revival and growth of vegetation as water becomes available from the melting snow caps. But the low nighttime temperatures and the small water supply would seem to exclude all but the most hardy of plants, such as the pale blue-green lichen that grows above the Arctic Circle on Earth.

In 1877 the Italian astronomer Giovanni Schiaparelli first observed markings on Mars that he called *canali,* which means "channels" in Italian, but this word was incorrectly translated into English as "canals." This unfortunate translation resulted in waves of speculation that have continued to this day and at times have reached astonishing heights of fancy. It was assumed that the Martian surface was laced with a vast network of canals for transporting water from the melting polar caps to the drier equatorial regions by a colossal, planet-wide irrigation project for the production of food. In any case, because of the $-100°F$ nighttime temperature on Mars, the canals would be solid ice canals rather than water canals. Recent pictures of parts of the Martian surface taken by orbiting satellites show no channels or canals. Instead, the Martian surface is pock-marked like the surface of our moon with meteor impact craters as shown in Figure 17.8.

What then did many reputable and outstanding astronomers see that appeared to be long straight channels on the Martian surface? There is a psychological tendency for the eye to connect objects such as craters to form straight features when they are near the limit of visibility. Although the question of canals has not been unequivocally resolved, there is certainly no direct evidence for their existence.

JUPITER

Of all the planets, certainly Jupiter is king. Not only is Jupiter more massive than any other planet; it is about two and one-half times more massive than all the other planets combined. Jupiter also has more satellites, twelve in all, than

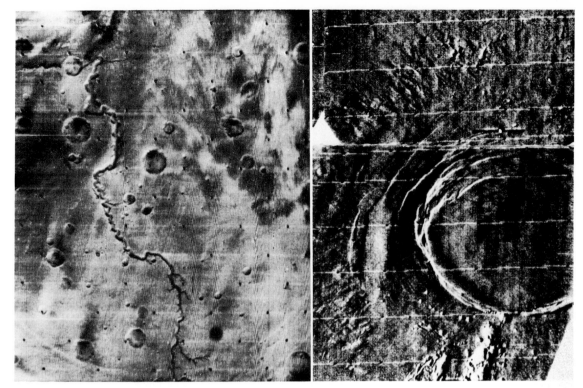

(Photographs Courtesy of NASA)

Fig. 17.8 Photographs of the pock-marked surface of Mars. One of the photographs shows what appears to be a water cut sinuous valley.

any other planet and has the distinction of having a moon, Ganymede, that is larger than Mercury. Jupiter was indeed aptly named after the Roman ruler of the gods. Jupiter's diameter, 87,000 miles, is approximately one-third the distance between Earth and its moon, and its mass is 318 times the mass of Earth. Because of this great mass, its surface gravity is 2.64 times greater than Earth's. A man who weighs 200 pounds on Earth would weigh 528 pounds on the surface of Jupiter.

One might intuitively feel that such a large planet would be rather sluggish in its movements and would spin slowly on its axis. However, Jupiter is the fastest spinning planet in the solar system and makes one rotation on its axis in nine hours and fifty-five minutes. As a consequence of the high speed of rotation, objects at Jupiter's equator are traveling at approximately 28,000 miles per hour (compare this with objects at Earth's equator that travel 1,000 miles per hour). This very fast spin results in considerable flattening at Jupiter's poles.

Whereas the inner planets have relatively high densities, from about 4 to 5.5 times as dense as water, Jupiter has a density of only 1.33 times the density of water. In order for Jupiter to have such a low density, it has been calculated

our solar family

Fig. 17.9 **Jupiter,
the largest planet
in the solar system.**

that the planet is about 75% hydrogen. Besides hydrogen, Jupiter's atmosphere also contains helium (He), methane (CH_4), and ammonia (NH_3). Although the planet's make-up certainly must include large quantities of water, the low temperatures (about $-130°C$ or $-202°F$) existing on a planet so far from the warmth of the sun would result in all water's being in the form of ice. The overall chemical composition of Jupiter, as well as Saturn, Uranus, and Neptune, is very similar to the chemical composition of the sun, other stars, and gaseous nebulae from which our solar system was formed. This similarity of composition is the result of the great masses of these planets (with the consequence of a large gravity) and their extreme distances from the sun (with its consequence of low temperatures). Because of its high gravitational field and very low temperature, Jupiter has not lost very much of its original composition.

The most spectacular feature on Jupiter is the Great Red Spot (see Fig. 17.9). This gigantic spot (30,000 miles long and 7,000 miles across) is large enough to contain eight planets the size of Earth with a few objects the size of our moon thrown in for full measure. The exact nature of this dull red blot on the surface of Jupiter is not known and remains one of the most intriguing mysteries of our solar system. The rest of the Jovian planet is encased by white, red, yellow, brown, and blue-green bands parallel to the equator. These bands, like the Great

358   **our solar family**

Red Spot, are permanent features of Jupiter's atmosphere, but their colors and band widths do slowly change with time.

Jupiter is surrounded by huge radiation belts similar to Earth's Van Allen radiation belts. Jupiter's radiation belts mean, of course, that Jupiter must have a very strong magnetic field that has trapped large quantities of charged particles. Since Jupiter does have a magnetic field surrounding it, one must assume that it also has a molten iron core. Radiotelescopes detect strong radio emissions from the region of Jupiter's radiation belts. These radio emissions are undoubtedly the result of the charged particles, trapped in the radiation belts, being accelerated as they spiral through the magnetic lines of force of Jupiter's magnetic field. Radio emission caused by the acceleration of charged particles in a magnetic field is known as *synchrotron radiation*.

Jupiter's satellite, Ganymede, has a thin atmosphere.

SATURN'S RINGS AND ROCHE'S LIMIT

Saturn is the most beautiful and spectacular telescopic object in the solar system mainly because of the unique and splendid set of radiant rings that encircle it. This "ringed planet" is the second largest planet in the solar system with a diameter of 72,000 miles. The outermost ring of Saturn, has a diameter of 171,000 miles, which is almost three-fourths the distance between Earth and the moon. While there are three concentric rings surrounding Saturn at its equator, the innermost ring of Saturn is 7,000 miles from the planet's surface and is 11,000 miles in width. After the inner ring there is a gap of 1,000 miles before the beginning of the second ring, which is 15,000 miles in width. The second ring is far brighter than the other two rings. Between the bright second ring and the outermost ring is a 3,000 mile gap. The outer ring is 10,000 miles in width. The complete ring system, then, is 40,000 miles in width and 171,000 miles in diameter. Amazingly, for such a gigantic ring system, the rings are only ten miles thick. Indeed, when Saturn is in a position that presents her rings edge-on, they are all but invisible and cannot be detected except by very powerful telescopes. In fact, in 1609, right after Galileo had built his first telescope, he saw Saturn's rings and published a paper describing his marvelous observations. About two years later, however, when he again decided to study Saturn through his telescope, the rings happened to be edge-wise to Earth and he was unable to see them. His failure to observe the rings the second time seriously confused and upset him because he thought that he might have simply imagined having previously seen Saturn's rings. Fortunately, many years later, the rings of Saturn were inclined so that he was able to see them again, and he, therefore, realized that he had not been tricked by some bizarre telescopic illusion. The rings of Saturn consist of billions of pebble and sand-grain sized fragments of rock and ice, each following its own orbit around Saturn just as the ten moons of Saturn follow their own individual orbits. As proof that the rings are composed of individual particles and are not solid, very bright distant stars can occasionally be seen dimly shining through the rings.

Fig. 17.10 Saturn, the "ringed" planet.

(Photograph Courtesy of New Mexico State University Observatory)

The existence of Saturn's rings puzzled astronomers for over two hundred years until in the middle of the nineteenth century calculations showed that if any satellite came closer to its mother planet than 2.4 times the planet's radius, the satellite would be completely disintegrated by the gravitational tidal forces of the planet. This critical distance, 2.4 radii from the center of a planet, is called the *Roche limit*. The outer edge of Saturn's ring is 85,500 miles from the center of the planet whereas the Roche limit distance is $2.4 \times 36,000$ miles or 86,400 miles. It is seen then that the outermost ring is a scant 1,000 miles within the Roche limit. Either the rings represent a situation in which the particles of the rings were not able to consolidate into a satellite, or they represent a satellite that at one time was beyond the Roche limit but strayed within the Roche limit and was summarily disintegrated into the rings we see today. Saturn's closest moon, Janus, is about 9,000 miles beyond the Roche limit and, therefore, quite safe.

Saturn has the lowest density of all the planets, 0.7 times the density of water. With this low density Saturn would quite easily float. Saturn's chemical composition and surface structure are quite similar to Jupiter's. Thus, Saturn's surface also has a banded appearance, although the colors of the bands are not as bright as Jupiter's. Since Saturn is about 900 million miles from the sun, its surface temperature is $-150°C$ ($-240°F$). This low temperature means that a large percentage of the methane that would normally be present in Saturn's atmosphere is in a liquid state, and a visitor to the planet's surface would find turbulent methane seas swept by violent storms of hydrogen winds.

Saturn rotates on its axis in ten hours and forty-five minutes.

Saturn was named after an ancient Roman Titan God of agriculture who was the father of Jupiter. Saturn's largest satellite, Titan has an atmosphere of methane.

URANUS

Uranus was the first planet to be discovered during recorded history and was named after the Greek god of heaven who was father of the Titans and, therefore, father of Saturn, and grandfather of Jupiter.

In general appearance and chemical makeup, Uranus is similar to Jupiter and Saturn. The planet is 30,000 miles in diameter and rotates on its axis in ten hours and forty-five minutes. Uranus' very low temperature, $-180°C\,(-292°F)$, allows ammonia and methane "snows" to fall from the hydrogen atmosphere onto methane seas.

The most unusual aspect of Uranus is that its axis of rotation lies almost in the plane of its orbit. The planet's north pole is 8° below the plane of its orbit.

Uranus has five satellites.

NEPTUNE

In the years that followed the discovery of Uranus, astronomers periodically noted its positions in the heavens and from this information were able to calculate the exact orbit that Uranus should take in its path around the sun. Before very many years had passed, it became obvious that Uranus was deviating from its calculated orbit. Many of the astronomers of that day began to wonder whether perhaps Newton's universal law of gravitation was invalid at such great distances or perhaps was only qualitatively correct rather than quantitatively correct. Some astronomers wondered whether Uranus was being pulled from its predicted path by the gravitational influence of an unknown planet that was circling the sun beyond the orbit of Uranus.

In 1845 a young English mathematician by the name of John Adams calculated exactly where an unknown planet should be in order to disturb Uranus' orbit by the observed amount. Adams sent his calculation to Sir George Airy, the Astronomer Royal of England. Sir Airy apparently did not have much faith in so young an astronomer and did not bother to have the telescopes at his command assigned the task of looking for the new planet. The following year, the French Astronomer, Leverrier, who did not know of Adam's work, also calculated the position of the unknown planet and sent his result to the astronomer Galle at the Berlin Observatory. Galle looked for and found the new planet on the night of the same day that he received Leverrier's letter. Today, both Adams and Leverrier equally share the honor of having predicted the position of the new planet. This new planet was named Neptune after the Roman god of the sea.

Neptune has a diameter of 28,000 miles and only two satellites. The planet's chemical composition and structure are similar to the composition and structure of Jupiter, Saturn, and Uranus. Since Neptune is so very far from the sun, about three billion miles, it requires 165 years to complete one orbit around the sun. Neptune has not had time to complete one orbit since its discovery in 1846!

PLUTO

The story of the discovery of Neptune is repeated in the discovery of Pluto. Before long it was realized that Neptune also did not follow its prescribed path around the sun, and scientists immediately suspected the existence of a planet still farther out in the solar system. In 1930 this outermost planet was discovered at the Lowell Observatory.

Pluto is something of a puzzle since it is quite unlike its neighbor planets and is actually more like the high density inner planets. Pluto is only 3,600 miles in diameter (compare Pluto's diameter with the 3162 miles diameter of Jupiter's satellite, Ganymede). Furthermore, part of Pluto's orbit actually falls within Neptune's orbit. For the above reasons, some scientists speculate that Pluto is an escaped satellite of the planet Neptune. Since Pluto is 3.7 billion miles from the sun, it takes Pluto 247 years to make one trip around the sun. It will not, therefore, complete its first revolution since its discovery until the year 2177.

From Pluto's unbelievably cold surface ($-240°C$ or $-400°F$), the distant sun would appear as a very small object in the sky and would be only one thousandth as bright as when seen from Earth. Surely this small airless planet at the outer rim of the sun's planetary system makes a very lonely voyage in its 247 year trip around the sun. Its name, taken from the Roman god of the underworld, is not meant to reflect its temperature but its condition of darkness and loneliness.

THE ASTEROIDS

Before the discoveries of Uranus, Neptune, and Pluto, astronomers wondered about the regularity of the spacing of planets about the sun. In 1772 the Director of the Berlin Observatory, Bode, published a paper in which he explained the work of Titius who had found an empirical relationship between the various distances from the sun of the solar system's planets. This empirical relationship, known as Bode's "Law," predicted that a planet should occupy an orbit between the orbits of Mars and Jupiter. However, no planet had been observed at this distance (2.8 A.U.) from the sun. In 1801, the Sicilian astronomer, Piazzi, found a planet-like object that was very close (2.9 A.U.) to the predicted orbit. Piazzi named the new object *Ceres* after a goddess of his native Sicily. A more careful investigation of the newly discovered object showed that it was only about 500 miles in diameter! Could an object only 500 miles in diameter be considered a planet when the smallest planet previously known in the solar system was 3030 miles in diameter? Surely not. This dilemma was eliminated when a year later a second diminutive planet was discovered with the same general orbit as Ceres and was named *Pallas* after Athena, the Greek goddess of wisdom. In 1804 a third object was found and named *Juno* after the wife of the ruling Roman god, Jupiter. Today, thousands of these "minor planets" or *asteroids* have been found that occupy orbits around the sun that correspond to the predicted orbit of the "missing planet." The known asteroids range in

TABLE 17.1
Some Data on Our Solar System.

Solar System Object	Diameter (Earth's Diameter = 1)	Mass (Earth's Mass = 1)	Distance From Sun (A.U.)	Number of Satellites	Revolution Period ("year")	Rotation Period ("day")	Escape Velocity (miles/sec.)	Surface Gravity (Earth's = 1)
Sun	109	333,000		9		~25^d	387	28
Moon	1/4	1/10	~1	none	27.3^d	27.3^d	1.5	1/6
Mercury	1/3	1/20	2/5	none	88^d	59^d	2.6	1/2
Venus	~1	4/5	3/4	none	225^d	*243^d	6.4	~1
Earth	1	1	1	1	1^y	1^d(23^{h}56^m)	7	1
Mars	1/2	1/9	1½	2	1.9^y	24.6^h	3.1	1/2
Jupiter	11.3	318	5	12	12^y	9.9^h	37.8	2.6
Saturn	9.5	95	9½	10	29.7^y	10.4^h	22.3	1.2
Uranus	3.9	15	19	5	83.7^y	10.8^h	13.7	~1
Neptune	3.5	17	30	2	165^y	15.7^h	15.3	1½
Pluto	1/2	1/5 (?)	39	none	248^y	6.4^d(?)	?	?

* Exhibits retrograde (reverse) rotation

our solar family

diameter from one mile to 500 miles and can, therefore, be considered "flying mountains" or "flying mountain ranges." There are probably untold thousands of asteroids, too small to be observed, that have diameters from one mile to fractions of an inch.

There is some speculation that billions of years ago a planet did exist in the orbits of the asteroids, but for some unknown reason it blew up into the fragments that we observe today. If this assumption is correct, the unfortunate planet would have been of extremely small size since it is estimated that all the asteroids combined account for no more than 1/80 the mass of the smallest planet, Mercury. Indeed, the assumed planet would not have been large enough to be considered a respectable satellite since it would have been only 1/20 the mass of our moon.

There is other speculation that the asteroids represent original solar system material that for some reason never congealed to form a planet. If the above assumption is true, the question we must ask ourselves is, "Why should the material existing between the orbits of Mars and Jupiter not have produced a planet by the process of accretion as happened for all the other planets?" Since the asteroids are far beyond the Sun's Roche limit, the tidal action of the sun is not the answer. Perhaps the massive Jupiter introduced tidal effects that prevented planet formation.

COMETS

From ancient times down to the Middle Ages, man considered the appearance of a comet as an apparition that portended grave disasters for individuals and for nations. Comets were thought to predict the defeat of armies and the coming of pestilence and plague to kingdoms. As an example, a comet appeared in 1453 when Constantinople fell to the Turks. People of that time were convinced that the comet had been sent to mark the end of the thousand-year reign of the Byzantine empire. Today enlightened people realize that comets are simply natural members of our solar system. As recently as 1910, however, considerable fear was generated when the Earth passed through the tail of Halley's comet. Scientists had pointed out that among other things comets contain methane, ammonia, and hydrogen cyanide. Knowing that the gases present in comets are deadly in high enough concentrations, many people prophesied that life on our world would end when we passed through the lethal gases present in the tail of Halley's comet. In fact, certain groups met on mountain tops to hold marathon religious meetings that were to extend up to the "doomsday" time when the inevitable end would come as Earth's orbit and the comet's orbit intersected. Other groups with a more materialistic and hedonistic outlook on life met in cellars to indulge in wild revelry and merrymaking. The basic tenet of the latter group seemed to be "make merry today for tomorrow we die."

Actually, nothing whatsoever happened when we passed through the comet's tail. The religious groups of doomsday prophets were delightfully surprised when the judgment failed to materialize, whereas the hedonistic groups

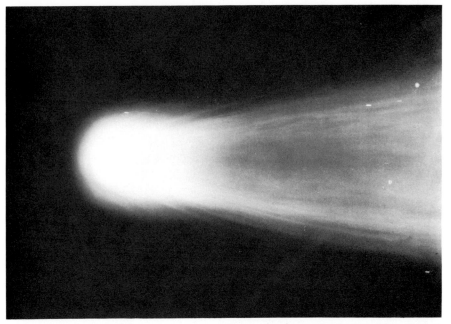

Fig. 17.11   Head of Halley's comet, May 8, 1910.

Fig. 17.12   Comet Arend-Roland, April 30, 1957.

probably felt that the "morning-after" effects of their nights of revelry meant that some sort of judgment as well as physical punishment had surely been meted out.

A really big comet is a beautiful thing to behold. The comet appears as an elongated filmy patch of soft light that may cover a larger portion of the night sky than any other astronomical object. The tenuous nature of the comet is obvious from the fact that background stars shine through the tail with very little diminution of their brightness. The name *comet* is very well chosen from the Greek meaning long hair. Thus, the word *comet* implies an apparition of soft, glowing heavenly hair.

Comets are most spectacular when they are at or near perihelion and consist of a head and a tail. The head, which can be from 10,000 miles to over 1½ million miles in diameter (1½ million miles in diameter is almost twice the diameter of the sun!), consists of a cloud of gas surrounding a collection of loose particles. The tails of comets have been known to reach almost 100 million miles, a distance which is farther than Earth is from the sun!

When a comet is far from the sun, it is mainly a collection of frozen water, ammonia, and methane in which is embedded rocklike particles. The diameters of comets under these conditions probably range from a few miles to a few thousand miles. As a comet approaches the sun, it becomes warmer and the frozen material melts and vaporizes to produce the head and tail of the comet. The tail grows in length, getting longer the closer the comet is to perihelion, and always points away from the sun (see Fig. 17.13). The growth of the tail as well as its direction is the direct result of the solar wind and radiation pressure of the sun. As the comet recedes from the sun, the tail decreases in length until it finally disappears, and the head gradually shrinks. The gaseous material of the head slowly condenses and solidifies until once again the comet is a frozen mass of material that is no longer visible even through our telescopes.

Whereas the planets all revolve around the sun in approximately the same plane, comets revolve around the sun at all possible angles. Thus, the orbits of some comets are in the same plane as the planets' orbits, other comets' orbits are perpendicular to the planets' orbits, and still other comets revolve around the sun in orbits between the above extremes. The orbits of comets are highly elliptical, and there is a great variety in the aphelion distances of various comets. Some comets take over a million years to complete one orbit around the sun and, therefore, have never been seen or seen only once by Homo sapiens. Other comets

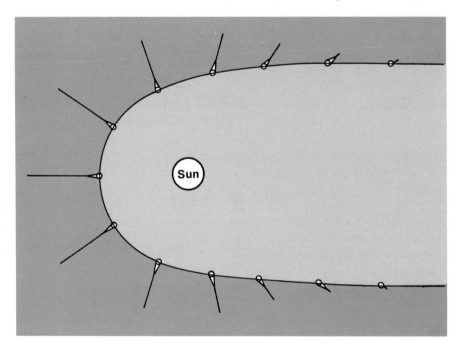

Fig.17.13 A comet's tail changes in length and direction as it approaches and recedes from the sun.

our solar family

have an aphelion distance from the sun fairly near the aphelion distance of the planet Jupiter, and actually are gravitationally influenced by Jupiter; these comets are called "Jupiter's family of comets." The most famous comet, and a member of Jupiter's family of comets, is Halley's comet, whose every passage around the sun has been faithfully recorded by man since 240 B.C. Halley's comet will next return in 1986.

Each time a comet makes a close approach to the sun, some of its gaseous material is swept away because of the action of radiation pressure and solar winds. Eventually, after hundreds of passes around the sun, comets lose their water, ammonia, and methane, and only small particles of rocks and metals remain. These small solid particles continue to circle the sun, and on occasion Earth's orbit intersects the orbit of one of these "dead comets" with the result of a beautiful and spectacular meteor shower. It would appear that as a result of comets' slowly dying because of their many passages around the sun, the time will come in the far, dim future when our solar system will no longer be blessed by the delicate beauty of our lonely comets.

METEORS

Who has not, as a child, lain on his back in the grass at night and counted "shooting stars" and wondered what they were and from where they came? These bright streaks of light that pierce the sky about ten times per hour are the result of meteors being heated to incandescence by the friction of their passage through the air. Most of the meteors that we see are no larger than the head of a pin. Most meteors in our atmosphere are too small to give any sign of their presence. It has been estimated that the Earth gains thousands of tons of mass each day by the collection of meteors. This rain of meteors upon the Earth amounts to a mass increase of about 400,000 tons per year and 20 thousand billion tons of material since Earth's beginnings as a planet. The total mass that Earth has collected in its four and one-half billion year lifetime, however, represents only three ten-thousands of one percent of Earth's present mass.

A few times each year there are an unusual number of meteors seen. These exceptional meteoric displays, called *meteor showers,* are the result of the Earth's passing through the orbit of a "dead" comet that has lost all but its rock and metal type of material (see Fig. 17.14).

Once a meteor has fallen to the ground, it is called a *meteorite.* Many hundreds of such meteorites have been collected, but most meteors are burned up in their passage through the air and never reach the Earth's surface. Of those meteors that do reach the Earth's surface, two-thirds of them will fall into the great oceans of our planet and never be seen by man. Most of the meteorites that have been found are relatively small, but a 50-ton meteorite has been found in South-West Africa, and a 35-ton meteorite, found in Greenland, is on display at the American Museum of Natural History. There is only one reliable report of a person's being struck by a falling meteor. A meteor crashed through the roof and into the living room of a house in Alabama and struck a woman on her hip.

Meteors are normally too small to make much of a crater on impact with the ground, but there have been some spectacular exceptions. In 1908 a meteor, estimated to have been one hundred thousand tons in mass, fell in Siberia. Fortunately, it fell in a desolate area and not in a highly populated region. The impact produced a shock wave that felled trees for twenty miles from the point of impact, and the air compression wave and the resultant earth tremor were recorded in Europe! The largest meteorite crater in the United States is at Winslow, Arizona, and is called the *Barringer Meteorite Crater*. The Barringer Crater is slightly over 4,000 feet in diameter and 600 feet deep.

CHAPTER SUMMARY

Our solar system consists of a central star, the sun, with its attendant nine planets and their thirty-two satellites or moons, tens of thousands of asteroids, innumerable comets, and myriads of meteors.

Our sun is an average star with a diameter of 864,000 miles, a "surface" temperature of 5700°C and a core temperature of about twenty million °C. *Sunspots*, which quite often appear on the surface of the sun, are areas that are of lower temperature than the rest of the sun's surface. Charged particles are emitted in all directions by the sun to produce the *solar wind* that permeates much of our solar system. The charged particles of the solar wind are trapped by the magnetic fields of Earth to produce the *Van Allen radiation belts*.

Outwards from the sun the planets of the solar system are Mercury, Venus, Earth, Mars, Jupiter, Saturn, Uranus, Neptune, and Pluto. Mercury is the smallest planet and Jupiter is the largest planet. Mercury, Venus, and Pluto have no satellites. Jupiter has the greatest number of satellites with its retinue of twelve moons. Venus is the only planet that rotates in a direction counter to the spin direction of all the other planets in the solar system. Saturn, the "ringed planet," has three concentric rings of rock and ice encircling it.

Between the orbits of Mars and Jupiter are extremely large numbers of orbiting rocky bodies known as *asteroids*, varying from 500 miles in diameter to the size of small pebbles.

Comets have relatively small masses and revolve around the sun in highly flattened orbits oriented in all possible directions to the plane of the orbits of the planets. While a comet is relatively near the sun, it forms a tail that always points away from the sun and may attain a length of millions of miles.

Meteors, commonly called "shooting stars" continually enter Earth's atmosphere. About 400,000 tons of meteors are captured by Earth each year.

QUESTIONS AND PROBLEMS

1. What are the names of the largest and the smallest planets of our solar system. (p. 363)
2. Which planet undergoes retrograde rotation? (p. 353)
3. What are some differences between the inner planets and the outer planets? (p. 352)
4. State Kepler's laws of planetary motion. (p. 346)
5. Why are sunspots dark? (p. 350)
6. Explain the existence of the aurora. (p. 373)
7. Even though Venus is farther from the sun than Mercury is, Venus has a higher daytime surface temperature. Why? (p. 354)
8. Which planet spins the fastest on its axis? (p. 363)
9. Why does Saturn have rings surrounding it? (p. 360)
10. Which planet was the first to be discovered in recorded history? (p. 361)
11. What would a comet look like at aphelion? (p. 366)
12. Why does a comet's tail always point away from the sun? (p. 366)

the planet earth

18

THE ENVELOPE OF EARTH
THE FOUCAULT PENDULUM EXPERIMENT
STELLAR PARALLAX
THE SEASONS
BLUE SKY AND RED SUN
SOLAR AND SIDEREAL TIME
PRECESSION OF EARTH'S AXIS
THE SHAPE OF EARTH
THE EARTH-MOON GRAVITY COUPLE
THE TIDES
SOLAR AND LUNAR ECLIPSES
THE FACE OF THE MOON
EARTHLIGHT
THE FUTURE OF EARTH

vocabulary

Barycenter	Precession (of Earth)
Sidereal Day	Solar Day
Stellar Parallax	Umbra

Although Jupiter's moon *Ganymede* is the largest satellite (3162 miles in dia.) in our solar system, Earth's moon is the largest moon in comparison to its mother planet. In fact, the Earth-Moon system is frequently referred to as the "double planet." Looking "down" on Earth from the surface of the moon, we see the most beautiful and colorful planet in the entire solar system. The brilliant multihued Earth hangs in the black airless sky of the moon in striking difference to the desolate landscape of contrasting dark black shadow and shining white light which mark the stark outlines of the deep gaping craters and the towering craggy mountains of the moon's surface.

Fig 18.1 A half-Earth as seen by Apollo 8 astronauts as they orbited the moon.

(Photograph Courtesy of NASA)

THE ENVELOPE OF EARTH

If we now leave the moon and travel toward Earth, we find at a distance of 15,000 miles from Earth the first of the *Van Allen radiation belts* that encircle our planet except at the poles. The belts contain charged particles, electrons and

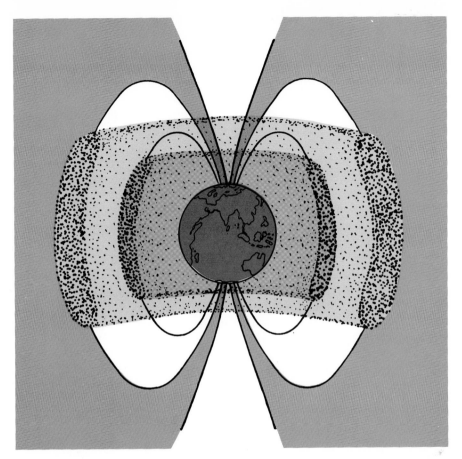

Fig. 18.2 **Earth's magnetic field and Van Allen radiation belts.**

protons, captured from the solar wind and held fast by Earth's magnetic field. At 5000 miles we encounter the second belt and, as we did with the first belt, quickly hurry through so as to expose ourselves for as short a time as possible to the x-rays generated by the high speed electrons and protons striking the walls of our space vehicle. As we approach Earth, we see soft, white clouds gently floating in an azure sky above the surface of a planet-encircling blue-green sea. At each pole are large, cold, white caps of ice and snow. The continents and islands which are scattered around the planet contain lush dark green forests, eye-searing white sand deserts, and lofty mountain ranges that, like a giant dinosaur's backbone, span whole continents.

As far out as a thousand miles from Earth, we notice faint and tenuous vestiges of Earth's atmosphere and from 6000 down to 100 miles we see a brilliant display of glowing curtains of light that silently move in a contorting dance of color and beauty. These are the auroras, the northern and southern lights. The auroras are caused by charged particles from the sun bombarding the atoms of the upper atmosphere in the polar regions to produce a glow similar to the glow of a neon light.

From the outer beginning of Earth's atmosphere down to approximately fifty miles above the surface, we find that the temperature of the air molecules decreases from about 2900°F (1600°C) to −60°F (−51°C) respectively. The high temperature existing at the higher altitudes does not mean that an object at this height would be hot. In fact, just the opposite would be true: the object would be quite cold! What is meant by saying that the temperature is about 2900°F is that the molecules of air in this rarefied portion of our atmosphere are traveling at very high speeds (the higher the temperature of a gas, the faster the molecules of the gas travel). There are so few molecules per cubic foot, however, that only a very few molecules would come in contact with an object, and, therefore, very little heat would be transferred from the molecules to the object. This portion of our atmosphere (from the outer fringe to fifty miles above the surface) is called the *ionosphere* since the radiation and the high speed particles from the sun strike the air molecules to produce ions. From about seventy to fifty miles above Earth the concentration of ions is relatively high and is responsible for reflecting radio waves back toward Earth and thus making possible long distance radio receiving. (See Fig. 18.3.)

From about fifty to ten miles above the surface, the atmospheric temperature is a relatively constant −60°F (−51°C). This layer is called the *stratosphere*. The most important portion of this layer is the ozone layer that exists between forty and thirty miles from the surface. Ozone, which is an oxygen

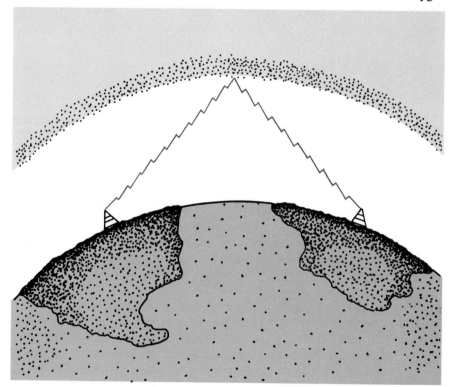

Fig. 18.3 The reflection of radio waves by the ionosphere.

molecule containing three oxygen atoms rather than the normal two atoms of oxygen, is an excellent absorber of ultraviolet light. This ozone layer is extremely important to life on our planet since most forms of life cannot exist if subjected to high ultraviolet concentrations. Indeed, it is doubtful that life could have ever emerged from the seas and invaded the land if it were not for the protective ozone layer that surrounds our planet.

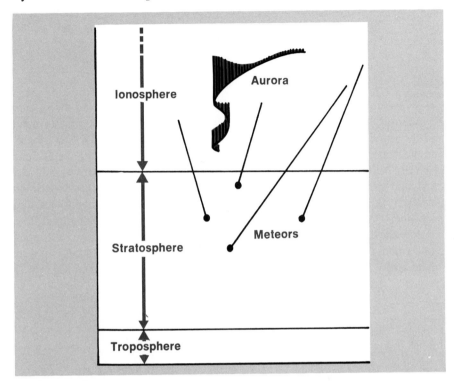

Fig. 18.4 **The various layers of Earth's atmosphere.**

From about ten miles out from Earth down to the surface is that portion of Earth's atmosphere called the *troposphere*. The troposphere is only five miles from the surface in the vicinity of the poles, however. The most striking fact about the troposphere is its temperature gradient. Thus, when we first enter the troposphere, the temperature is approximately −100°F (−73°C), and the temperature increases 19°F for every mile closer to the surface so that by the time we reach the surface of Earth we encounter a temperature of +90°F, assuming a balmy summer day.

During our trip from the outer reaches of the atmosphere down to the surface, we noticed a continuous increase in the concentration of air molecules with an accompanying increase in pressure. The gravitational attraction of Earth for the air molecules pulls them toward the surface. If this gravitational attraction were the only force operating on the air molecules, they would eventually settle in a thin, hardpacked layer encasing Earth. However, since the molecules have a certain kinetic energy which results in a constant random motion, the air molecules are closer together nearer the surface and increasingly farther apart at increasingly greater distances from Earth. As a result of the opposing forces of gravity and molecular motion, 90% of all of Earth's atmosphere is contained within ten miles of the surface, and 50% is contained within three and one-half miles of the surface.

Analysis of Earth's atmosphere reveals that it consists of 78% nitrogen, 21% oxygen, 0.9% argon, 0.03% carbon dioxide, 0.01% water vapor, and traces of helium, neon, hydrogen, methane, and other gases. This analysis does not include the smog and noxious gases that prevail around large cities!

THE FOUCAULT PENDULUM EXPERIMENT

How do we know Earth rotates on its axis? For centuries the ancient Egyptians, Greeks, Babylonians, and Romans wondered whether Earth rotated and the stars, sun, and moon were stationary or whether Earth was stationary and the heavens moved. Though there were advocates on both sides, the ancients generally believed that Earth stood still and the heavens moved. It was not until 1851 that this question was finally settled beyond all doubt by an experiment devised by the French physicist Jean Foucault. The famous "Foucault pendulum experiment" was performed in Paris, but it will be easier to understand if we consider the experiment's being done at one of Earth's poles. Consider setting up a tall tripod, the center of which is right over the north pole. From the top of the tripod a fifty pound steel ball is suspended by a fine steel wire (see Fig. 18.5).

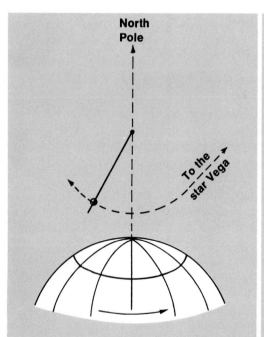

Fig. 18.5 A Foucault pendulum experiment at the North Pole.

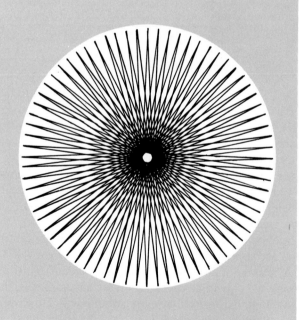

Fig. 18.6 The type of pattern that would be obtained if the Foucault pendulum experiment were carried out at the North Pole.

At the bottom of the steel ball is a steel point of such length that as the ball swings like a pendulum the point will trace the ball's path in a layer of sand. If the pendulum is put in motion, we find that the point does not make the same line in the sand over and over. Instead, after a few hours, many marks are made by the pointer, all passing through a common point. The explanation of the result of the above experiment is that the pendulum, once set in motion, continues to swing in the same direction while Earth rotates underneath it. Indeed, if the experiment had been carried out at night and if the pendulum had been started swinging toward and away from the star Vega in the constellation Lyra, the pendulum would be observed to continue swinging toward and away from Vega throughout the whole experiment. Obviously, Earth must be turning beneath the swinging pendulum, and a line made by the pointer is displaced a small amount by the rotation of Earth before the pendulum can retrace its previous mark. If we continued the experiment for twenty-three hours and fifty-six minutes (for length of the sidereal day see page 380), we would see a pattern as shown in Figure 18.6 in which, eventually, the first trace made by the pendulum's pointer would be retraced because Earth would have completed one rotation on its axis. Foucault did not obtain a perfect pattern in his experiment because he was at Paris, France, instead of at one of the poles. As this experiment is performed at different places on a line from one of Earth's poles to the equator, the circular pattern takes longer and longer to complete until at the equator only a single straight line is obtained since the part of the Earth directly underneath the pendulum is moving in a single direction. At one of the poles, however, Earth is rotating 360° beneath the pendulum.

One thing that may bother the student at this point is that since the tripod of this experiment is attached to Earth and the steel ball is attached to the tripod by a wire, why doesn't the direction of swing of the pendulum change as Earth rotates. Rest assured that it does not. You can verify this fact by swinging an object on the end of a string. If you twist the string between your fingers while it is swinging, the direction of swing will not change.

STELLAR PARALLAX

As early as about 315 B.C. and as late as 1600 A.D. the philosopher Aristotle and the astronomer Tycho Brahe respectively dismissed the possibility that Earth revolves around the sun because they could observe no shifting back and forth of nearby stars with respect to very distant stars during a year. They assumed correctly that if we did revolve around the sun our motion would cause the position of some stars to shift in relation to other stars as Earth speeds along in its orbit (see Fig. 18.7) in the same fashion that a nearby telephone pole is seen to block out different distant objects as one moves past the pole. Since they were not able to observe any shifting of position of stars, called *stellar parallax*, they concluded that either all stars are unimaginably distant from Earth, and, therefore, the parallactic shift is too small to see or else Earth is stationary.

the planet earth   **377**

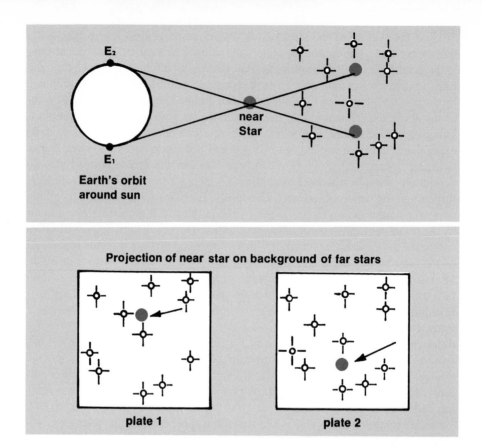

Fig. 18.7 **The observation of stellar parallax. When Earth is at E₁ the near star would appear to be positioned as shown in plate one in relation to background stars. Six months later, when Earth is at E₂, the near star would appear to be positioned as shown in plate two in relation to background stars.**

The largest parallax of any star is only three-fourths of one second of arc (60 seconds of arc in 1 minute, and 60 minutes in 1 degree of arc), and since the eye is not capable of resolving anything less than one minute of arc, the parallactic shift of the nearest stars (Alpha Centuri system) is eighty times too small to be observed with the naked eye. Small wonder then that Aristotle and Brahe rejected the heliocentric concept.

THE SEASONS

Earth follows an elliptical path in its annual journey around the sun so that at times it is closer to the sun and at other times it is farther from the sun. Actually Earth is closest to the sun (perihelion) during the northern hemisphere's winter and farthest from the sun (aphelion) during summer. Obviously then, we must seek an explanation for the seasons elsewhere. Earth's axis, determined by

Earth's rotation, makes an angle with the plane of Earth's revolution around the sun of 66½ degrees—more commonly expressed as 23½ degrees from a line perpendicular to the plane of the Earth's orbit. Because of this 23½° "slant," the northern hemisphere receives more sunshine than the rest of the planet during part of the year to produce the northern hemisphere's summer.

As shown in Figure 18.8 and Figure 18.9, the sun's rays strike the northern hemisphere almost perpendicularly during summer and the southern hemisphere at an extreme slant. This difference means that more sunlight falls on each square foot of Earth in the northern hemisphere than in the southern hemisphere. Thus, while the northern hemisphere is enjoying summer, the southern hemisphere is suffering through winter. Six months later the situation is reversed and the southern hemisphere receives the greatest amount of sunlight per square foot. On June 21, the *summer solstice,* the sun is at zenith (directly overhead) at 23½° north latitude, and this latitude is called the *Tropic of Cancer.* On the same day the sun shines 23½° past the north pole toward the equator or to 66½° north latitude. This latitude is called the *Arctic Circle,* and on June 21 the Arctic circle basks in twenty-four hours of daylight. At the same time, the sun reaches only to 66½° south latitude (23½° from the south pole), and this latitude is called the *Antarctic Circle.* On June 21, people above the Antarctic Circle face the gloomy prospects of what to do during twenty-four hours of total darkness. Six months later, on December 21, the *winter solstice,* the sun is

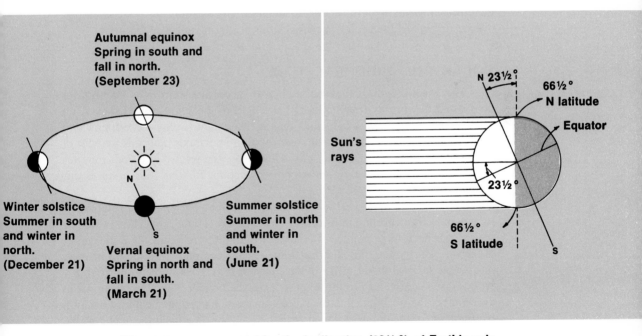

Fig. 18.8 **The seasons are caused by the inclination (23½°) of Earth's axis of rotation to the plane of Earth's orbit.**

Fig. 18.9 **Earth during the summer solstice (June 21). The sun is at zenith (directly overhead) at 23½° north latitude. This latitude is called the Tropic of Cancer. The sun's rays reach 23½° beyond the north pole or to 66½° north latitude and this latitude is called the Arctic Circle. Note that the sun's rays are more slanted or oblique in the southern hemisphere than in the northern hemisphere.**

at zenith of 23½° south latitude, and this latitude is called the *Tropic of Capricorn*. On March 21 and September 23, Earth is half way between the winter solstice and the summer solstice and is said to be at the equinoxes (from the Latin meaning equal light). During these two days Earth's axis neither slants toward nor away from the sun but instead is "parallel" to the sun as shown in Figure 18.8. Consequently, Earth receives twelve hours of sunshine and would experience twelve hours of night were it not for the fact that diffraction and scattering of the sun's rays considerably extend the period of daylight. March 21 is the *vernal equinox,* the spring equinox, and September 23 is the *autumnal equinox.* At the equinoxes the sun is at zenith on the equator.

BLUE SKY AND RED SUN

As the light from the sun passes through our atmosphere, the gaseous air molecules scatter the light. This scattering process is wavelength or color selective since blue light is very highly scattered but red light is little affected. As a result of this scattering, the blue wavelengths of light are scattered in all possible directions to produce our blue sky but do not reach us directly as do the red wavelengths of the sun's light which pass relatively unchanged through the atmosphere to Earth's surface. This selective scattering has considerable aesthetic consequences for man since, as just stated, it produces our beautiful blue sky and fiery red sun. Even the moon when it is very low on the horizon, and, therefore, when its light must pass through a longer path of air than when it is overhead, has a golden color because of this scattering effect.

SOLAR AND SIDEREAL TIME

The exact length of time between two successive sunrises is twenty-four hours, and this twenty-four hours is the length of the *solar day.* However, an observer in outer space would find that it takes only twenty-three hours and fifty-six minutes for Earth to make one rotation on its axis, and this amount of time is the length of the *sidereal day.* Obviously, the solar day is four minutes longer than the sidereal day. How can we account for this four minutes difference?

As shown in Figure 18.10, while Earth is making one complete rotation on its axis, it is also traveling in its orbit about the sun (about 1,600,000 miles per day). Consider a large pointer so constructed that it points directly at the center of the sun. During the time that Earth makes one complete rotation on its axis, that is, one sidereal day, Earth will have moved from position E_1 to position E_2. At position E_2, after one rotation on its axis, the pointer will be pointing not at the center of the sun but in the same direction that it had pointed while at position E_1. In order for the pointer to once again be pointing directly at the center of the sun, Earth must turn one degree more on its axis. It takes Earth about 4 minutes to rotate one degree. Since Earth must rotate 361° for a solar day but only 360° for a sidereal day, the solar day is four minutes longer than a sidereal day.

PRECESSION OF EARTH'S AXIS

During our youth we all saw a spinning top, and we noticed that as the top spun on its axis, it did not stand upright or perpendicular to the floor but, instead, had a tilted motion which caused its axis to sweep out a cone in space as shown in Figure 18.11.

This conical motion of the axis of a spinning object is called *precession*. The precession of a spinning top is caused by the gravitational attraction of Earth for the spinning top. If a top were held rigidly perpendicular to the surface of Earth, the center of gravitational attraction on the top would be along the top's axis and no precession would occur. If the top wobbles, however, the direction of the force of Earth's attraction for the top is not in line with the axis of the spinning top. This nonalignment of the direction of gravitational force with the axis of the spinning top produces the observed precession of the axis of a spinning top.

Consider Earth as a huge spinning top. The sun exerts a gravitational attraction upon Earth which is not in line with Earth's spin axis. Consequently, Earth's axis precesses like the axis of a toy top. Earth's axis maintains its $23\frac{1}{2}°$ tilt from a line perpendicular to the plane of Earth's orbit around the sun, but the axis precesses and cuts out a cone in space. It takes 26,000 years for Earth's axis to make one complete precession.

Earth's axis today is pointing approximately to the star Polaris, our "North Star." Because of precession, however, Earth's axis is slowly pointing farther and farther away from Polaris, and in the future Polaris will not be our polar or north star. Indeed, for a thousand years before and well over a thousand years after the birth of Christ, Earth had no North Star. As shown in Figure 18.12,

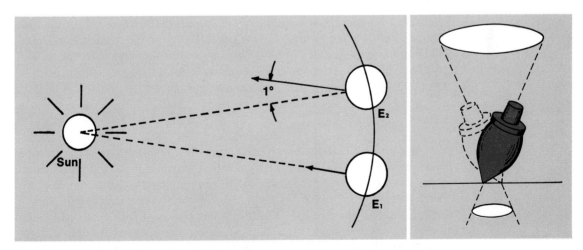

Fig. 18.10 **In one sidereal day Earth makes one complete rotation (360°) on its axis. Earth must rotate 361°, however, in order to complete one solar day.**

Fig. 18.11 **Precession of the axis of a top.**

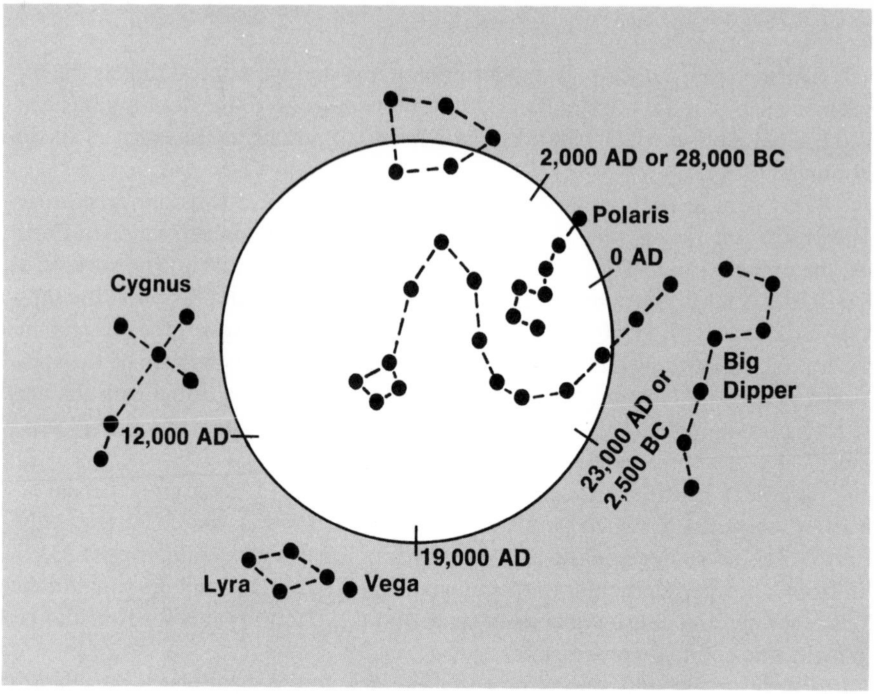

Fig. 18.12 The path among the stars followed by Earth's north axis during a 26,000 year period. This path is the result of the precession of Earth's axis.

Labels in figure: 2,000 AD or 28,000 BC; Polaris; 0 AD; Cygnus; Big Dipper; 23,000 AD or 2,500 BC; 12,000 AD; 19,000 AD; Lyra; Vega

by the year 3000 A.D. there will be no star close enough to the direction in which Earth's axis points to be called a pole star. By the year 11,500 A.D., however, a star in the constellation Cygnus will be fairly close to where Polaris is now. By the year 14,000 A.D. the bright star, Vega, in the constellation Lyra, will be our pole star. Then man must wait about 9,000 years before there will be another pole star, for it will not be until 23,500 A.D. that a star in the constellation Draco will be in the approximate position. This star will be the same pole star seen in 2,500 B.C. by the Babylonians and Egyptians. Then, 26,000 years from now, in about 27,973 A.D., Polaris will once again be our pole star.

THE SHAPE OF EARTH

As early as the sixth century B.C., some men realized that Earth must be round. Aristotle (384–332 B.C.) reasoned that the world must be round because as one travels to the north or to the south, new stars and constellations become visible that were not visible before, and known stars and constellations slowly disappear below the horizon. Much later in history, man developed sailing ships with very tall masts. As a ship sailed into the distance, the ship did not simply appear smaller and smaller until it disappeared but instead, the hull of the ship soon disappeared over the curve of Earth while the top of the masts was still visible. The manner in which a ship disappears over the horizon is another indication that Earth is spherical.

Earth, however, is not a perfect sphere but an oblate spheroid—which means that the diameter of Earth at the equator is greater than it is at the poles. (If Earth were longer through the poles than through the equator it would be a prolate spheroid.) This flattening at the poles, with the subsequent bulging at the equator, is the result of Earth's rotation on its axis. As a result of Earth's rotation, an object at the equator is traveling approximately 1,000 miles per hour! The difference in Earth's equatorial diameter (7927 miles) and its polar diameter (7900 miles) is quite small, only 27 miles; consequently, as an astronaut approaches Earth from outer space it is impossible for him to see Earth as anything but a perfect sphere.

Since the outward force exerted on a body at the equator, due to Earth's rotation, opposes the force of gravitational attraction for the body, there is a slight decrease in the weight of the body. At the poles, where the force due to rotation is zero, the weight of a body is exactly what would be expected by the attractive force of gravity. Thus a man at one of Earth's poles who weighs 300 pounds would weigh only 299 at the equator. Actually, there is a second factor that affects the weight of objects at the equator and at the poles. Since Earth's rotation flattens Earth in the polar regions, objects at the poles are 27 miles closer to the center of Earth than are objects at the equator. Since the farther away an object is from Earth's center the less it weighs, objects at the equator weigh less than objects at the poles. It is both effects, then, Earth's rotation and Earth's oblate shape, that are responsible for the decrease in the weight of objects at the equator as compared to the weight of objects at the poles. Since the force of gravity is decreased by one-fourth as the distance from an object to the center of Earth is doubled, the change in weight of an object as it gets farther from Earth is rather dramatic. For instance, a man who would weigh 200 pounds on the surface of Earth would weigh only fifty pounds at a distance of 3950 miles from Earth's surface.

THE EARTH-MOON GRAVITY COUPLE

We have all said or thought that the moon revolves around Earth, and although such appears to be the case, it is not the full story. Actually, Earth and the moon revolve around a common point on a line connecting their centers. The point around which the Earth-moon system revolves is called the center of gravity or *barycenter* of the system.

If the bodies are of equal mass, such as is the case in some binary star systems, the two bodies revolve around their barycenter which is on a line half way between the centers of each body. What about a system such as the Earth-moon system in which Earth is considerably more massive than the moon? If a very large boy and a very small boy want to seesaw, the large boy must sit, perhaps, in the center of his half of the seesaw whereas the small boy must sit as close to the outer edge of his half of the seesaw as possible in order for the seesaw to be balanced. The Earth-moon system must balance itself in an analogous manner. Because of the much greater mass of Earth, the barycenter is very close to Earth's center, and, indeed, it lies one thousand miles below the surface of Earth. The moon and Earth both revolve around the center of gravity one thousand miles within Earth. If one were looking "down" on the solar system from the direction of Polaris, one would see Earth "wobble" as it speeds along on its yearly journey around the sun. This wobble would be the result of Earth's revolving around the Earth-moon barycenter.

THE TIDES

Many ancient societies worshiped that delicate, soft globe of light that we call the moon. Because of its great beauty (or perhaps in spite of its beauty) many people thought that the moon directly affected their lives as well as other animate and inanimate things on Earth. Some thought that the waxing moon gave them strength. Others thought that the moon, during certain phases, could give them wisdom. Of course, even today, most women know that the gentle glow of moonlight gives them a certain beauty that the harsh light of day can never equal. However, whereas the moon does not directly affect our lives, it does cause physical changes in the atmosphere, oceans, and land masses of our globe.

Anyone who has spent at least five to six hours at the seashore has noticed that the sea level rises and falls in a rhythmic pattern. This "coming in" and "going out" of the tides can make variations in sea level from a few feet to over fifty feet. The gravitational pull of the moon causes the water of the oceans to be heaped up at certain places and to be lowered at other places. Even the huge, seemingly rigid continental masses cannot resist this unrelenting, inexorable, attractive force of the moon, and they develop tidal bulges up to eight inches. The moon also causes tides in our atmosphere.

Even though the sun is much more massive than the moon (its attraction for Earth is 150 times greater than that of the moon's), the sun is so much farther away that its tide-raising force is only about one-third that of the moon's. As seen in Figure 18.13, during the full moon and the new moon, the moon and sun are lined up so that their tide-raising forces are additive and they produce tides that are higher than normal. These highest tides are called *spring tides,* a term that has absolutely nothing to do with the spring season. When the moon is in its first and third quarters, the pull of the moon and the pull of the sun are at right angles to each other, and the tides produced are correspondingly lower. These lower tides are called *neap tides.*

A complete explanation of all the forces that are involved in the production of Earth's tides is quite complex, but Figure 18.14 gives an explanation adequate for our purposes. In Figure 18.14, the black arrows represent the forces due to the moon's gravitational attraction. The white arrows represent the centrifugal forces due to Earth's movement around the Earth-moon center of gravity (barycenter) as well as the centrifugal forces caused by Earth's axial rotation. The resultants of the gravitational forces and the centrifugal forces are the causes of the tides. These resultant forces cause the waters of the oceans to flow toward a

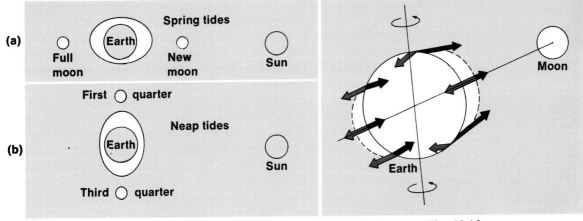

Fig. 18.13

Fig. 18.14

Fig. 18.13 **Different types of tides produced by different positions of the moon. (a) Spring tides where the moon is either full or new. (b) Neap tides where the moon is either in its first or third quarter.**

Fig. 18.14 **The forces that cause the formation of lunar tides on Earth.**

point on Earth's surface that is directly between Earth and the moon on the side toward the moon and to a point on the opposite side of Earth that is in line with the moon.

The tidal bulge does not point toward the moon as one might expect but instead points ahead of the moon in an easterly direction. The tidal bulge is pulled ahead of the moon by the eastward rotation of Earth on its axis since Earth rotates faster on its axis than the moon revolves around Earth (see Fig. 18.15).

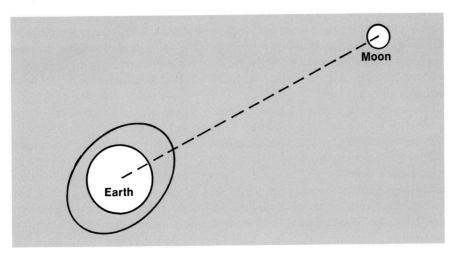

Fig. 18.15 **The tidal bulge caused by the moon actually points ahead, or east, of the moon because Earth's rotation eastward pulls the bulge ahead of the moon.**

the planet earth

There is considerable friction produced as the tides roll over the sea floors and particularly where huge quantities of water are funneled through very shallow and narrow passages. Energy must be used to overcome these frictional forces, and energy is made available at the expense of Earth's rotation on its axis. The consequence of this loss of rotational energy is that Earth is slowly decreasing its speed of rotation. This decrease means that the length of our day is correspondingly increasing. Even though the increase in the length of the day is very small, approximately 1/1000 of a second per century, this increase can be an appreciable length of time if one considers the many eons that our planet has existed. It is not known exactly how long Earth's rotation has been slowing down, though one would assume that this slowing effect has been in operation for as long as our planet has had oceans. It is known, however, that this slowing down has been going on at least since the Devonian period, about 400 million years ago. Geological evidence shows that during the Devonian period Earth's year consisted of 400 days which means that each Devonian day was only twenty-one hours long instead of the twenty-four hours that is the length of our modern day.

If the moon causes the crust of our planet to bulge out as much as eight inches, what must be the tidal effect of the much more massive Earth on the moon's crust? It is probable that the surface of the moon is bulged out three to four feet above normal by the tidal action of Earth.

SOLAR AND LUNAR ECLIPSES

As the moon slowly comes between Earth and the sun and, therefore, covers the face of the sun, the sky slowly darkens and a chill pervades the air. When the sun's golden disk is completely blotted out by the moon, birds fly to their nest to roost and stars can be seen shining brightly at midday. It is small wonder that the ancients and even more recent primitive societies were clutched in a state of fear by the awesome spectacle of a solar eclipse. Their life-and-warmth-giving sun was apparently being swallowed up by some god of darkness that could turn day into night. During an eclipse the ancient Chinese would beat on the ground and on houses with sticks in an effort to make enough noise to scare away the devil-god that had swallowed their sun. Although lunar eclipses last much longer than solar eclipses, they are not as impressive as solar eclipses and consequently did not capture the imagination of primitive man as did solar eclipses.

As shown in Figure 18.16, Earth casts a long conical shadow, called the *umbra,* within which the sun cannot be seen and there is complete darkness. Surrounding the umbra is a less dark region, called the *penumbra,* in which only part of the sun is visible.

Earth's umbra is 850,000 miles in length. Since the moon is only 238,000 miles from Earth, we would expect the moon to pass within Earth's shadow and be eclipsed. From the above discussion one might quite reasonably expect that the moon should be eclipsed once each twenty-nine and one-half days (from new moon back to new moon) as would indeed be the case if the plane of the

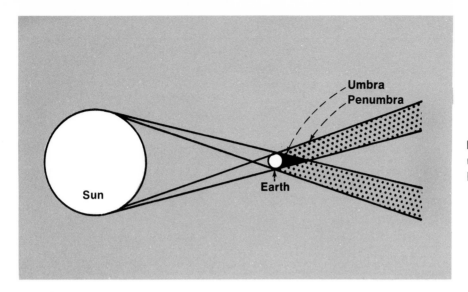

Fig. 18.16 **Earth's umbra and penumbra.**

moon's orbit around Earth coincided with the plane of Earth's orbit around the sun. The plane of the moon's orbit is inclined 5° away from the plane of Earth's orbit. Consequently, an eclipse can occur only where the two planes intersect and only when the intersection includes the moon, Earth, and the sun in this order. This special arrangement of moon, Earth, and sun occurs only twice a year, and, therefore, lunar eclipses are seen only twice a year.

Like lunar eclipses, solar eclipses can occur only when Earth, the moon, and the sun are lined up in that order along the intersection of the plane of the moon's orbit and the plane of Earth's orbit. Consequently, a solar eclipse can occur only about every six months.

Interestingly, it is pure happenstance that we on Earth can observe a total solar eclipse. In the first place, if the moon were slightly farther from Earth, the moon's shadow, or umbra, would not be long enough to fall upon Earth's surface. In fact, there are many times when the alignment of Earth, the moon, and the sun is correct for a total eclipse, but the moon's umbra is not long enough to reach Earth. The average length of the moon's umbra is only 232,500 miles, whereas the mean distance of the moon from Earth is 238,800 miles. Therefore, it is only when the moon is near perigee (closest approach to Earth) that a total solar eclipse of the sun can occur, since near perigee the moon is only 219,500 miles from Earth. Secondly, it is pure coincidence that the apparent size of the moon's disk is slightly larger than the apparent size of the sun's disk so that the sun can be completely covered by the moon. Although the sun has a diameter that is 400 times larger than the diameter of the moon, the sun is approximately 400 times farther from Earth than is the moon. As a result, the apparent size of the moon's disk is large enough to cover the sun. Within our solar system there are many moons that are close enough to their mother planet to produce eclipses. Shown on the following page is a solar eclipse occurring on the planet, Jupiter.

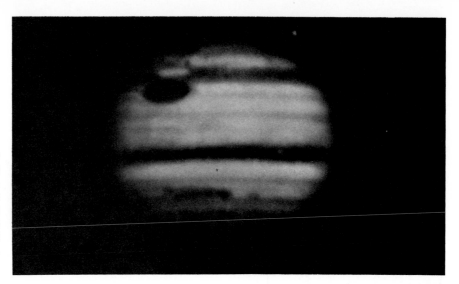

Fig. 18.17 **A solar eclipse on the planet Jupiter. Jupiter's satellite Ganymede casts its shadow above Jupiter's great red spot.**

THE FACE OF THE MOON

From Earth the moon appears as a magnificent celestial object that bathes the night Earth in a soft, pale light that has prompted poets and lovers to speak eloquently of its exquisite beauty. Because the moon always keeps the same side toward Earth, its general features are well known, and some people fancy that its gross features resemble the smiling face of a man—thus, the man in the moon. All the above comments, however, belie the severe conditions that actually exist on the surface of the moon where a harsh, scorching sun blazes down on a very inhospitable landscape. During the day the surface temperature soars to above the boiling point of water and during the night plunges to −283°F (−175°C).

Only recently has man photographed and observed the far side of the moon, and until space travel made these events possible, man had to be content with seeing only one side of the moon. The reason that the moon keeps one side perpetually facing Earth is that the rotation of the moon on its axis is synchronized with the revolution of the moon about Earth. The moon has not always had its rotation and revolution synchronized, however. The tidal bulge on the moon, created by the Earth's gravitational pull, has slowly reduced the speed of the moon's axial rotation to its present value. The moon's present speed of rotation is now locked by gravitational forces to its revolution around Earth.

EARTHLIGHT

When the moon is at crescent phase, we see the sun shining on only a thin crescent of the moon's surface whereas the rest of the moon that we faintly see is the nightside. The light of the sun does not touch this nightside, yet we see the nightside faintly illuminated. What is the source of this illumination? It is "earth-light," Earth's equivalent of moonlight. The nightside of the crescent moon is facing the daylight side of Earth with the result that earthlight shines upon the nightside of the moon in the same manner that moonlight shines upon the night-side of Earth. An astronaut on the nightside of the moon would be bathed in "earthlight" from a "full Earth" that would be five times brighter than the moon-light we see on Earth from a full moon.

Fig. 18.18 **This gibbous Earth lights up the nightside of the moon with "earthlight" that is about five times brighter than the nightside of Earth bathed in moon-light from a gibbous moon.**

(Photograph Courtesy of NASA)

THE FUTURE OF EARTH

Earth has existed for four and one-half billion years and will continue much as it is now for another four and one-half to five billion years, assuming of course that man does not completely alter Earth's surface by pollution and war. After another five billion years, however, our sun will enter the red giant stage and expand until its outer edges engulf the planet Mercury. Then the surface of our planet will be so hot that the water of the oceans and the ice of the polar caps will be turned into super heated steam that will completely enshroud our planet in white clouds much as Venus is today. At this time, life will cease to exist on the third planet of our sun. Eventually, the sun will begin to shrink and will continue to shrink until it reaches the stage of a white dwarf. As the sun reaches the white dwarf stage, Earth will become a bare, frozen rock, encased in continental ice sheets and frozen oceans. Ultimately, after many tens of billions of years, the nuclear fires of our sun will be completely exhausted and the sun will become a cold, burned-out cinder which, with its retinue of dead planets, will continue its lonely journey through space.

CHAPTER SUMMARY

From the surface of Earth to about ten miles out is the *troposphere,* where most of Earth's weather occurs. From about ten miles to about fifty miles above Earth's surface is the portion of the atmosphere called the *stratosphere.* The stratosphere has an ozone layer that prevents excessive amounts of ultraviolet radiation from reaching Earth. From about fifty miles above the surface to the outer fringes of Earth's atmosphere is the *ionosphere.* A layer of ions in the ionosphere reflects radio waves.

The Foucault pendulum experiment proves that Earth rotates on its axis.

On June 21, the *summer solstice,* the sun is directly overhead on the *Tropic of Cancer,* 23½° N latitude. Then land above the *Arctic Circle,* 66½° N latitude, has twenty-four hours of sunshine and land below the *Antarctic Circle* has twenty-four hours of darkness. On December 21, the *winter solstice,* the sun is directly overhead on the *Tropic of Capricorn,* 23½° S latitude. On the same day land above the Arctic Circle has twenty-four hours of darkness and land below the Antarctic Circle has twenty-four hours of sunshine.

On March 21, the *vernal equinox,* and on September 23, the *autumnal equinox,* the sun is overhead on the equator.

Because Earth is a spinning object influenced by the gravitation field of the sun and the moon, Earth's spin axis *precesses* once every twenty-six thousand years.

An eclipse of the sun occurs whenever part of Earth passes through the *umbra* or shadow of the moon. An eclipse of the moon occurs whenever the moon passes into the umbra of Earth.

The gravitational attraction of the sun and the moon for Earth produces tides in Earth's atmosphere, oceans, and solid crust. Only the tides produced in the oceans and seas of Earth are easily discernible. The tides in the crust of Earth do not exceed more than about four or five inches.

QUESTIONS AND PROBLEMS

1. Why does Earth have radiation belts? (p. 372)
2. What experiment proves that Earth rotates on its axis? (p. 377)
3. What proof is there that Earth revolves around the sun? (p. 377)
4. Where on Earth will the sun be directly overhead (Zenith) on December 21? (p. 379)
5. On which two days of each year will there be an equal number of hours of sunlight and no sunlight? (p. 379)
6. Why does Earth precess? (p. 381)
7. Why is the speed of rotation of Earth very slowly decreasing? How does the decrease of rotational speed affect the moon? (p. 386)
8. What causes our sky to be blue? (p. 380)
9. What is the cause of the aurora? (p. 373)
10. What would happen if the ozone layer of Earth disappeared? (p. 375)

the birth
of
earth

19

vocabulary

Basalt	Crust (of Earth)
Granite	Mantle (of Earth)
Nebula	Peridotite

Flung throughout the vast reaches of our galaxy are immense areas of gas and dust particles. Some sections of the spiral arms of our galaxy, The Milky Way, contain relatively high concentrations of these gases and solid particles. The distribution of the dust particles in our galaxy is not uniform, however, with the result that some regions are more dense than others and have a "cloud-like" appearance. These shadowy patches, extending over enormous regions, are called *nebulae* from the Latin meaning *cloud*.

Even though the concentration of material in the nebulae is thousands of times higher than in interstellar space in general, the nebulae, consisting of very sparse dust particles, constitute a better vacuum than man can make in the laboratory.

There are two types of nebulae that man can see in the vast lonely stretches of the universe: dark nebulae and bright nebulae. Though both the dark and the bright nebulae are composed of essentially the same material, the bright nebulae are close enough to large, hot stars to be illuminated by reflection and fluorescence. On the other hand, the dark nebulae lie so far away in the remote depths of interstellar space that no appreciable amount of starlight is reflected by them. The light that shines from stars and bright nebulae is sometimes obscured by the material of the dark nebulae. This absorption of light by dark nebulae is what makes them appear as dark patches blotting out background stars. Two of the most famous dark nebulae are the so-called "Great Dark Rift" that appears to split the Milky Way and the "Horsehead" that conceals part of the bright nebula in the constellation Orion. Although the density of nebulae is very low, some extend over such vast distances that the total mass of material contained in them is unimaginably large.

Fig. 19.1 The "Great Dark Rift" in the Milky Way.

(Photograph Courtesy of the Hale Observatories)

the birth of earth

How were these nebulae formed? Their formation is part of the endless cycle of cosmic events that includes the continuing and cyclic birth, life, and death of stars in our galaxy and throughout the universe. Nebulae gain their life from the violent explosions of unstable stars that for some as yet unknown reason suddenly and explosively eject part of their mass as a rapidly expanding shell of superhot gaseous material. These unstable stars hurl into the far reaches of the space between them part of themselves. This ejected "star dust" accumulates over eons of time to form the nebulae of our universe.

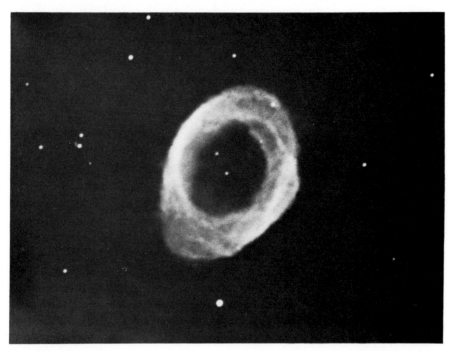

Fig. 19.2 **The Ring Nebula in Lyra. The star in the center of the "halo" ejected the expanding gas.**

(Photograph Courtesy of the Hale Observatories)

What is the composition of this matter that stars suicidally eject into space? The answer to this question depends a great deal on the size and age of the star. If the star is of reasonable size and age, it will have slowly converted by nuclear fusion reactions parts of its hydrogen into helium and heavier elements. The material ejected by a star contains hydrogen, helium, and varying amounts of other elements. Of course, the different elements can chemically combine once they escape from the fantastically high temperatures of a star. Theoretically, one would expect all the elements possible in nature to be present in the debris that exists in the space between stars, but not all the elements have been detected. Some of the various elements and compounds found in various nebulae are hydrogen, helium, oxygen, nitrogen, sodium, calcium, sulfur, argon, titantium, iron, methane, water, ammonia, hydrogen cyanide, formaldehyde, methyl alcohol, and chlorophyll-type compounds. It is these types of elements and compounds then that constitute the gases and dust particles of nebulae.

As we shall soon find out, the condensation of the material of these nebulae gives birth to second-and-later generation stars. Then, these second-and-later generation stars eventually become unstable and in turn eject part of their matter into space from which it came, and in the process give birth to new nebulae in a never-ending cycle of birth, life, death, and rebirth.

IN THE BEGINNING

Our solar system consisting of one star, nine planets, thirty-three satellites or moons attending six of the planets, thousands of asteroids, countless numbers of comets, and myriads of meteors originated from just such a nebulosity of gas and dust. About five billion years ago the tremendous formless cloud, flung across three thousand billion miles of space, would have started to contract due to the mutual gravitational attraction of the atoms, molecules, and particles present. Each particle would exert a small gravitational attraction on each of its neighbors, causing all the particles to be pulled closer together. As this condensation occurred throughout the vast cloud, a gradual contraction toward the center resulted. The contraction would have been a very slow process at first. Indeed, it probably took millions of years for the cloud to contract from a diameter of three thousand billion miles to a diameter of ten billion miles. During this initial contraction the particles were relatively so far apart that the mutual gravitational attraction was extremely small. As a result, the particles would have "fallen" very slowly toward the center of the cloud.

After the diameter of the cloud reached approximately ten billion miles, the speed of the particles as they moved toward the center of the cloud was greatly accelerated until eventually the particles attained speeds of thousands of miles per hour. As the primeval nebula continued to contract and the particles moved even faster toward the center, the deep interior of the cloud began to heat up as the potential gravitational energy was slowly changed into heat energy. During the process of contraction, the whole cloud rotated slowly in a direction that would appear counterclockwise to a stellar observer in the vicinity of the North Star. As the mass of material increased in the center, various eddy currents or whirlpools of matter present in the swirling cloud began to build up much smaller concentrations at varying distances from the center of the contracting cloud. It is these smaller contracting masses produced by the eddy currents in the rotating cloud that eventually became the planets.

LET THERE BE LIGHT

As the heat liberated in the center of the mass continued to increase, the temperature rose until the dense center became incandescent, and the newly created star reigned over our infant solar system in lovely splendor. This brilliant birth-shine of the new star heralded the beginning of life for our sun. However, although our sun had begun its life-task of pouring out untold quantities of warmth, light, and other forms of energy, it was still not assured of a long life.

This long life, probably about ten billion years, could only be assured if the internal temperature of the new star continued to increase until the temperature reached ten million degrees Centigrade (eighteen million degrees Fahrenheit).

Approximately ten million years after the star began to shine, the temperature of the sun's interior rose to the critical value of ten million degrees Centigrade by virtue of its continued contraction and conversion of potential gravitational energy into heat energy. When this temperature was reached, the nuclear fires spontaneously ignited.

At this temperature hydrogen (H) nuclei, hydrogen atoms stripped of their electrons, have sufficient energy because of their high speeds to fuse together when they collide. On collision they form the isotope deuterium (D), sometimes called heavy hydrogen, and release an electron (e^+).

$$\,_1^1H + \,_1^1H \longrightarrow \,_1^2D + e^+ + energy$$

In turn, the new deuterium nuclei, themselves traveling at awesome speeds, collide and fuse with other hydrogen nuclei until eventually nuclei of the element helium (He) are produced.

$$\,_1^2D + \,_1^1H \longrightarrow \,_2^3He + energy$$

Then nuclei of the element helium crash together to produce new helium nuclei plus hydrogen nuclei.

$$\,_2^3He + \,_2^3He \longrightarrow \,_2^4He + 2\,_1^1H + energy$$

So in essence, four hydrogen nuclei by successive collisions fuse into a nucleus of the element helium.

$$4H \longrightarrow 1\,He + energy$$

If one could weigh the four hydrogen nuclei that fuse together to produce one nucleus of helium, he would find that the helium nucleus weighs less than the four hydrogen nuclei that produce it. The mass lost in this process is changed into energy according to Einstein's relationship $E = mc^2$, which can be stated as the energy released is equal to the mass that disappears times the square of the speed of light. This nuclear process whereby hydrogen nuclei are fused into helium is essentially the same process that man uses on a very puny scale in his so-called hydrogen or fusion bombs. Although the amount of mass that is converted to energy when each helium nucleus is created is extremely small, the total number of helium nuclei being produced in the sun in a short period of time must be considered. The amount of hydrogen being converted into helium every second by the sun is so large that the total amount of mass converted into energy per second staggers the imagination. Thus, every second, the sun produces 559 million tons of helium from 563 million tons of hydrogen. This means that the sun converts 4 million tons of matter into energy every second. To get some idea of the unbelievable quantity of energy the sun is releasing every second, one would have to burn 22 million pounds of coal to produce the energy released when only one ounce of matter is changed into energy! Even at this prodigious rate of matter conversion, the sun has enough "fuel" to keep the nuclear fire burning for at least ten billion years!

FORMATION OF THE PLANETS

Now that we feel fairly comfortable about our sun's future, let us turn our attention once again to the struggle for existence that our Earth and the other planets of our solar system must have made in order to become the worlds that we know today.

The eddy currents present in the contracting, slowly rotating cloud inexorably pulled more and more material toward their own smaller centers at the expense of the material rushing toward the central part of the cloud which would eventually become our sun. In the same fashion that the mass of material was collected to produce our sun, material on a less grand scale slowly collected in these whirlpools of dust and gas at different distances from the main cloud and eventually formed dense bodies of material slowly circling around the new sun. We must keep in mind that like the material consolidating to produce the sun, the material condensing to eventually become planets was mostly hydrogen, a small amount of helium, and a relatively small percentage of all the heavier elements and compounds present in the original cloud. Eventually, numerous comparatively small bodies of material congealed in the slowly rotating cloud. These "protoplanets" were distinguishable as the beginnings of future planets. Their volumes were very large in terms of present-day planetary volumes and probably encompassed a volume of space thousands of times larger than their present diameters. At the same time and in the same fashion that the planets were formed, the satellites or moons of the planets were forming as still smaller condensations circling their parent planets.

By the time that our new sun became sufficiently hot to reach incandescence, the material forming the planets was fairly well consolidated. Even so, the denser, heavier material was concentrated in the center of the mass forming the protoplanets, whereas the extremely light hydrogen and helium, forming by and large the greater bulk (90%) of the total protoplanet masses, enveloped the denser centers in a cocoon extending outward to great distances. The mass of protoearth must have been a thousand times greater than Earth today, and its diameter must have approached ten million miles.

As the new star began to shine more and more brilliantly, giving off more and more heat, the hydrogen and helium molecules surrounding those protoplanets closest to the new star began to boil away. Also, the stream of ionized particles, mainly electrons and protons, that are continually pouring out from the sun to produce the so-called "solar winds" physically bombarded and pushed away the light hydrogen and helium molecules. In the same fashion that the sun utilizes light pressure and solar wind today to "push away" the tails of comets so that the tail always points away from the sun (see Chapter 17), the protoplanets were denuded of their hydrogen and helium atmospheres.

Fig. 19.3 **An artist's conception of the solar system toward the end of its formation.**

The closest protoplanet, Mercury, being so near the sun, received the largest quantity of heat and solar particles and consequently was the first protoplanet to be stripped of its envelope of hydrogen and helium. What was left was only a small fraction of the total mass of material that had once been the protoplanet Mercury. On the other hand, what was left was the dense, heavy material that we are familiar with as constituting and making up the mass of the present-day "inner" planets—Mercury, Venus, Earth, and Mars. These four planets all have similar densities (five times as dense as water), as one would expect if indeed they were all formed by similar means. Also the inner planets, because of their relatively small sizes, exerted correspondingly modest gravitational attractions on the hydrogen and helium gases surrounding them.

However, those planets farther removed from the sun, such as Jupiter, Saturn, Uranus, and Neptune, were so far away and, because of their huge masses, exerted such large gravitational attractions for the gases that surrounded them that even the prodigious outpouring of heat and charged particles from the sun was not sufficient to heat up and brush away more than a fraction of the vast quantities of hydrogen and helium that initially condensed and encircled them. Consequently, today these "outer planets" have comparatively very low densities (from 0.7 to 2.3 g/cm^3; together their densities are only one-third that of the inner planets), and a major portion of their masses is due to the relatively light weight hydrogen and helium remaining from the initial "creation."

Saturn, for example, is only 95 times heavier than Earth but is 769 times larger in volume than Earth. Indeed, Saturn's density is less than the density of water. If there were an ocean large enough to hold Saturn, the entire planet would float!

As the slowly rotating original cloud condensed and its radius shrank, the cloud began to spin faster and faster. Why the speed of rotation of the cloud must have increased as the cloud shrank is easily illustrated by the example of an ice skater. Everyone has noticed that as an ice skater starts spinning, her arms are outstretched, but as she pulls her arms in closer to her body, she spins more rapidly. This is an example of the conservation of angular momentum. In order for the angular momentum of a contracting nebula to remain constant, as it must, the speed of rotation of the nebula increases.

As a consequence of the increased speed of rotation of the contracting cloud, the new-born sun rotated on its axis approximately fifty times faster than it does today. How then can we explain the present speed of rotation of our sun (one rotation in 26 Earth days) in view of its original rotation once in twelve Earth hours? As the rapidly rotating new star shone upon its brood of infant protoplanets, the intense bombardment of charged particles and light from the star ionized the envelope of gases relatively near the sun. Interaction took place between the sun's magnetic field and the rotating, now-ionized envelope of gaseous material. This interaction resulted in a decrease of the rotation of our sun in the course of its evolution and an increase in the speed of rotation of the rest of the solar system material.

As the powerful radiation from the sun pushed away the miscellaneous gases and materials left from the formation of the sun and the planets of our solar system, not all the material was completely ejected from the solar system into interstellar space. At distances beyond the orbit of Pluto, some of the debris condensed into relatively small bodies that circle the sun at the very outposts of our solar system—probably up to thousands of billions of miles from the sun. Occasionally, these cold, dark objects came close enough to the sun to shine by reflection and fluorescence and to become the brilliant, spectacular objects we know today as comets.

THE INFANT EARTH

Now that we have some idea of how the solar system in general was formed, let's look in greater detail at the formation of our Earth and its moon.

After the excess hydrogen and helium surrounding protoearth had been swept away, there was left a bleak, shrunken core of material held together by gravity. This aggregate of atoms, molecules, compounds, and chunks of material contained all the elements that now exist on Earth.

As the Earth was forming, so was Earth's lone satellite, the moon. The two bodies were formed together as gravitationally bound twin condensations during the solar system creation. Even from the beginning, the moon was bathing the desolate, bare new Earth in its soft, gentle light.

There still remained in the general vicinity of Earth, as well as in the vicinity of the other planets, debris from the creation. This debris ranged in size from dust particles to huge mountains of rock-like material circling the sun in the general orbit of Earth. With the passage of perhaps a billion years, these myriads of debris were slowly captured by the more massive Earth until eventually Earth's path around the sun was almost swept clean.

Until this time Earth had been generally cool, but the overall process of gravitational contraction had caused warming of the infant planet. Also, as Earth attracted the debris in its vicinity, the fragments of material would crash into Earth at very high speeds, and the kinetic energies of the fragments would be converted into heat. This continual release of heat energy, produced by constant bombardment of Earth, resulted in further warming of the planet. In addition, another heat-releasing process was taking place. Uranium, thorium, potassium, and other unstable elements were slowly and inexorably liberating heat from radioactive processes of decay that are an intrinsic part of the life processes of radioactive elements. Over millions and millions of years, these radioactive elements liberated vast quantities of heat. Between the heat liberated by debris crashing into Earth and by radioactive decay, Earth became very hot. It probably never became hot enough to completely melt but did reach a sufficiently high temperature to allow various elements and compounds to segregate and migrate according to their chemical and physical similarities and affinities.

It was during this period, and perhaps extending up to the present day, that the very dense iron from various parts of Earth slowly separated and migrated under the influence of gravity toward the center of the planet. Since certain types of meteorites, which can be considered the fossil relics of the solar system creation, contain iron mixed with 5 to 15% nickel, it is reasonable to assume that nickel also migrated with the iron.

Eventually, the iron-nickel core of Earth melted. Although some scientists do not agree, it is possible that the outer portion of Earth surrounding the core may have melted also. One reason for this assumption is that although Earth is at least four and one-half billion years old, no one has so far found any rocks that are over three and one-half billion years old. Of course, rocks older than three and one-half billion may be found in the future.

The iron-nickel alloy that makes up Earth's core started melting when the temperature of the core reached about 1500°C (iron melts at 1535°C), but today it is estimated that the core has a temperature of approximately 3500°C.

It is because of the molten, iron-nickel core that Earth has a strong external magnetic field with the results that we have north and south magnetic poles as well as the Van Allen radiation belts that surround our planet from two to five thousand miles from the planet's surface. Mercury, Venus, and Mars have no appreciable magnetic poles.

But since Earth is inhabited by man, it is fortunate that our planet has a liquid core with its resultant strong north and south magnetic poles. Without the compass which our magnetic poles control, man could not have ventured so far out to sea many hundreds of years ago in search of sea commerce and new lands. But with the discovery of the magnetic compass needle by the Chinese in about 300 A.D., mariners were able to venture boldly forth on relatively long ocean voyages that previously would have been quite perilous.

Earth's magnetic field is produced by Earth's rotation, which sets up eddies that circle from west to east within the liquid core. The result is an electric current moving in the same direction. Just as a magnetic field surrounds any conductor carrying an electric current, the electric current in the core acts as a huge electromagnet that produces a magnetic field external to our planet with lines of force extending north and south. Note that Earth's magnetic field is oriented at right angles to the core's electric field and roughly parallel to the axis of Earth's rotation.

Besides the elements and compounds initially present in the primeval cloud, many other substances were formed during the formation of the solar system. On protoearth the oxygen present reacted with hydrogen, silicon, magnesium, iron, aluminum, calcium, carbon, sodium, potassium, etc. to produce their corresponding oxides: H_2O, SiO_2, MgO, FeO, and Fe_2O_3, Al_2O_3, CaO, CO and CO_2, Na_2O, K_2O, etc. Some of these oxides in turn reacted with each other in the presence of differing amounts of heat to produce, among many other minerals, the following silicate minerals (a mineral is a naturally occurring inorganic substance having distinctive chemical and physical properties). For convenience, the minerals listed below have been divided into sial and sima materials. The sial minerals are rich in silicon and aluminum, whereas the sima minerals are rich in silicon and magnesium.

Mineral Class and Name	Chemical Formula
Sial	
Quartz	SiO_2
Orthoclase Feldspar	$KAlSi_3O_8$
Plagioclase Feldspar	$(Ca_xNa_y)\,Si_3O_8$
Muscovite Mica	$KAl_3Si_3O_{10}(OH)_2$
Sima	
Augite	$Ca(MgFeAl)Si_2O_6$
Biotite Mica	$K(MgFe)_3AlSi_3O_{10}(OH)_2$
Hornblende	$NaCa_2(MgFeAl)_5Si_8Al_8O_{22}(OH)_2$
Olivine	$(MgFe)_2SiO_4$

These minerals, if one assumes the necessary temperatures and pressures, would have produced the following types of rocks that are present on Earth today:

Granite—consisting of 2 parts orthoclase feldspar + 1 part plagioclase feldspar + 1 part quartz + small quantity of sima minerals.
Basalt—consisting of 1 part plagioclase feldspar + 1 part sima minerals.
Peridotite—consisting of 1 part plagioclase feldspar + 3 part sima minerals.

THE STRUCTURE AND COMPOSITION OF THE EARTH

After the new sun swept away the mist surrounding protoearth, there remained the rock-mass that was infant Earth. The infant Earth now had a diameter of 7910 miles (1.3×10^9 cm), a volume of 1.1×10^{27} cm³, a weight of 6.6×10^{21} tons (6.6 billion trillion tons or 6×10^{27} g), and a density of 5.5 g/cm³ (5.5 times the density of water). It is difficult to comment on the overall inner structure of the infant Earth, but it might be valuable to our discussions ahead to assume that after a period of a few hundred million years the structure of infant Earth was essentially similar to its structure today. The interior of present Earth is structured into zones as shown in Figure 19.4.

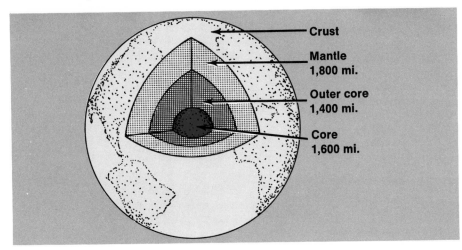

Fig. 19.4 The interior of Earth.

Man's knowledge of the different zones within Earth (crust, mantle, core—see Fig. 19.4) is deduced largely from a study of the speed of earthquake waves as they pass through different portions of our planet. In a manner of speaking, man uses earthquakes to "x-ray" Earth. It is a fine example of man's ingenuity that he has been able to put to a good purpose the life-and-property-destroying fury of earthquakes.

The boundary between the crust and the mantle was discovered by recording the speed with which earthquake waves travel through Earth. It has been observed that these waves attain a speed of five miles per second when traveling through the upper part of the mantle, but the speed drops abruptly to approximately four miles per second when the waves enter the crust. The boundary between the two zones is called the Mohorovicic discontinuity, after its discoverer, or the Moho for short. This change in speed is accounted for by assuming that the rocks of the mantle and the crust are of different composition.

Actually, there are two different kinds of earthquake waves: primary (P waves) and secondary (S waves). The P waves can travel through solids or liquids, but the S waves can travel through only solid material.

A truly startling thing happens to S waves from an earthquake when they reach a distance of 1,800 miles below the surface, which is about halfway to the center of Earth. They are stopped cold! This behavior could mean only that the S waves have reached a liquid zone which must be molten rock.

Note from Figure 19.5 that the center portion of Earth consists of an inner core and an outer core. Although the outer core is liquid, the inner core is solid. The solidity of the inner core is implied because P waves travel through the outer liquid core at a set speed (slower than their speed through a solid) but at a distance of 800 miles from Earth's center the P waves suddenly speed up again to the speed at which they travel through solid material.

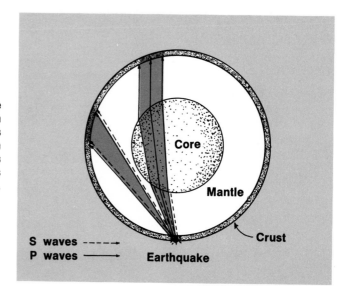

Fig. 19.5 **The bending of certain earthquake waves give information about the various zones of Earth's interior.**

S waves - - - - →
P waves ———→   Earthquake

the birth of earth

It is estimated that the chemical composition of our planet is as follows:

Section of Earth	Substance	Percentage
Crust	SiO₂ plus miscellaneous	1–2
Mantle	SiO₂	32
	MgO	23
	FeO / Fe₂O₃	6–7
	Al₂O₃	2
	CaO	2
	Na₂O	1
Core	Fe	24
	Ni	4
	Si	4

These percentages are very close to the percentages obtained for the chemical composition of certain meteorites. This similarity is not unexpected since we assume a common origin for planets and meteorites.

THE CORE

The core accounts for 32% of the mass of our planet but only 22% of its volume. Obviously, the core must be made of very heavy material. The core's density is calculated to be 15 g/cm³ in order for Earth as a whole to have an overall density of 5.5 g/cm³. The assumption that the core is an iron-nickel alloy is consistent with a density of 15 g/cm³.

It has already been pointed out that the molten iron-nickel outer core is responsible for Earth's magnetic fields. It is interesting to note that for some as yet unexplained reason, Earth reverses the polarity of its magnetic poles with fair regularity. The north and south magnetic poles are reversed approximately every one million years. The present magnetic field polarity has lasted for the past 700,000 years.

THE MANTLE

Between the core and the crust is a zone approximately 1,800 miles in thickness called the mantle. The mantle, which contains slightly more than two-thirds of Earth's mass, must be solid since it transmits S waves during an earthquake.

It appears that the mantle probably is composed of peridotite, a rock made up mainly of the iron-and-magnesium-rich material olivine, e.g.: olivine = 2 parts magnesium + 2 parts iron + 1 part silicon + 4 parts oxygen.

Although the mantle is definitely solid, the high temperatures and high pressures that exist in the mantle make it slightly plastic in nature. Consequently, there are huge convection currents operating below the crust whose slow but inexorable movements greatly affect the surface features of our planet.

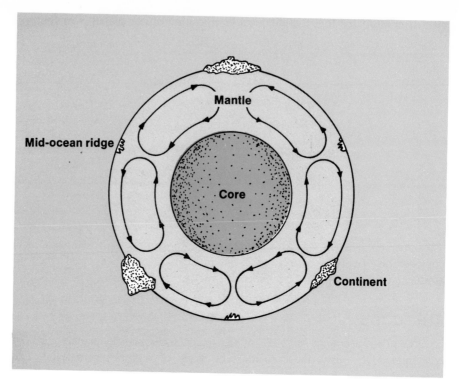

Fig. 19.6
Convection
currents within
the mantle and
their effect on
surface features
of the crust.

ORIGIN OF THE OCEANS

The infant Earth must have had only a moderate amount of atmosphere and a correspondingly modest amount of water on its surface. A considerable amount of the water that exists today in our great ocean basins was produced by "outgassing" (gases and liquids being squeezed out by high pressures, high temperatures, and crystallization of minerals) from the mantle and crust of our planet. This outgassing took place through volcanoes, hot springs, and fumaroles over eons of time to produce, along with the water initially present, the only "water" planet in our solar system. This outgassing yielded not only water but carbon dioxide that resulted in a fairly constant rate of deposition of limestone (calcium carbonate) in the seas of the changing Earth.

THE CRUST

The thin outermost shell of Earth, which accounts for only about 1% of its mass, is called the crust. This is the portion of Earth that man has lived on and is most familiar with and is, therefore, that portion of our planet that we have the most information about. The crust varies in thickness from about three miles to twenty-five miles, being thickest under the continents and thinnest under the oceans. In most places on our planet the crust can be divided into an upper and lower part. The lower part has properties similar to the rock layer underlying the oceans and consists mostly of basalt. Chemically, the crust is mostly oxygen and silicon, as is shown below:

Composition of the Earth's Crust

Element	Percentage	
Oxygen	46.6	} 74.4%
Silicon	27.7	
Aluminum	8.1	
Iron	5.0	
Calcium	3.6	
Sodium	2.8	
Potassium	2.6	
Magnesium	2.1	
All other elements	1.5	

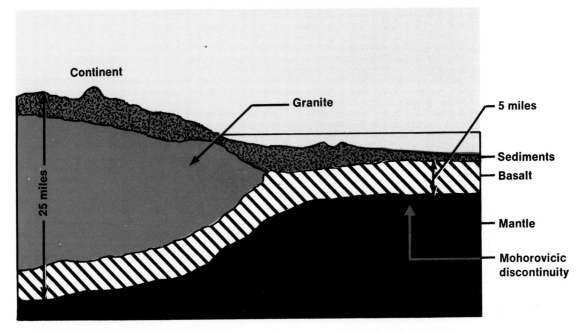

Fig. 19.7 The crust of Earth showing the basalt layer under the oceans and under the granite of the continents.

Interestingly, only eight elements make up approximately 98.6% of the crust, and two, oxygen and silicon, constitute 74.4% of the total crust! Incidentally, though it might seem that a lot of gold has been mined from the crust, the percentage of gold present in the crust is estimated to be only 0.0000001%.

the birth of earth

THE PLASTIC MANTLE AND ISOSTASY

We have all been awed by the lofty majestic mountains that exist on Earth. We would be even more intimidated if we realized that nine-tenths of the mass of a mountain exists beneath the surface crust that we stand upon. Indeed, an apt analogy of mountains is that of an iceberg floating in the ocean. Approximately nine-tenths of the mass of an iceberg is below the surface of the water. This means, of course, that nine-tenths of the iceberg must displace its weight of water to give the necessary buoyance to float the whole mass. The same requirement is necessary for our immense mountains. Some mountains have "roots" that extend down into the mantle for fifty miles and there float in the more dense mantle. As the massive mountains are eroded by wind, water, ice, and earthquake action, the buoyance of the monolithic mass slowly pushes the mountains up so that they are maintained in towering splendor much longer than would otherwise be possible.

This upward movement, in response to the decreased weight of a mountain (caused by erosional agents) is similar to the response of a heavily ladened ship that sits very low in the water but while being unloaded slowly rises in the water in response to its decreased weight.

An immediate question that springs to mind is how can the mantle of our Earth be sufficiently fluid to "float" an immensely heavy mountain range? Of course, the concept that is so startling is that if the mantle is rigid, as it must be in order to transmit S waves from earthquakes, how then can it be fluid enough to act as a liquid and buoy up massive structures such as mountains in the same fashion as the ocean floats icebergs? In order to understand this apparent anomaly of physical properties that the mantle must have, one can consider the common candle that is used in the home on birthdays and on nights when society's sophisticated electrical generating capacity fails, and we must revert to candles for light as if we were once again in the dark ages. A candle will most certainly break if struck a sharp blow, but a long candle supported by one end for a long period of time will slowly but inexorably bend under the force of its own weight. The same type of physical characteristics is exhibited by the mantle of our Earth. The mantle is sufficiently rigid to break under stresses and strains to produce earthquakes and transmit S waves, but under tremendous pressures and temperatures over long periods of time the mantle acts as a plastic "fluid" material that will float huge masses existing on the crust of our planet. This condition of equilibrium whereby tall, heavy land masses supported by deep roots on the denser plastic mantle is known as isostasy from two Greek words *iso*-equal and *stasis*-standing still.

CONTINENTS THAT DRIFT

Many times in the past while you were looking at maps of the world you probably made the casual observation to yourself that the opposing shorelines of the continents of South America and Africa would fit as two pieces of a

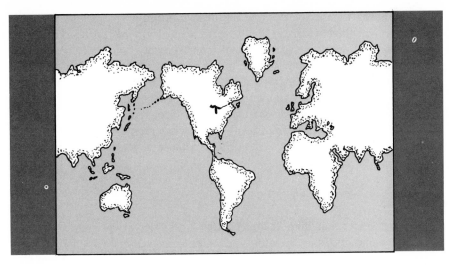

jigsaw puzzle if pushed together. The surprising fact is that not only will they fit together very nicely but at one time in the distant past these two continents, as well as many others, were together! They slowly, over hundreds of millions of years, drifted apart to reach their present positions.

Approximately two hundred million years ago when dinosaurs reigned supreme over our Earth, the continents of this planet formed one immense solitary continental island, called Pangaea, sitting alone in a world-encircling sea. This primeval planetary land mass broke into two parts. The southern half was called Gondwanaland and the northern half was called Laurasia. At this time the Atlantic Ocean did not exist since the North and South American continents were joined to the African and European land masses respectively.

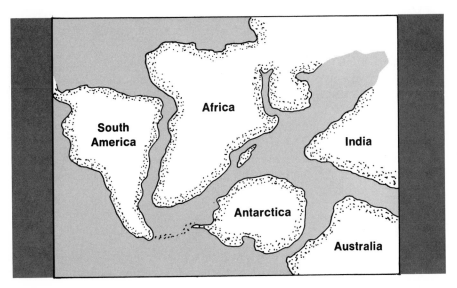

Fig. 19.9 **Gondwanaland soon after the great land mass became fragmented into present day continents.**

About 150 million years ago, ruptures occurred in the huge land masses, and the fragments that formed slowly drifted away from the rest of Laurasia and Gondwanaland at an average speed of approximately two-thirds of an inch a year. The North and South American continents drifted westward, and India broke away from Antarctica and drifted northward a considerable distance to eventually crash into the Asian continent. This violent collision resulted in the formation of the Himalaya Mountains between Tibet and India. Australia drifted eastward and Antarctica drifted southward to its present position at the south pole. The continents of our planet drifted because of sea floor spreading (see below).

TROPICAL FORESTS AND DINOSAURS IN ANTARCTICA

What evidence is there to suggest that Antarctica at one time was part of Gondwanaland and was, therefore, attached to the present African continent? Geologic studies of the Antarctic continent reveal extensive coal beds and fossil remains of a dinosaur-type animal. It turns out that during the Permian period large portions of the Antarctic continents were covered by lush tropical forests and swamps, and Lystrosaurus, an early dinosaur type, roamed Antarctica at will. How could this situation have occurred if Antarctica had always been in its present position at the frigid glacier-producing south pole? The answer, of course, is that it has not always been in its present position, but 200 million years ago Antarctica was attached to Africa only 15 to 30° south of the equator. It was during this period of time that the Lystrosaurus and the hot moist climate conducive to flourishing, lush vegetation existed. Moreover, the same types of plants and animals found in Antarctica and shown to have lived about 225 million years ago are found in the southern parts of South America, Africa, and India. Since Lystrosaurus, as well as the flora present during the same period of time, could not have survived an ocean trip between these continents, one must assume that Antarctica, South America, Africa, and India were joined together during the Permian period.

These "continental islands," floating on a sea of heavy mantle materials, drifted to the various corners of the globe. The North and South Atlantic Oceans were slowly formed by the westward drift of the North and South American continents respectively until today a distance of approximately 1,500 miles separates these two continents from their original positions against Europe and Africa.

The obvious question arises: "Are the continents of today still drifting?" The answer is yes! The continued movement of continents is being measured and studied by scientists in many nations today.

SEA FLOOR SPREADING

Figure 19.10 shows the mid-ocean ridges encircling the world to form the largest single geologic structure on our planet. It is at these ridges that material

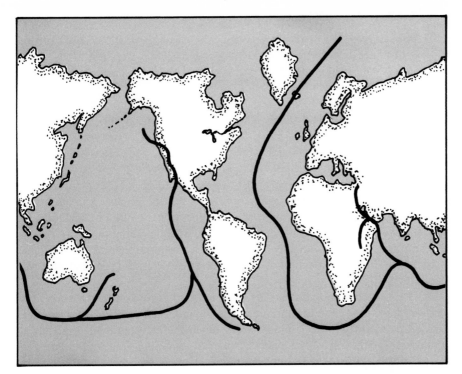

Fig. 19.10 **The world encircling mid-ocean ridges show as heavy lines. It is at these ridges that sea-floor spreading occurs.**

from the hot mantle of our Earth issues forth and spreads out along both sides of the ridges. This outpouring produces spreading of the sea floors. The upwelling of mantle material which spreads the ocean floors, as illustrated in Figure 19.6, shows the convection currents in the mantle.

The mid-ocean ridges are the scars remaining from deep planetary wounds from which molten mantle material constantly issues forth. It is the same wounds that opened up under the primitive behemoth mother continent, Pangaea, to cause the monolithic structure to fragment into smaller continental masses.

From Figure 19.6 it can be seen that new sea floor is being continuously formed by submarine lava outflowing from the mid-ocean ridges. Hot mantle material, being lighter than cooler mantle material, rises to produce upwelling lava. This spreading at the mid-ocean ridges causes the whole ocean floor to move away from the ridges and toward the continents. If a continent is attached to a sea floor that is spreading, the continent will move with the sea floor. Thus, sea floor spreading is the agent responsible for continental drift. Near the continental slopes the ocean floor dips down under the continents and is mixed with and melted by the hot plastic mantle.

An excellent example of a sea presently being formed by the spreading of two land masses is the Red Sea. The width of the Red Sea is slowly increasing as the Arabian peninsula gradually separates from Ethiopia, the Sudan, and the United Arab Republic. One can see an excellent fit between the opposing coast lines of the Red Sea land masses.

GEOLOGICAL EVOLUTION

The composition of the atmosphere, the contents of the oceans, and the variety of minerals present on primitive Earth certainly were not what we find today. Our planet has undergone, and is still undergoing, geological as well as biological and chemical evolution.

The atmosphere of our Earth has changed from one containing mostly hydrogen and helium with lesser amounts of methane, water, ammonia, and oxides of carbon to an atmosphere of 78% nitrogen, 21% oxygen, and small amounts of argon, carbon dioxide, and water. Our seas and oceans have changed from almost fresh water bodies containing a thin organic soup to a 3.5% solution of inorganic salts that supports a staggering tonnage of life in an unbelievable diversity of forms. The lithosphere or solid portion of our Earth has undergone changes in the variety of minerals and types of rocks present, including the amount of vulcanism and earthquake activity. Also the entire outer layer of the crust of this planet is continually being carved and sculptured into ever-changing patterns by the erosion action of wind, water, ice, earthquakes, and lava outflows.

There is a continuing struggle between the forces within Earth's crust that push up huge masses to form mountains and plateaus and the restless forces of erosion that slowly but relentlessly wear down the elevated land in a vain attempt to make the land surface of our globe one continuous sea level plain.

Fortunately for the esthetic sensibilities of man, the struggle between land uplift and erosion gives our planet its multitude of surface features such as mountains, valleys, hills, lakes, plains, and plateaus.

Indeed, our planet is unique in its beauty. We are the only planet with large quantities of water on its surface and, therefore, the only planet with lush vegetation. These two features give us the blue of our oceans and the green of our hills. The other planets in our solar system present stark and bleak appearances. It is easy to understand, therefore, the feelings of Rhysling, the space troubadour in Robert Heinlein's story, *The Green Hills of Earth,* when he sings:

> "I pray for one last landing
>   On the globe that gave me birth;
>   Let me rest my eyes on the fleecy skies
>   And the cool, green hills of Earth."

Copyright Saturday Evening Post, 1946

CHAPTER SUMMARY

Our solar system was born out of a slightly rotating giant cloud of gas and dust called a *nebula*. Mutual gravitational attraction existing between the particles of material in the cloud caused the material to contract. During the contraction, gravitational potential energy was converted into heat energy. The material coalescing at the center of the contracting cloud became hotter and hotter until it started giving off heat and light to become a newly born star—our sun.

During the time that the material of the nebula was contracting, eddy currents surrounding the center caused the formation of protoplanets and their satellites.

The nebula increased its speed of rotation during the contraction process. By the time the new star and its brood of planets had been born out of the star dust of space, the planets were revolving around the star and rotating on their axes.

Not too many years after a star begins to shine, the internal temperature of the star reaches the critical value of about ten million degrees Celsius. At this temperature the nuclear fusion reactions spontaneously ignite, and four hydrogen nuclei are fused into one helium nucleus with the liberation of energy. The difference between the mass of the four hydrogen nuclei and one helium nucleus is converted into energy, and it is this fusion reaction that is the main energy source of stars such as our sun.

The outermost layer of rock on Earth, the *crust*, is composed mainly of oxygen and silicon. The crust varies in thickness from about three miles under the oceans to about twenty miles on the continents.

The *core* of Earth has a diameter of about four thousand miles and, except for the central portion of the core, consists of molten iron and nickel. The small *inner core* is solid iron and nickel. The section of Earth between the core and the crust is the *mantle*. The mantle is solid but has plastic characteristics.

The continents of Earth float on the plastic mantle. About 200 million years ago, all the continents of Earth were linked together in one huge mother continent called *Pangaea*. Pangaea broke into two halves called Gondwanaland and Laurasia. Both Gondwanaland and Laurasia in turn broke up into smaller sections that drifted in various directions to produce the continental masses we know today. The drifting of the continents is the result of sea floor spreading that occurs at the *mid-ocean ridges*. At the mid-ocean ridges, hot mantle material issues forth to cause spreading of the sea floors and, consequently, movement of the continents.

QUESTIONS AND ANSWERS

1. Why are some nebulae dark and some light? (p. 394)
2. What is the chemical composition of the gases and dust particles in nebulae? (p. 395)
3. Earth's atmosphere makes the sun appear redder than it actually is and makes the sky appear blue. By analogy, how should star light passing through a gaseous nebula be affected? (p. 380)
4. How could the pressure of light and particles from the sun be used to propel a space ship?
5. How much gasoline would have to be burned in order to release as much energy as the sun releases in one second? (p. 397)

the birth of earth

6. What would happen to Earth's magnetic field if Earth stopped rotating on its axis? (p. 402)

7. From the mass and volume of Earth given on page 403, verify the value given for Earth's density. (p. 403)

8. Why do scientists believe that Earth's core is liquid? (p. 403)

9. Earthquake waves indicate that Earth's mantle is solid, but it is thought that continents "float" on the mantle. Explain. (p. 408)

10. In approximately what year is Earth's magnetic polarity expected to reverse itself again? (p. 405)

11. What evidence suggests that Antarctica was at one time connected to the African continent? (p. 410)

12. Explain the presence of so much salt in our oceans. (p. 406)

the chemical origin of life

20

THE CHANGE IN THE ATMOSPHERE OF PRIMITIVE EARTH
THE BEGINNING OF THE OCEANS
ENERGY FOR SYNTHESIZING PRELIFE
BUILDING BLOCKS FOR LIVING ORGANISMS
CHEMICAL EVOLUTION
BIOLOGICAL EVOLUTION
LIFE: ACCIDENTAL OR INEVITABLE?
LIFE ON OTHER WORLDS?

vocabulary

Bacteriophage | Catalysts
Photosynthesis

At the end of the formation of our planet, Earth circled its mother star with its eight companion planets. Its one-time overwhelming mass of atmosphere had been bombarded by the sun's outpouring of heat and charged particles until only a remnant of its initial gaseous envelope remained. The atmosphere that remained consisted mainly of hydrogen (H), ammonia (NH_3), and methane (CH_4), with smaller amounts of carbon monoxide and dioxide (CO and CO_2), water (H_2O), helium (He), hydrogen cyanide (HCN), and formaldehyde (H_2CO). Notice the conspicuous absence of free oxygen (O) from the list of constituents of the primeval atmosphere. This absence is certainly not due to any lack of oxygen in Earth itself since the crust alone is made up of minerals whose weight is approximately 45% combined oxygen. This high percentage indicated what happened to the original free oxygen; it reacted chemically during the early stages of Earth formation to form the compounds and minerals that make up the crust and mantle of our planet. There just wasn't any appreciable amount of "free" oxygen left.

In support of the above-assumed chemical composition of Earth's primitive atmosphere are the facts that nebulae, like the one that formed our solar system, have been shown to contain the compounds listed above, and comets, the leftover debris from our solar system's birth, are made of methane, ammonia, and water among other compounds.

THE CHANGE IN THE ATMOSPHERE OF PRIMITIVE EARTH

What happened to change the atmosphere of Earth from a primitive mixture of hydrogen, helium, ammonia, and methane to the present-day nitrogen and oxygen mixture? With the passage of eons and eons of time, the hydrogen and helium escaped into interplanetary space while the ammonia was oxidized to nitrogen and water and the methane was oxidized to carbon dioxide and water. The carbon dioxide produced was slowly incorporated into deposits of limestone.

From our knowledge of the launching speeds for space probes, we know that any object that travels "straight up" from our Earth with a speed of approximately seven miles per second will be lost from Earth. When anything attains this speed, it is capable of leaving forever Earth's gravitational influence. Let's consider for a minute what the speeds of the various atoms or molecules of hydrogen and helium were when these elements were abundant in the atmosphere of our planet.

the chemical origin of life

Consider a hypothetical situation where we have one hundred molecules of hydrogen in a box. Assume the box and its molecules are at room temperature. If we had a magic microscope with which to see the molecules in the box, we would observe that most of the hydrogen molecules are traveling at a speed of about one mile per second but that approximately one-fourth of the molecules would be traveling five times faster than this and approximately one fourth would be traveling five times more slowly. If we looked more carefully, we would see that 5 or 6% are hurtling along at seven miles per second, a speed which is fast enough for them to escape from Earth's gravitational field. Also, if the molecules traveling at seven miles per second are removed from the box or if their counterparts in Earth's primeval atmosphere are allowed to escape from Earth, the remaining molecules immediately redistribute their speeds so that, assuming the temperature remains the same, the group of molecules remaining still has the same small percentage with speeds of seven miles per second. This means then that given sufficient time all the molecules of hydrogen in Earth's primitive atmosphere would be lost into space. The same considerations are true for the slightly heavier helium atoms, which would also leak away into space but would do so at a slower rate because of their greater mass.

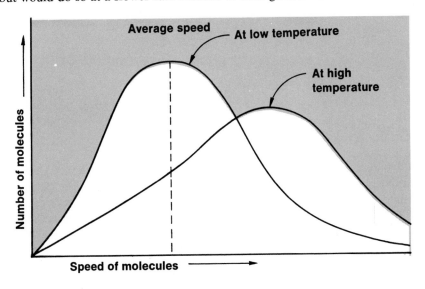

Fig. 20.1 A distribution of speeds of molecules at two different temperatures.

It has been calculated that molecules that have an average speed of only 1.4 miles per second will be completely lost into space in a period of a few hundred million years. Conversely, in order for any particular gaseous substance to have been retained by Earth during its four and one-half billion years of existence, the molecules of the gas must be heavy enough so that at room temperature the average speed of the molecules is less than one-sixth of the escape velocity of Earth. This is the case for the heavier gases of Earth in the past as well as today. Shown on the following page is a distribution of speeds of oxygen molecules at 0°C.

the chemical origin of life

Range of speed in miles per hour	Percent of molecules having that speed
0–224	1.3
224–447	8.1
447–671	16.7
671–895	21.5
895–1119	20.3
1119–1342	15.2
1342–1658	9.2
1658–above	7.7

The argument for the loss of original hydrogen and helium in Earth's atmosphere because of Earth's relatively small gravitational field is born out when one studies the present atmospheres of the other planets of our solar system. The massive planets, Jupiter, Saturn, Uranus, and Neptune, have large gravitational fields and consequently high escape velocities:

TABLE 20.1

Relationship of Planetary Mass to Escape Velocity

	Mercury	Venus	Earth	Mars	Jupiter	Saturn	Uranus	Neptune
Mass in terms of Earth's Mass (approximate values)	1/20	4/5	1	1/8	318	95	14	17
Escape velocity in miles/sec.	2.6	6.4	6.9	3.1	37.8	22.9	13.6	15.5

They are also very cold planets and consequently the molecules of gases in their atmospheres have a lower average molecular speed and, therefore, a smaller percentage of molecules have high enough speeds to escape the planet's gravitational field. As a consequence of high gravitational fields and low temperatures, these outer planets have retained their hydrogen, helium, and methane atmospheres. Water has been frozen out of the atmospheres of all four, and ammonia has been frozen out of the atmospheres of Uranus and Neptune. In the case of Mercury, Venus, Earth, and Mars, their comparatively lower gravitational field and higher temperatures have allowed all the hydrogen and helium originally present in their atmospheres to escape. Obviously, therefore, even if Earth hundreds of millions of years after its formation was not at a higher temperature than it is today, Earth still would have lost almost all of its hydrogen and helium molecules within the first 500 million years after the sun began to shine.

As we shall see later, it is quite probable that all the organic compounds necessary for the emergence of life on our planet were synthesized in the atmosphere of our world during the first one-half billion years after the birth of our sun.

THE BEGINNING OF THE OCEANS

It was probably during the earlier part of the 500 million year period when Earth still had a large amount of hydrogen in its atmosphere that the temperature of Earth reached its highest point. Toward the end of this period the temperature of Earth had dropped below the boiling point of water. Once the temperature of the land fell below the boiling point of water, the torrential rains that were falling upon Earth collected in depressions and crevices and over a long period of time slowly and inexorably filled the depressions to produce oceans, seas, lakes, ponds, and rivers. The rains falling upon the hot land were in turn warmed, and the warm waters flowed by the force of gravity to the now existing bodies of water. During the time that the water was upon the land, it dissolved small quantities of various chemicals, mostly inorganic salt-type substances. The unending rains continued to leach out these salt minerals from the land and carry them in solution to the seas. Thus, the buildup of the salt of the seas and oceans had begun.

When the dissolved chemicals and suspended particles carried by swift streams were carried into the still waters of ponds, lakes, and oceans, there resulted an unceasing drizzle of particles of various compounds through the waters that would finally rest on the sides and bottoms of the lakes and seas. Some of the compounds were natural chemical reaction catalysts of various types.

Catalysts participate in a reaction in such a manner that the reaction proceeds at a speed several to tens of thousands of times faster than it would if no catalyst were present. Catalysts are effective in very small amounts since they are not used up in a chemical reaction. We know, of course, that catalysts will not cause any reaction to occur that would not occur in the absence of a catalyst, but simply allow the reaction to be completed much more quickly.

ENERGY FOR SYNTHESIZING PRELIFE

The desolate and barren Earth continued to receive endless rains. The blazing, brilliant sun, occasionally obscured by dark, thunderous clouds of methane and ammonia, bathed Earth in the warmth of infrared rays, the light of visual rays, and the searing energy of ultraviolet rays (UV light). Lightning flashed brokenly across the energy-filled skies in great electrical discharges that ruptured chemical bonds of molecules present in the atmosphere and caused new bonds to be formed. Water, passing over the subterranean volcanic lava flows, rose to the surface to produce hot springs and pools containing water at all possible temperatures.

In most places on the planet the land seas were warm, but there were at higher altitudes land and water much cooler than the average. Also, from the subterranean fires deep within Earth there were occasional flaming and brilliant displays of erupting volcanoes spewing their red fiery molten lava skyward in awesome exhibitions of the primeval release of raw energy. Consequently, at this

the chemical origin of life

stage of development of Earth, water and land could be found that had temperatures ranging from below the freezing point of water to the temperature of molten lava. Radioactive elements, existing as compounds dissolved in the waters and deposited on the land, silently and incessantly emitted high energy electrons, neutrons, and gamma rays. Debris from the solar system formation crashed into Earth at high speeds to produce shock waves.

Now our picture of the primitive Earth is as follows:

(a) Various compounds are present in the atmosphere such as H, NH_3, CH_4, H_2O, CO_2, etc.

(b) Many energy sources are available for the breaking of existing chemical bonds. These energy sources are electric discharge (lightning), ultraviolet radiation from the sun, high energy particles from radioactive substances, heat, and shock waves.

(c) Many different catalysts are present on the land and in the waters.

(d) Temperatures vary from below the freezing point of water to thousands of degrees Celsius. (Of course, as the temperature of a certain reaction increases, the speed of the reaction increases also.)

What will be the inevitable consequence of all the materials, catalysts, and energy forms coexisting on our infant planet? Laboratory experiments and theoretical discussions by scientists such as Oparin, Urey, Miller, Calvin, Fox, Ponnameruma, Orio and many others have given some answers to the above question. The answer is that the beginnings of primitive life forms were produced, which in turn evolved into self-duplicating, life-sustaining cells.

BUILDING BLOCKS FOR LIVING ORGANISMS

When a mixture of methane, ammonia, and water is subjected to electric discharges simulating lightning in the primitive atmosphere, or subjected to high energy electrons acting as the natural radioactivity present on the infant earth, or simply heated together at various temperatures, a generous mixture of amino acids is created.

$$CH_4 + NH_3 + H_2O \xrightarrow{\text{various energy forms}} \text{amino acids}$$

Simple heating of the amino acids formed produces protein materials similar to natural protein which form cell membranes.

When hydrogen is added to the above mixture and the same conditions exist, other compounds as well as amino acids are formed.

$$H_2 + CH_4 + NH_3 + H_2O \xrightarrow{\text{various energy forms}} \text{aldehydes, amino acids, cyanides, hydrocarbons, etc.}$$

From these relatively simple organic structures, many substances of biochemical significance to life forms can be produced by the same energy forces that created the amino acids and other substances.

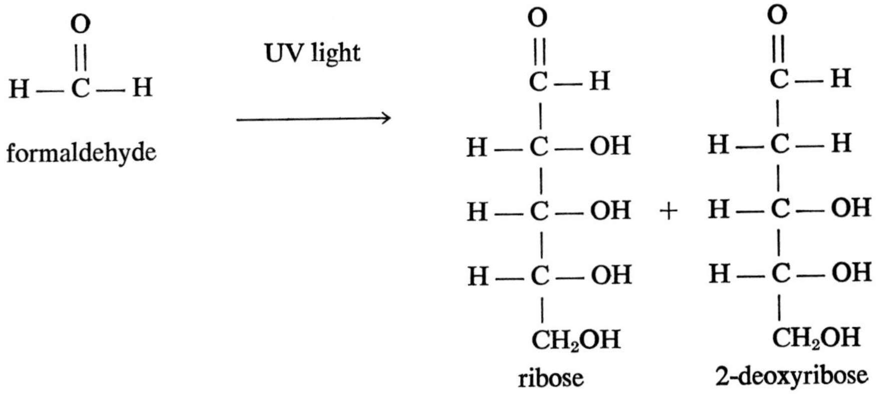

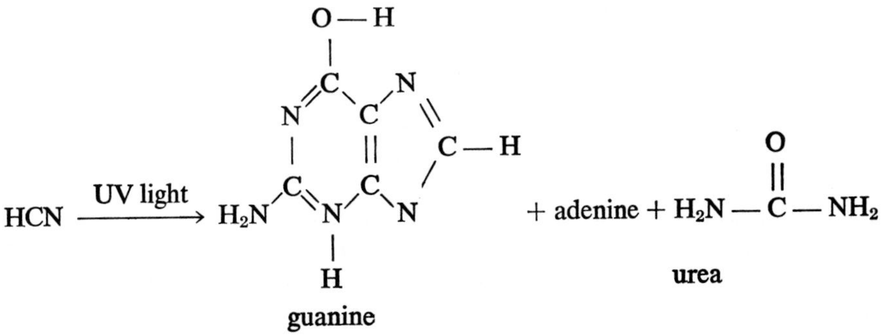

formaldehyde

ribose

2-deoxyribose

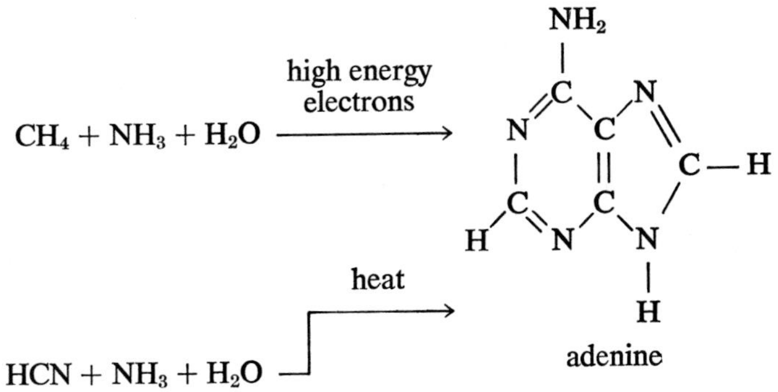

$CH_4 + NH_3 + H_2O \xrightarrow{\text{high energy electrons}}$

$HCN + NH_3 + H_2O \xrightarrow{\text{heat}}$

adenine

$HCN \xrightarrow{\text{UV light}}$

guanine

+ adenine + $H_2N - C - NH_2$

urea

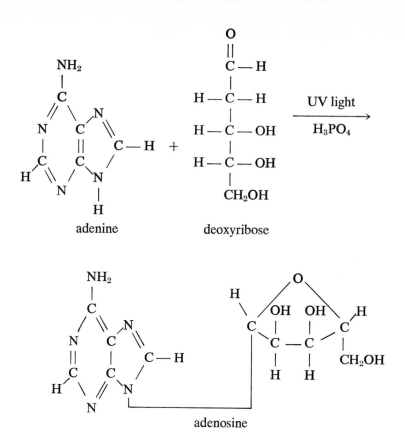

adenosine

Adenine, guanine, and adenosine are present in deoxyribonucleic acid (DNA), the genetic material that carries the genetic code in the form of a double helix. Different sites on the DNA double helix control the production of different

enzymes (chemical catalysts in cells). The enzymes produced by the DNA in turn direct the construction of carbohydrates, proteins, and fats in the cells, tissues, and organs that make up living organisms.

The amino acids, proteins, sugars, and DNA precursors rained into and were washed into the ponds, lakes, and oceans of our new world to produce a warm dilute organic "soup" that contained the chemical building blocks of life itself.

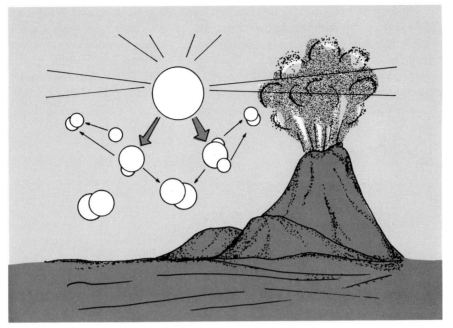

Fig. 20.3 A representation of the formation of a dilute organic "soup" on primitive Earth.

CHEMICAL EVOLUTION

Many scientists theorize that during the passage of untold eons, large protein molecules were formed. Because of the large number of sites of unequal charge distribution in protein molecules, they align themselves together in water so as to form geometrical patterns similar to the structures in present-day living cells. Dr. Sidney Fox and others have produced microspheres of synthetic protein material that are membrane-type spherules reminiscent of living cells. Photomicrographs of the proteinoid microspheres show that the spheres increase in number by "budding" and increase in size in concentrated protein solutions. Furthermore, these protocells exhibit enzyme activity similar to enzymes from living systems. It is not too difficult to visualize a natural chemical evolution of the microspheres to cells that can undergo life process.

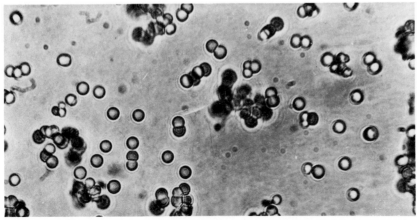

Fig. 20.4 Microspheres of synthetic protein material.

(Photograph Courtesy of Dr. Sidney Fox, Institute of Molecular and Cellular Evolution, University of Miami)

Growth of living matter involves the duplication of structures containing DNA. One of the simplest known examples of this duplication involves the viruses that infect bacteria: the bacteriophage. The bacteriophage consists essentially of DNA packed in a protein cover. General composition of the bacteriophage is approximately 60% protein and 40% DNA. In view of what we have already discussed, it seems certain that the formation of protein sub-

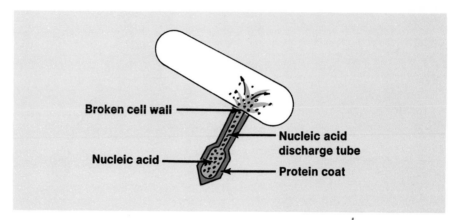

Fig. 20.5 **A bacteriophage attacking a bacterium.**

stances and of DNA must have occurred rather early on primitive Earth. The bacteriophage reproduces itself by attaching to a host bacterium, breaking through the bacterium membrane, and injecting its DNA molecules into the bacterium. The virus DNA multiplies and induces the formation of new protein coats from the bacterium cell material and then the newly formed DNA is packed into the new protein coats. The above process results in exact duplicates of the original bacteriophage. The cell wall of the victim bacterium ruptures and the new bacteriophages, about 200, are released to attack other cells.

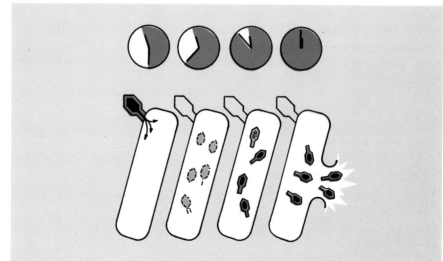

Fig. 20.6 Reproduction of bacteriophages.

BIOLOGICAL EVOLUTION

Now that we have assumed the transition from the membrane-like spherules or protocells to regular-reproducing primordial life single cell forms, what did the new life eat? Fortunately, there was a whole ocean of delicious organic compounds available of the exact type needed by the new organism as food. If the primordial cells had evolved no further, however, the newly emerged life forms would have had a placid indolent life consuming the thin organic soup that surrounded them and multiplying until eventually all the organic material produced through hundreds of millions of years would have been consumed. When all the organic "food" had disappeared, the so-recently-formed life on Earth would have perished of starvation, and nature's most noble experiment would have died in birth. New life could rise again only after the long period of time necessary for the production and buildup in the seas of the organic materials necessary as precursors of life forms.

An oganism's acquiring its needed energy by the simple breakdown of organic compounds present in its environment is a very inefficient way to obtain energy. However, there are many different primitive organisms that use this process even today since this is essentially the process of fermentation. If, however, life is to continue to evolve, a more efficient way of obtaining energy from its surroundings is absolutely imperative. The manner in which the original life forms managed this transition was the evolution into organisms that could acquire their energy from the sun in a direct manner. We are talking, of course, about the process of photosynthesis.

Photosynthesis depends on the three dimensional molecular structure of chlorophyll which holds an ion of magnesium in a central position in the structure. This particular arrangement of organic molecule and inorganic metal ion

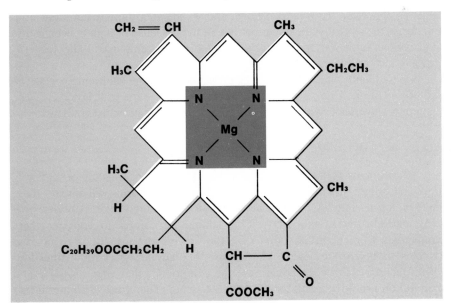

Fig. 20.7 **The structure of chlorophyll a.**

allows an organism to trap the energy of sunlight directly and use that energy to synthesize from carbon dioxide, water, and trace amounts of other materials the food that it requires for its growth and reproduction. The process of photosynthesis, while consuming carbon dioxide and water, releases free oxygen. Therefore, with the advent of photosynthesis by living organisms, oxygen was slowly released into the atmosphere of Earth.

$$6CO_2 + 6H_2O + energy \longrightarrow C_6H_{12}O_6 + 6O_2$$

There already were very small quantities of oxygen in the atmosphere, produced by the photodissociation of water vapor into hydrogen and oxygen.

As life utilizing photosynthesis flourished in the waters of our planet, the concentration of oxygen released into the atmosphere continuously increased. As the amount of oxygen slowly built up, the strong radiation of the sun changed some of the oxygen into molecules of ozone (the triatomic molecule of oxygen).

$$3O_2 \xrightarrow{\text{energy}} 2O_3 \text{ (ozone)}$$

We are all familiar with ozone, at least with the odor of ozone, since this is the pungent odor one detects in the vicinity of electric motors and other electrical devices that produce electric discharges. The electric discharges convert small quantities of oxygen in the air into ozone. The ozone was formed about ten miles up in the atmosphere so that after eons had passed, a layer of ozone completely surrounded our planet, as it still does today. This relatively thin layer of ozone very efficiently absorbs a large percentage of the ultraviolet radiation that our sun continuously showers upon us. After the ozone layer was formed, the amount of ultraviolet radiation reaching the surface of our planet was greatly reduced. As it turned out, this development was an absolutely necessary requirement for life to emerge from the waters of Earth and invade and multiply upon the land. Life forms, as we know them, cannot live in the presence of high concentration of ultraviolet radiation. Indeed, ultraviolet radiation is employed today to sterilize food and equipment.

Consequently, until the ozone layer was formed, the land remained sterile and devoid of life. Life was able to continue to thrive in the seas, however, since the upper layers of water absorbed much of the ultraviolet radiation.

LIFE: ACCIDENTAL OR INEVITABLE?

Experiments verify the fact that in an atmosphere consisting of methane, ammonia, water, and oxides of carbon, various energy forms will convert these small molecules into the building blocks of life: e.g. amino acids, peptides, proteins, carbohydrates, and DNA. One should be careful to consider these compounds from a certain point of view, however. Thus, life does not use proteins, carbohydrates and DNA, because they are necessary for various life-form functions, but because these are the compounds that were available and plentiful on primitive Earth for life to evolve from. Therefore, it is natural that life systems are composed of and do utilize these particular building blocks as food-energy sources.

Actually then, given a water-type planet with an atmosphere of compounds similar to those that existed on primitive Earth, and given normal energy forms such as ultraviolet radiation and electric discharge by lightning, many scientists theorize that it is inevitable that life systems will be produced if sufficient time lapses. In other words, a primitive planet with the proper conditions of temperature, atmosphere, etc. will evolve life because of the natural "law" of the chemical evolution of life that exists in our universe.

As evidence of the fact that amino acids, for instance, have been produced on solar system bodies other than Earth, moon samples and certain types of meteorites, called chondrites, have incorporated in their material many of the amino acids that are essential to life forms that we know today. It is assumed that these amino acids are not the result of decomposition of life systems that existed on the moon or the meteorite bodies at one time, but that the amino acids were formed abiotically in the same fashion that they were formed on primitive Earth.

LIFE ON OTHER WORLDS?

If one accepts the hypothesis that life systems will "spontaneously" be created if the necessary compounds and conditions exist, then the question arises, "Is there life on planets in other solar systems in our galaxy and in other galaxies of the universe?" In order to answer this question, we must have knowledge of the existence of other solar systems. In 1968 Dr. Peter van de Kamp analyzed data collected over a period of fifty-one years on the motion of the star called Barnard's star. The results of his study prove that Barnard's star has at least two planets as a companion.

Barnard's star may have many planets revolving around it as our sun does. At the present stage of technology, however, man's instruments are not sensitive enough to detect the presence of small planets with the mass of, say, our Earth. The planets that Dr. van de Kamp detected are about the size of the planet Uranus in our system. Other stars in our galaxy have also been shown to have planets. Actually, it was assumed in the past that stars with planets were a very rare occurrence in nature, but now scientists believe that it may be the rule for stars to have planets rather than the exception.

Other conditions than the mere existence of planets must be met, however, in order for life to exist. A few such conditions are listed below:

1. The star (sun) must not fluctuate to any appreciable extent in energy output.
2. Only planets at certain distances from their sun sustain life. If they are too far from their sun, the planets are so cold that water would stay frozen, and life as we know it must use liquid water in its cells. If the planets are too close to their sun, the resulting high temperatures would destroy life. In our solar system only Earth and Mars have surface temperatures between the freezing and boiling points of water.

3. The planets' orbits must not be too eccentric or they will receive too much heat when at perihelion (closest approach to the sun) and too little heat at aphelion (farthest distance of planetary orbit from sun). Obviously, a nearly circular orbit is the safest for life forms.
4. The planets must be "water" planets so that the protoplasm of cells can function.
5. The primitive atmosphere of the planets must contain the type of atmosphere that Earth once enjoyed. This is the easiest requirement to fulfill since it is probably a universally existing condition; remember that amino acids are present on the moon and in some meteorites.
6. The planets' rotation on their axes must not be synchronized with their revolution around their suns. If synchronization occurs, then the planets would keep the same face to the sun and the same hemisphere away from the sun with the result that all water on the planet would eventually be collected and frozen on the dark side, and the lighted side would become a searing, inhospitable desert.
7. The planets must have a minimum mass. If the mass is too small, all the planetary atmosphere would leak away.
8. The planets' speed of rotation should not be too fast in relation to their mass. For instance, if our planet rotated on its axis about eighteen times faster than it does, the centrifugal force at our equator would slightly exceed the gravitational force and our atmosphere would all escape in a short time. It might be pointed out that people and things near the equator would also drift off into space!

If our solar system is a reliable gauge, then one in nine planets can readily support life.

In view of these restrictions on life's ability to use certain planets, let's assume that only one star in ten million has a planet with the prerequisites to produce and sustain life. Since there are 100 billion stars in our galaxy then there should be $\frac{10^{11}}{10^7} = 10^4$ or 10 thousand stars with planets that probably have life forms present! Even more stunning in its implications is the fact that since there are 10^{21} (ten hundred thousand million billion) observable stars in the universe, there must be $\frac{10^{21}}{10^7} = 10^{14}$ (one hundred thousand billion) stars in the universe that we can see that have planets on which life exists!

CHAPTER SUMMARY

The original atmosphere of Earth contained mainly hydrogen, helium, methane, ammonia, and water. Various forms of energy available on primitive Earth caused the elements and compounds present in Earth's atmosphere to react and produce amino acids, sugar, organic nitrogen bases, and many other organic compounds. The amino acids in turn reacted to form proteins. One of the sugars, deoxyribose, the organic nitrogen bases, and phosphoric acid further reacted in the presence of catalysts to form molecules of deoxyribonucleic acid (DNA), the molecule that carries the genetic code of life. Further reactions of proteins, DNA, and other compounds produced the first primordial life cells.

The cells reproduced for untold generations and eventually developed the ability to utilize chlorophyll to obtain their energy directly from the sun to synthesize their needed food.

As a result of the photosynthetic process using chlorophyll, life slowly changed the primitive atmosphere of Earth by releasing oxygen from photosyntheses into the air. Over untold eons of time the oxygen reacted with the compounds in the air of Earth until the present atmosphere of nitrogen and oxygen prevailed.

Darwinian selection and evolution operated on the simple life forms of Earth to eventually produce the plants and animals we know today.

QUESTIONS AND PROBLEMS

1. How was our present atmosphere formed? (p. 418)
2. How could one calculate an age for Earth based on the amount of salt in today's oceans?
3. Do you think amino acids will be found on Mars? Why? (p. 429)
4. What would happen to the civilizations on Earth if Earth's rotation and revolution were synchronized? (p. 430)
5. How were our oceans produced? (p. 406)
6. If all life on Earth were destroyed, could life again originate on Earth with Earth's present atmosphere? (p. 422)

the record of life in the rocks

21

RADIOACTIVE "CLOCKS"
TELLING TIME BY ROCK LAYERS
ANCIENT LIFE – LAYER BY LAYER
THE SUCCESSION OF LIFE ON THE PLANET
EARTH
THE GEOLOGIC TIME SCALE
PRECAMBRIAN LIFE
THE PALEOZOIC ERA
THE MESOZOIC ERA
THE CENOZOIC ERA AND THE RISE OF MAN

vocabulary

Cenozoic era	Fauna
Flora	Fossil
Mesozoic era	Paleozoic era

In the Chapter "The Chemical Origin of Life," we traced the various processes by which the chemicals constituting Earth's primitive atmosphere were converted into the complex organic compounds that were the precursors of primordial life on our planet. These organic compounds, over untold millennia of time, slowly increased in complexity until, by chemical evolution, the first self-duplicating primordial form of life was evolved.

From a human's point of view, this surely was the most magnificent and incredible act of nature that had yet occurred in our universe. The importance of this event is that this lowly original life form, born in the misty ages of the past, was destined to evolve into ever increasingly more sophisticated forms of life until eventually Earth gave birth to the first sentient beings to live upon this planet. These sentient beings eventually became, through continued biological evolution, modern *Homo sapiens* who, for better or for worse, rule over all other life forms on the planet Earth.

What the far-distant future holds for man in terms of continued evolution man does not know, and surely it is best that he doesn't. According to Greek mythology, Prometheus, the Titan who made and loved mankind, took away man's knowledge of the future and in its place gave man hope. It is this hope that has always sustained the individual and the race in the face of disasters and ill fortunes. Man stands now, not at the end of nature's evolutionary path, but at the beginning of the evolution of intelligent, thinking men. It is quite possible that man may be evolutionarily changed in the future in a way we cannot now dream of even in our wildest imagination.

RADIOACTIVE "CLOCKS"

We are told that Earth is four and one-half billion years old, that dinosaurs reigned supreme on our planet 150 million years ago, and that primitive animals and plants left the security of their traditional home in the seas and invaded the lands of Earth 400 million years ago. How can scientists (paleontologists) that give us this information possibly know when various events occurred or when certain animals lived in the dim, distant past? In order to answer this question, we must first understand the principles and results of the radioactive decay of certain elements. Let's consider the radioactive decay of the element ^{238}U. Uranium spontaneously transmutes into other elements until eventually ^{238}U becomes the element lead (^{206}Pb). There is no way to tell exactly when a specific atom of uranium will decay. The atom may decay instantaneously, or it may wait several million years before decaying. Although the time when a specific atom of uranium will decay is not known, the length of time it takes for a certain fraction of a large group of atoms to decay can be calculated with great precision. In a similar manner, it is impossible for an insurance company to say when a certain person will die, but the company can very accurately calculate exactly how many deaths will occur each year in a very large group of people.

As shown in Figure 21.1, if one pound of ^{238}U had been isolated at the initial formation of Earth, today (four and one-half billion years later) there would be one-half pound of ^{238}U left. Thus, one half of all the uranium atoms initially present would have commenced the series of transformations that end when they become lead atoms. Four and a half billion years from now (nine billion years from when the pound of uranium existed) one half of the present day one-half pound (or one-fourth pound) of uranium would still remain and so on.

In other words with the passage of each four and one-half billion years' unit of time, the amount of uranium present will decrease by one half. The length of time it takes for one half of a radioactive substance to decay is called the half-life of the substance. If a rock containing uranium is analyzed and the ratio of uranium to lead is determined, then one can calculate how long it must have taken for that particular ratio of uranium to lead to be produced. Thus, the approximate age of rocks that contain uranium can be found.

There are many other radioactive elements besides uranium that are used to date rocks and buried bones. It is by the use of these various radioactive "clocks" that man can tell when in the past certain animals lived. If the bones of a saber-toothed tiger are found buried in layers of rocks that, from uranium-lead dating, are shown to be twenty-five million years old, then the saber-toothed tiger must have died and been buried twenty-five million years ago.

TELLING TIME BY ROCK LAYERS

Unfortunately, not all the layers of rock that exist in the crust of our planet contain the various radioactive materials necessary to determine the rocks. ages. Fortunately, there are other methods that can be used to give, if not absolute ages of formation, at least the relative ages of rock formation. For instance, as shown in Figure 21.2, if layer A has been dated by radioactive clock techniques as having been formed 100 million years ago and if layer E has been dated as having been formed 80 million years ago, then layers B through D must have been formed between 80 million and 100 million years ago. One can further state that since layer B is below layer C, layer B must be older than layer C since layer B had to be deposited before layer C could be deposited so as to rest on top of layer B. The above conclusion can be stated in a general fashion by saying that each layer or bed of rock is younger than the beds below it and older than the beds above it. The above is a statement of the *principle of superposition*. Thus, in reference to Figure 21.2, although we cannot tell the absolute age of each of the beds B through D, we do know that beds B through D are between 80 million and 100 million years old and that bed D is closer to 80 million years older than the other beds and that bed B is closer to 100 million years older than the other beds. Bed C must be intermediate in age between beds D and B.

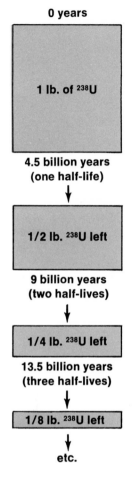

0 years

1 lb. of ^{238}U

4.5 billion years (one half-life)

1/2 lb. ^{238}U left

9 billion years (two half-lives)

1/4 lb. ^{238}U left

13.5 billion years (three half-lives)

1/8 lb. ^{238}U left

etc.

Fig. 21.1
Radioactive decay of uranium −238 into lead −206 in terms of half-life units of time.

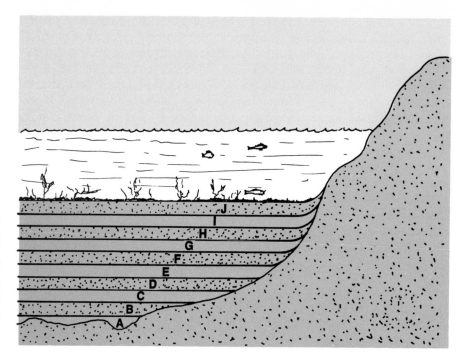

Fig. 21.2 An illustration of the principle of superposition. Layer A is older than layer B which is older than layer C, etc.

ANCIENT LIFE—LAYER BY LAYER

As early as about 500 B.C., the Greek philosopher Xenophanes postulated that the rocks that were formed in the bottom of seas and lakes were later made part of the dry land. He reasoned that the transition from sea bottom to dry land was the result of the sea's receding or of the sea bottom's being pushed up above the surrounding water. His reasoning was based on the fact that shells of certain mollusks were found great distances inland and even in rocks high up in mountains. Further, he had observed in broken rocks impressions of many different kinds of fishes and other marine life. The indications of past life that Xenophanes observed are today called *fossils*. A fossil may be defined as any indication of past life. By convention, indications of life that existed since the beginning of recorded history are not included. These indications of past life may be the bones or shells of once-living creatures or replicas of bones or shells in which the original bones or shells have been replaced by various minerals in the same manner in which petrified forests are produced. Fossils also may be only impressions of past life such as the imprint of a leaf or the footprint of a dinosaur.

Sometimes, though rarely, complete flesh and blood carcasses of ancient animals are preserved. Occasionally, insects are found imbedded in amber, and more than once the frozen carcasses of huge wooly mammoths have been found. The mammoths apparently had fallen into glacial crevasses and were hurriedly deep frozen. After more than 20,000 years those portions of glaciers melted to give forth the undecayed flesh of those majestic ancient animals.

When fossils are found imbedded in layers of rock, it is possible to use the principle of superposition to determine which layer of rock contains the oldest fossils and which layer of rock contains the youngest fossils. Thus, fossils present in different layers of rock present a record of the succession in time of different life forms. We know that the fossils in the uppermost layer are a record of more recent life than the fossils in lower layers. The fossils in the succession of rock layers from the bottom to the top show that the animals that left the fossils in the bottom layer were the oldest; the animals that left the fossils in the next higher layer were more recent than the fossils in the bottom layer, but older than the fossils in the layer immediately above them, and so on. This orderly succession of ancient through recent life forms recorded in the rocks is referred to as the *law of faunal succession*. The word *fauna* actually is a term meaning animals but it is to be understood that the law of faunal succession applies to plant fossils as well as to animal fossils.

THE SUCCESSION OF LIFE ON PLANET EARTH

As one studies the fossils that exist in rock layers, one is impressed by the gradual increase in the complexity of life as one goes from the lower, more ancient beds to the higher, more recent beds. In fact, one observes a picture book of the succession of life from lower to higher forms that gives considerable validation to Darwin's theory of evolution.

This slow climb up the evolutionary ladder by life was world-wide, and evolutionary changes that were taking place in one part of the world were, in general, also taking place in other parts of the world. Consequently, if certain types of fossils are found in a particular bed in Europe, then a bed in the United States that contains the same fossils should be of the same general age as the European bed. In fact, the ages of different beds are actually dated in this manner by paleontologists.

Even the environment in which animals and plants lived in past geologic time can be deduced from the fossils present and the mineral composition of the rocks in which the fossils are embedded. If a geologist finds oyster shells, coral, or fossilized fish in a bed, it is safe to assume that the bed was formed at the bottom of an ancient sea. On the other hand, if the fossils are of four legged animals, one would postulate that the rock was formed on land. If a bed containing fossils also contains salt, then the fossils probably lived in a desert climate which had been responsible for the evaporation of a shallow sea to produce the salt deposit. A bed of coal implies a very warm climate with a generous rainfall that would produce the lush vegetation necessary for the eventual formation of coal.

THE GEOLOGIC TIME SCALE

The Geologic Time Scale, shown in Table 21.1, shows some of the main events that heralded the changes in life forms on Earth with the passage of time.

The time scale is divided into four main time segments called *eras*. These eras are as follows:

1. *Cenozoic*—from 70 million years ago to the present. Cenozoic means recent life.

2. *Mesozoic*—from 230 million years ago to 70 million years ago. Mesozoic means middle life.

3. *Paleozoic*—from 600 million years ago to 230 million years ago. Paleozoic means ancient life.

4. *Precambrian*—from the time of formation of Earth to 600 million years ago. The term Precambrian is used to indicate any life forms that existed before the beginning of the Paleozoic era.

In turn, each era, with the exception of the Precambrian, is subdivided into smaller segments of time called periods. The periods in turn are subdivided into even smaller time segments called *epochs*. In the geologic time scale shown in Table 21.1 the epochs of only the periods of the Cenozoic era are indicated.

TABLE 21.1.
The Geologic Time Scale.

Millions of Years Ago	Era	Period	Epoch	Duration in Millions of Years	Major Evolutionary Changes	Age of
	CENOZOIC	Quaternary	Recent	2	Man dominates Earth,	Man
			Pleistocene	0.01	Ice Ages	
		Tertiary	Pliocene	10	Flowering plants widespread	Mammals
			Miocene	15	Great diversity of mammals	
			Oligocene	13	Grasses abundant; First saber-toothed tigers	
			Eocene	20	Horses, rhinoceroses, whales appear	
			Paleocene	10	First primates	
70	MESOZOIC	Cretaceous		60	Flowering plants appear; dinosaurs become extinct	Reptiles
		Jurassic		50	Dinosaurs rule Earth; First mammals, first toothed birds	
		Triassic		50	First dinosaurs; many mammal-	

the record of life in the rocks

the record of life in the rocks

Era	Period	(value)	Events	Life forms
PALEOZOIC	Permian	50	Rise of reptiles	
	Pennsylvanian	30	Extensive coal-forming swamps; reptiles appear; giant insects evolve wings	Amphibians
	Mississippian	35	Great diversity of fish; insects	Fishes
	Devonian	60	Amphibians appear on land; first insects	
	Silurian	20	First air breathing animals; First land plants	
	Ordovician	75	First vertebrates appear—primitive fish	Invertebrates
	Cambrian	100	Marine invertebrate animals and algae abundant	
PRECAMBRIAN	Precambrian		Variety of simple plants and soft bodied animals (From the Ediacara formation—Australia)	
			Diverse multicellular plant life forms (From the Gunflint formation—Canada)	
			Single cell photosynthetic organisms similar to present day blue-green algae. (From the Fig Tree formation—Africa)	Photosynthesis
			First primordial life form	Life
			Complex organic compounds present in earth's seas	Chemical Evolution
			Formation of the Earth	

Time scale (millions of years): 230, 600, 1000(?), 2000, 3200, 3600(?), 4000(?), 4500

PRECAMBRIAN LIFE

The very ancient Precambrian rocks contain few fossils in comparison to the great number and variety of fossils found in the Paleozoic, Mesozoic, and Cenozoic eras. The scarcity of Precambrian fossils is mainly due to the following:

1. Most of the Precambrian animal life were soft-bodied organisms not having hard parts easily preserved as fossils.
2. Life was not as abundant and widespread as in later geologic eras.
3. Most Precambrian rocks have been subjected to intense heat and pressure with the result that fossil remains which may have been present at one time have been destroyed.
4. Lack of study of Precambrian rocks by paleontologists.

In any case, Precambrian fossils are relatively scarce. There have been, however, some very notable fossil finds in Precambrian rocks in the last ten years. Enough information has now been collected to allow man to sketchily reconstruct the evolution of life from earliest Precambrian times.

Surely the most momentous event in life's evolutionary climb during the Precambrian era was the first step—the appearance of the first primordial self-replicating life system. This primordial form slowly evolved over untold eons into single cell organisms similar to present-day blue-green algae. Fossil remains of these single cell organisms have been found in South Africa by Drs. Elso Barghoorn and William Schopf in rock that is 3.2 billion years old. Interestingly, Africa is the same continent that gave birth to man almost two million years ago.

Fig. 21.3 Fossil remains of microorganism (left portion of picture) 3.2 billion years old from the Precambrian era of South Africa.

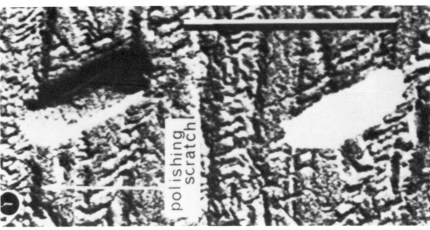

(Photograph Courtesy of Professor E. S. Barghoorn of Harvard University)

These 3.2 billion-year-old organisms had developed photosynthesis. The ability of life to obtain much of its energy by the photosynthetic process was a great leap forward up the evolutionary ladder. Prior to the development of photosynthetic ability, cells had to acquire nourishment from the organic nutrients of the shallow seas in which they lived. This non-photosynthetic method of acquiring energy is very inefficient and, besides, the organic nutrients available at this stage of life development would have been quite limited.

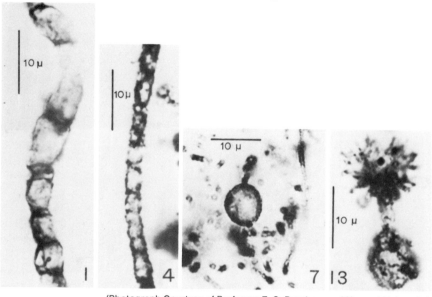

Fig. 21.4 Fossil remains of photosynthetic multicellular organisms that are two billion years old. Found in Ontario, Canada.

(Photograph Courtesy of Professor E. S. Barghoorn of Harvard University)

The next major step in the evolution of life was the development of a cell containing a nucleus. With the appearance of nucleated cells, sexual reproduction became possible. Sexual reproduction opened the door to genetic variability which resulted, by natural selection techniques, in increased complexity of multicellular organisms. Eventually a single cell life form, similar to the present-day Mastigophora, evolved that was a plant-animal. This Mastigophora type of orga-

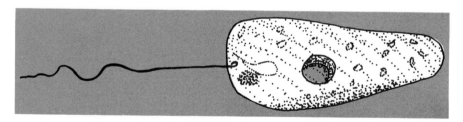

Fig 21.5
A drawing of present day Mastigophora.

nism could live strictly as a plant by obtaining its energy by photosynthesis or could live strictly as an animal and in the absence of sunlight obtain its energy by ingestion of organic nutrients from plants.

By at least two billion years ago multicellular forms had developed in which different cells of the organism had different specific functions necessary for the organism's normal life operations. The relatively simple multicellular organisms

the record of life in the rocks

evolved, in less than a billion years, into the more complex plants and soft-bodied animals represented by the fossils found in the Ediacara formation in Australia.

Fig. 21.6 Fossil remains of organisms that lived 900 million to one billion years ago. These fossils were found in Ediacara formation in Australia. *Spriggina floundersi, Tribrachidium meraldicum, Dickinsonia costata.*

(Photographs Courtesy of Professor M. F. Glaessner of the University of Adelaide, Adelaide, Australia)

the record of life in the rocks

THE PALEOZOIC ERA

At the close of the Precambrian era, and during the whole of the Paleozoic era, great inland seas inundated the continents of Earth. These ancient intracontinental marine seas teamed with life. The ancient life left its fossil remains deep within the interior of continents for man to see and ponder many hundreds of millions of years later.

By the Cambrian period of the Paleozoic era, 600 million years ago, the shallow seas of Earth swarmed with invertebrate animal life; that is, they did not have internal skeletons but had hard external shells. Some of the soft-bodied animals present in the latter part of the Precambrian era had acquired body armor that protected them from predators and harmful doses of ultraviolet radiation. The dominant animal in the Cambrian seas was the trilobite that existed in a great variety of shapes and patterns. Other important animals present in the Cambrian seas were brachiopods, archeocyathids (which formed great limestone reefs), sponges, snails, worms, and jellyfish. (See Fig. 21.7.) Plant life consisted mostly of blue-green algae. Although life flourished in the Cambrian seas, the stark desolation of the Cambrian land masses was mute testimony to the fact that evolution had not yet succeeded in producing life that was viable enough to live out of the life-protective seas.

The Ordovician period is remarkable for the great diversity as well as the great abundance of marine animal life. As an indication of the profuse animal sea life at that time, the oldest and largest deposits of oil and gas come from chemically changed animal bodies from this geologic period. The most important development of the Ordovician period was the appearance of the first vertebrates, which were primitive fish. The largest animals that had yet existed on

(Photograph Courtesy of the American Museum of Natural History)

Fig. 21.7 Life in the Cambrian seas.

Fig. 21.8 Life in the Ordovician seas. On the right is a large nautiloid cephalopod and on the left is a trilobite.

(Photograph Courtesy of the American Museum of Natural History)

Fig. 21.9 Life in the Silurian seas.

(Photograph Courtesy of the American Museum of Natural History)

the record of life in the rocks

Earth were the cephalopods of the Ordovician period. Some of the cephalopods attained a length of fifteen feet and were ten inches in diameter at their widest part.

The Silurian period marks a double milestone in the history of life on our planet. The first appearance of land plants and of air-breathing animals occurred in the Silurian period.

The first invasion of the desolate land by plants occurred during the Silurian period. This invasion surely was a long and difficult struggle. Before plants could successfully survive on land, they had to develop a protective outer layer to prevent loss of water. Furthermore, the plants had to evolve a vascular system for transporting water from the soil to the rest of the plant.

The first air-breathing animals were apparently sea scorpions, some of which attained a length of seven feet.

The outstanding events of the Devonian period were the transition from fish that used gills to fish that used lungs and the appearance of insects. The lung fishes, descendants of which have survived up to the present time, evolved into amphibians that slowly developed legs and are represented today by salamanders. Thus did the land vertebrates appear.

Insects first made their appearance during the Devonian period and flourishing coral formed great reefs throughout the world.

(Photograph Courtesy of the American Museum of Natural History)

Fig. 21.10   Life in the Dovonian seas.

the record of life in the rocks

Fig. 21.11 **Life in the Mississippian seas.**

Forests began to develop as the result of natural selection. Those land plants that could rise above competing plants and spread their leaves for greater sun absorption had a considerable advantage in the struggle for survival.

During the Mississippian, Pennsylvanian, and Permian periods of the Paleozoic era the climate of Earth was mildly tropical and, therefore, very favorable to plant growth. Consequently, the latter part of the Paleozoic era saw the development throughout the world of great forests and lush vegetation that yield today most of the world's coal. Because coal formed during the Mississippian and Pennsylvanian periods is so plentiful and widespread, the Mississippian and Pennsylvanian periods are often referred to collectively as the *Carboniferous period*.

During the Mississippian period insects developed wings, and there was a great diversity of fishes. During the Pennsylvania period giant insects had evolved and the swish and rustle of wings measuring to thirty inches filled the air above the luxuriant tropical vegetation. Of all the animals that have ever lived on Earth, the insects and the fishes have been more carefully tailored by nature to survive capricious environmental changes than all other types of animals. Insects and fishes have flourished with only modest change for about 400 million years!

Fig. 21.12   A coal forest during the Carboniferous period.

The amphibians of the Pennsylvanian period slowly evolved legs and began laying eggs with a tough outer shell that prevented dehydration. This adaptation allowed the animals to lay their eggs on the land rather than in the water. This transition from marine animals to land animals resulted in the first reptiles. The reptiles, who continued to evolve during the Permian period, must have been well suited to their environment because they established a dynasty that ruled the Earth, uncontested, for 200 million years!

THE MESOZOIC ERA

During the early part of the Mesozoic era, the South American continent was still attached to Africa as part of Gondwanaland. Since the oldest rocks in the floor of the Indian Ocean were formed no later than the Cretaceous period, the land mass of India must have also been attached to Africa in early Mesozoic time. About the middle of the Mesozoic era South America and India broke away from the African continent and began their slow journeys to their present positions.

The North American continent during the early Mesozoic era was mostly free of the great inland seas that had been present during the latter part of the

the record of life in the rocks

Paleozoic era. Toward the middle of the Mesozoic era, however, the seas began once again to encroach upon the land, and by the Cretaceous period a great and extensive marine sea extended from the Arctic Ocean to the Gulf of Mexico. The middle of the North American continent was thus completely inundated, and North America appeared as two large elongated islands.

The Mesozoic era is called the Age of Reptiles since during this era the reptiles were by far the most dominant animals on this planet. During the Triassic period of the Mesozoic era, the most famous of all ancient life forms appeared—the dinosaurs. The word *dinosaur* comes from the Greek meaning terrible lizard. The first dinosaurs were modest-sized animals, but by the Jurassic and Cretaceous periods evolution had produced great reptilian monarchs such as *Brontosaurus, Triceratops,* and *Tyrannosaurus rex*. Brontosaurus, the largest land animal ever to walk the Earth, attained a length of eighty feet and a weight of 100,000 pounds.

Even while the great reptilian lords were thundering over the lands of Earth, however, the animals that would one day usurp their sovereign rule were scurrying about almost unnoticed. During the Triassic period mammals had slowly evolved from small reptiles. The mouse-sized unpretentious appearance of these first mammals belied their future greatness.

The Jurassic period saw winged reptiles invade the skies and eventually become larger than any animal that ever learned the secrets of flight. The Pteranodon developed a wingspread of twenty-seven feet.

Fig. 21.13 Feeding Brontosaurus of the Mesozoic era.

(Photograph Courtesy of the American Museum of Natural History)

the record of life in the rocks

During the Cretaceous period flowering plants appeared. At the close of the Cretaceous period, and, therefore, the close of the Mesozoic era, dinosaurs became extinct throughout the world. The ending of the reptilian reign appropriately marks the end of an era.

THE CENOZOIC ERA AND THE RISE OF MAN

The Cenozoic era opens with the appearance of the first primates (any mammal of the Order Primates including man, apes, monkeys, etc.). It was these initial primates that were destined to give birth, approximately sixty-eight million years later, to *Homo sapiens*. Mammals had been transformed from rather inconspicuous animals thinly scattered over Earth to the dominant animal life forms.

During the Tertiary period, a great diversity of mammals occurred. Some mammals, bats, evolved wings and sought their food in the air. Other mammals, whales and porpoises, reentered the seas and oceans that had been the home of their ancient ancestors.

During the middle of the Tertiary period grass lands became widespread with the result that grazing animals evolved. Herds of horses, camels, and rhinoceroses were constantly preyed upon by fierce saber-toothed tigers. Flightless birds of prey eight feet tall roamed the continents in search of food. The largest land mammal that ever existed flourished during the middle of the Tertiary period. This mammal, Baluchiterium, stood eighteen feet high at the shoulders and was thirty feet long.

Fig. 21.14
A saber-toothed tiger (Smilodon) that roamed North America during the Tertiary period.

(Photograph Courtesy of the American Museum of Natural History)

the record of life in the rocks

The Pleistocene epoch of the Quaternary period of the Cenozoic era is usually called the Ice Age. During the Pleistocene Ice Age, great continental glaciers covered as much as one-fourth of the land. Since the water used to produce these massive ice sheets came from evaporation of ocean water, the level of the world's oceans was 350 feet lower than they are today.

Four times did the vast continental glaciers spread over much of North America, Europe, and Asia, and four times did the climate warm up and the glaciers retreat. The glaciers did not completely melt away, however, since remnants of the Pleistocene glaciers still exist as the Greenland and Antarctic ice sheets.

During the early Pleistocene a creature that weighed no more than a hundred pounds and that was no taller than five feet roamed southern Africa. These creatures were only slightly more man-like than they were ape-like. These men-apes, called *Australopithecus africanus* (southern apes of Africa), did not make tools or weapons. They did, however, use bone and wooden clubs that by happenstance had the desired shape for weapons. They did not modify these weapons but used them as they found them. At the same time that the southern apes of Africa were roaming their domain, there appeared in southeast Africa a slightly more man-like ape creature called *Homo habilis* (man with ability). *Homo habilis,* who lived at least 1¾ million years ago, was a pygmy by our standards and was very ape-like in appearance, but he was a true ancestor of modern man. He walked erect and fashioned crude tools.

About a half million years ago, the ape Man of Java (*Homo erectus*) and the Peking man (*Homo pekinensis*) walked the lands of Java and China respectively. Both *Homo erectus* and *Homo pekinensis,* descendants of *Homo habilis,* made crude tools and were more advanced in culture than was *Homo habilis,* but they were still quite ape-like in physical appearance.

(Photograph Courtesy of the American Museum of Natural History)

Fig. 21.15 Neanderthal man with family as he lived in France about 130,000 years ago.

About 130,000 years ago, the most famous of all extinct species of men was spread throughout Europe and the Near East. He was Neanderthal man (*Homo neanderthalensis*) who looked more like present man than his predecessors. Even so, he was brutish in appearance with prominent brow ridges, slightly bowed legs, and a great barrel chest. He did not have a double curve to his spine as modern man has and, therefore, could not hold his head up as high as we can. Neanderthal man averaged only five feet four inches in height and covered himself from the cold of the glacial period with animal fur. We should be careful not to assume a condescending attitude toward Neanderthal man, however, because his brain case was slightly larger than the average brain case of modern man and his cunning and strength allowed him to kill the great mammoths of his day. A great advantage that Neanderthal man had over his predecessors was his use of fire and his better-made weapons. The use of fire, along with his better tools, gave him a decided advantage in survival. These advantages allowed Neanderthal man to rule Europe and the Near East for 100,000 years.

It is not known what caused the decline of Neanderthal man. Perhaps it was the scarcity of food brought about by the advance and retreat of glaciers. In any case he became extinct or was assimilated by Cro-Magnon man about 35,000 years ago.

During the last glacial age, almost 35,000 years ago, Cro-Magnon man appeared. Modern man had finally, after approximately 3½ billion years of evolution of life forms, arrived to claim his destiny. Cro-Magnon man was a magnificent human specimen. He stood slightly over six feet tall and because he had a double curvature of the spine, he could hold his head high. He had a high forehead, a highly developed chin, and an absence of heavy brow ridges.

(Photograph Courtesy of the American Museum of Natural History)

Fig. 21.16. Cro-Magnon artists painting a hunting scene.

The tools that Cro-Magnon man made were considerably more sophisticated than the tools of Neanderthal man. Also, Cro-Magnon man made a greater variety of tools and weapons. Besides the more common stone axes, knives, clubs, and spears, Cro-Magnon man produced bone needles, fishhooks, dart throwers, and, eventually, bows and arrows. He cooked his food and wore fur clothes which he sewed together with his bone needles.

Cro-Magnon man, from whom we are descendants, was quite intellectually and culturally superior to Neanderthal man. In fact, Cro-Magnon culture was advanced to the point that within a particular community there were skilled craftsmen, cooks, painters, and religious leaders that were at least partially supported by a highly organized work force. The people of a Cro-Magnon community decorated themselves with beads and bracelets made from ivory and sea shells much as people do today. Cro-Magnon man buried his dead which may hint at a belief in a life after death.

Perhaps the greatest accomplishment of Cro-Magnon man was his art. The skilled Cro-Magnon painters produced superb paintings in outstanding style and color that clearly speak of a sensitivity and elegance of soul that is unquestionably a quantum jump above Neanderthal man. Reasonably delicate carvings and sculpture further attest to a civilization whose members were not only able to recognize the beauty and wonder of nature but who spent time and effort trying to create their own beauty in a manner that would express some of their sensitive inner feelings.

If the first primordial life forms did appear on Earth three billion six hundred million years ago, it required about three billion five hundred and ninety

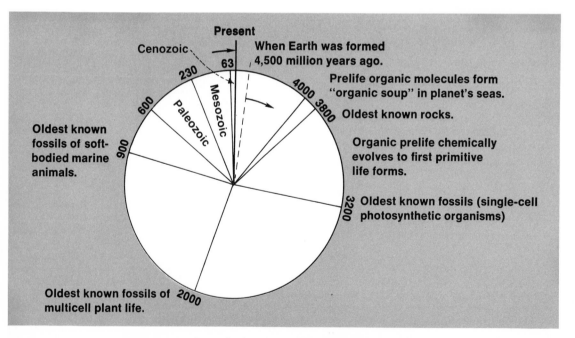

Fig. 21.17 A geologic clock. Man has been on Earth for such a relatively short period that it is impossible to show his existence on the above geologic clock. Numbers around clock are in millions of years.

eight million years for the evolutionary process to produce man. Man, then, has been an animal on this planet for only a very, very brief part of the time that life has existed on Earth.

CHAPTER SUMMARY

Comparing the ratio of a radioactive substance to its decay products present in a rock allows one to determine the rock's age and the age of fossils present in the rock.

The order in which fossils occur in rock layers allows one to deduce the sequence in which various ancient life forms occurred on Earth. It is possible, therefore, to know the chronology of life forms from the simplest single cell organisms of billions of years ago through many intermediate life forms to the eventual evolution of man on our planet.

Geologic time is divided into the following time segments called *eras*: *Cenozoic* (from 70 million years ago to the present), *Mesozoic* (from 230 million years ago to 70 million years ago), *Paleozoic* (from 600 million years ago to 230 million years ago), *Precambrian* (from the time of Earth's formation to 600 million years ago). Eras are subdivided into time spans called *periods,* and periods are further subdivided into *epochs.*

Life existed on Earth at least 3.2 billion years ago, when single cell organisms lived in the shallow seas of Earth. At least two billion years ago multicelled organisms developed, and by approximately one billion years ago complex plants and soft-bodied animals were gathering food in the seas of our planet. About 600 million years ago the seas of Earth swarmed with thriving invertebrate animal life, many with hard external shells.

Life forms continued to change until the evolutionary process culminated in *Cro-Magnon man,* from whom we are descendants.

QUESTIONS AND PROBLEMS

1. How long would it take for one pound of uranium −238 to decay until only one-sixteenth of a pound would be left? (p. 435)
2. What fraction of the total time life has existed on Earth has man been present? (p. 452)
3. How long was life present in the seas of Earth before life invaded the land? (p. 445)
4. What reasons can you think of that might explain why the dinosaurs disappeared from Earth?
5. How are fossils used to correlate rock layers separated by great distances? (p. 437)
6. Why are soft-bodied animals such as jellyfish usually not preserved as fossils? (p. 440)
7. How can the climatic conditions present many millions of years ago be deduced by the types of fossils found in an area? (p. 437)

our physical earth

22

UNIFORMITARIANISM
THE GEOLOGIC PROCESSES
MINERALS
ROCKS IN GENERAL
IGNEOUS ROCKS
SEDIMENTARY ROCKS
METAMORPHIC ROCKS
THE PHYSICAL LAWS OF GEOLOGY
OUR MINERAL RESOURCES

vocabulary

Chemical Change	Diastrophism
Geosyncline	Igneous Rocks
Lithification	Magma
Metamorphic Rocks	Mineral
Physical Change	Sedimentary Rocks
Specific Gravity	Unconformity
Uniformitarianism	Vulcanism

The origin of Earth is of considerable interest to geologists; however, geology usually concentrates upon the study of Earth after solidification of the surface occurred. Many of the various details about Earth in regard to its composition, shape, and size have been previously discussed, but a brief reconsideration of these items helps set the stage for the topic at hand.

Earth is considered an oblate spheroid—almost spherical but somewhat flattened at the poles and bulging along the equator. The diameter of this planet through the equator is some twenty-seven miles greater than the same measure through the poles. The difference between the two measures amounts to about 0.5%, a consideration of the flattening of Earth.

The surface of Earth is somewhat irregular, but a comparison of its greatest surface irregularity with that of an orange reveals that Earth is proportionally smoother. The highest point on Earth's surface is Mt. Everest, 29,141 feet above sea level. The lowest point is located on the ocean floor near the Philippine Island of Mindanao, 36,560 feet below sea level. The total relief of twelve miles amounts to a variation of only 0.3% of Earth's radius whereas the variation on an orange's surface is greater when the two radii are compared. The continents average about 0.5 miles above sea level. Asia, some 3200 feet above sea level, is the highest, and Europe, elevated slightly less than 1000 feet, is the lowest. The North American continent averages about 2400 feet above sea level.

UNIFORMITARIANISM

The outer layer of Earth, several miles thick, is called the *crust*. This layer is the only part of Earth accessible for our direct observation, though only about 30% of it is available because of the amount of the surface covered by the oceans. Classical geology developed from the study of the continents, which represent the preponderance of Earth not covered by water. One of the most fundamental observations which concerns visible Earth is that of *uniformitarianism*. This concept suggests that the present is the key to the past. Any structure in old rocks must have been formed by processes identical to those presently occurring on Earth. We note the process of erosion, soil deposition, and the effects of earthquakes and realize that geologic development, while uniformitarian, recognizes differences in the speed and intensity by which geologic processes occur. We see a deep valley such as the Grand Canyon and can realize how early geologists believed that it was created by great earthquakes. The fact that water flows in the valley was readily explained by earlier scholars of geology through comparison of the low area with the higher surrounding land. We now take a closer look at nature and note the material carried by muddy rivers and streams and conclude from the amount of accumulated sediment that rivers eroded the deep valleys and formed the present beds. The main process occurring today on Earth's surface is clearly erosion as evidenced by the sedimentary deposits located at the mouths of rivers and streams. Earth is constantly being worn down and theoretically all topography could be removed in forty-four million years. However, erosion has been going on for billions of years. The work of the rivers, easily the most important erosion agent, has its work interrupted through the uplift of the continents. Thus, much of the science of geology is centered around the struggle between the forces of erosion and the forces of uplift, including earthquakes, volcanoes, and the folding of rock layers.

THE GEOLOGIC PROCESSES

The geologic processes which operate upon and within the crust of Earth are often grouped under three main headings: *gradation, diastrophism,* and *vulcanism.* A fourth heading, *earthquakes,* could possibly be included; however, earthquakes are generally interrelated as an effect of a combination of the three processes.

Gradation includes that group of processes which tend to lower or smooth Earth's surface. The term encompasses the opposing processes of *degradation,* erosional processes which involve the wearing down of rocks by water, wind, and ice; and *aggradation,* the building up of low spots in rock layers by the accumulation of sediment which is deposited by the action of water, wind, and ice.

The second process, called diastrophism, includes mountain building processes which result from displacement (faulting) or deformation (folding) of Earth's crust. The crust of Earth is considered solid, but rocks which compose it are plastic under great pressure. When the pressure and the accompanying high temperature exceed the limit that the rock layers can withstand, the rocks are prone to bend. Slight bending over a long area is called *warping,* a process directly involved in early stages of mountain building. *Folding* is much like warping but is more intensive. It is generally caused by forces which act mostly horizontally in Earth's crust. Much of the knowledge about Earth is derived from the mountains that rise majestically above Earth's surface, since some of Earth's lower crust is exposed through uplift. Folded mountains occur in systems of ranges that appear as a series of ridges which alternate with valleys. This type of mountain range extends far across the continents and under the oceans, such as those formed in the East Indies, and extends throughout the Philippines and Japan.

Each of the world's major ranges of mountains has resulted from deformed *geosynclines,* elongated regions of shallow ocean basins that are still sinking and being filled with sediment. Such a subsiding trough is illustrated in Figure 22.1.

Sediment has formed rock layers which are about 40,000 feet thick in New York and decrease to about 4000 feet in thickness in the West. Other sediment has formed rocks which are 30,000 to 50,000 feet thick throughout the world. Rocks which represent the same time span are considerably thinner in the plains

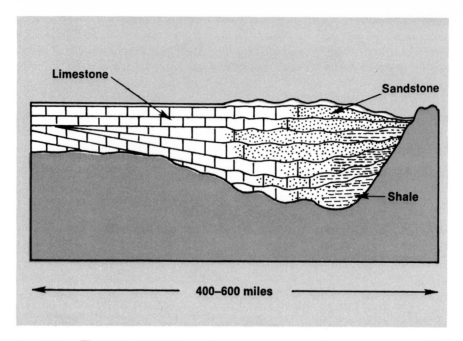

Fig. 22.1 **The geosyncline is a common occurrence in various geographic areas. The weight of the sediment that is deposited causes the geosyncline to sink continuously.**

areas that adjoin the regions of thick sediment. A detailed study of the variations in thickness shows that the sediment accumulated in shallow water. This observation is revealed by fossils, ripple marks, and wave markings that are imprinted in the layers. How is the accumulation in shallow water possible? A generally accepted theory is that the sea floor sank due to the weight of the sediment. Also possible is the hypothesis that the rate by which the sea level rose was equal to the rate that the sediment was laid down. However, available evidence discounts this theory. The conclusion that geosynclines are the result of the sinking of Earth's crust under the great weight of sediment indicates a stress on the layers of Earth caused by external forces. Mountain building, then, is caused by external stresses as well as forces within Earth.

As the geosyncline continues to form under pressure from the weight of the sediment and depresses to a depth of 20,000 feet or more, compressional forces present in the inner crust cause the adjacent layers to move upward and lap over onto the trough that is formed. The lower layer of sediment heats up from the warm interior of Earth and melts, forming *magma*. The folding and splitting of the upper layers permit the intrusion of the magma, and it rises toward the surface in the form of batholiths and other igneous bodies or perhaps reaches the surface in the form of volcanoes. Erosion eats away at the newly formed mountain, whose height maintains equilibrium as the geosynclinal cycle continues. The sequence of events is known as an *orogeny,* an important mountain building process. The massive lateral crushing, the fracturing, and the slipping continue today along the elevated regions that surround the Gulf and East coasts of North America.

The third geologic process is known as *vulcanism*. This term refers to all movements of molten rock and the formation of solid rock layers from the molten state. This igneous activity is divided into two categories: volcanism, surface deposition or lava flow; and plutonism, deep-seated activity which takes place several thousand feet or deeper inside Earth. The molten rock which cools under the surface of Earth and produces magma is mainly composed of silicates. About 11% of its composition is steam and other gases dissolved under pressure in addition to crystals that were previously formed in the inner regions. The temperature of magma ranges from 900°F to 2500°F, according to the chemical composition of the magma, and produces such rocks as granite and basalt or separates from a homogeneous solution into regions of unlike kinds by a process known as magmatic differentiation. This latter process is responsible for the forming of the different kinds of igneous rocks. Magma may also undergo other changes as chemical reactions occur between it and other rocks with which it comes in contact, including other magmas.

Magma is often lighter and naturally more fluid than the solid rock which surrounds it. As a result of lesser density it tends to rise in the crust of Earth and is aided in this venture by the pressures which act upon it. During the formation of mountains, magma is squeezed from its deep reservoirs upward into areas of lower pressure. It expands and releases some of its corrosive gases which eat upward into surrounding rock layers. As the hot fluid reaches shallow regions where cracks may exist, it moves more rapidly upward and emerges as *lava*. The molten rock cools rapidly on contact with the air and water and maintains the small void spaces from where the trapped gases are released.

The molten rock which cools deep within Earth takes on an entirely different appearance. The structure of the cooled magma is that of tiny discrete particles, mostly crystalline in nature. The size of the crystals that are formed varies but generally becomes larger if the rate of cooling is slow.

Igneous rocks, those formed from magma, fit into two categories. First, those rocks which formed from cooling lava are known as *extrusive* rocks. Second, *intrusive* igneous rocks are those that formed within the crust of Earth and are surrounded by older rock which they have entered as intruders, or invaders. The classification of intrusive bodies is accomplished according to their geologic occurrence, which involves their shape, size, position, and relation to the rock which encloses them.

Several of the bodies are common enough to warrant a brief discussion. If an igneous intrusive body of rock is so large that more than forty square miles of its surface are exposed on Earth's surface, geologists refer to it as a *batholith*. This body undoubtedly supplies the magma that forms smaller intrusive bodies at various levels in the crust. Batholiths originate during an orogeny, the sequence of events that produced Earth's continents and ocean basins, as intense forces present cause folding of the layers of surrounding rock layers. Because of their extremely large size, these intrusive bodies cool very slowly, a feature which produces coarse-textured rock.

Those intrusive bodies whose surface exposure is less than forty square miles are known as *stocks.* Other flows of magma which are partially exposed by erosion pass downward continuously and form pipe-like bodies called volcanic *necks.* Other flows fill cracks or fissures in the outer crust and form *dikes. Sills,* still another class of deposit, resemble dikes in that they are flat and thin, but they intrude upon other parallel rock layers. Sills do not have to be horizontal but can be vertical like the dike since the rock in which they intrude may be folded. Each of the types of igneous intrusions discussed is illustrated in Figure 22.2.

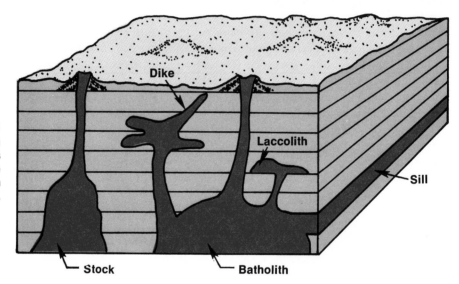

Fig. 22.2 Common intrusions which often make their way through Earth's crust.

A vast amount of knowledge can be obtained about igneous processes from the volcano. It offers the only direct evidence about the existence of magma within the crust, for it is a central vent by which heated rock and magma emerge to the surface. Most of the main volatiles in magma are nothing other than water that escapes in the form of steam along with some carbon dioxide and sulfur gases. If the magma erupts very suddenly, it may form a frothy substance which solidifies into a light rock that actually will float in water. The formation of *pumice,* as the light rock is known, is much the same as the formation of froth produced by opening a warm bottle of soft drink. As the lava is deposited on Earth's surface, it forms a conical hill or mountain with a funnel-shaped center called a *crater.* On the other hand, the fluid release of magma along with super-heated steam and gases which result causes total destruction over large areas since the eruptions may be quite violent. For example, in 1902 an eruption of Mt. Pelée on Martinque Island destroyed the city of St. Pierre and killed all but one of the city's 30,000 inhabitants. The survivor reportedly was a prisoner who was locked behind the strong, thick walls of the prison.

our physical earth

Many mountains formed by volcanoes are things of beauty as they lie dormant and are exposed to nature. Nevertheless, the threat of reeruption is always present, the towns which surround the quiet mountains are always on the alert for the telltale earthquakes and tremors which usually accompany the awakening movement of magma deep within Earth's crust. Volcanoes comprise only an estimated 0.1% of the total volume of known igneous rocks, but they are very significant in that they are a great source of knowledge and of fertile soil.

Great tremors in Earth's inner region and across its surface are by far the most terrifying naturally occurring phenomena to which man, of ancient and modern times, has been exposed. As fearful as Earth tremors are, their occurrence has led scientists to many clues about the composition and structure of Earth's interior.

The earthquake, as it is felt by mankind on Earth's surface, is the result of some fundamental geological process, never the actual cause. Locally, the quake may set off landslides, avalanches, explosions, and, among the worst, severance of water lines along with the inextinguishable fires that always accompany such major calamities. The actual cause may be the collapse of a layer of rock into an underground pocket produced by some erosional process, an eruption of a volcano and the resultant explosive tremors that the vast release of energy produces, or, as is typical, a *fault,* a fracture in the rock layers and the displacement of the layers that follows.

The destruction caused by an earthquake depends on many factors: duration, frequency, acceleration, velocity, origin, and, of course, intensity. The more violent quakes cause significant changes in Earth's topography. For instance, giant fissures or cracks open Earth's surface and gigantic landslides and avalanches engulf widespread areas. Also the great compressions that occur internally cause great ridges to appear or vast depressions in Earth's surface to occur where before was level land. At other times, vertical faults several stories high occur that were created by uplifted fractures. Gigantic icebergs are set free as the tremors tear them from the glaciers to which they were attached. Most vivid of all is the manner in which great rivers and streams are caused to flow backwards and thus to create great lakes in regions which before were bypassed by the direction of river flow.

Most of the widespread destruction from earthquakes results from the spectacular tidal waves (called tsunamis) caused by disturbances under the seas. The long, low waves which the quake causes travel at great speeds and constantly break on the shores of islands as well as on mainlands. Vast destruction accompanies the repeated poundings caused by the waves, such as those which struck Japan in 1703 and 1896 and the Aleutians at the close of World War II.

The frequency of earthquakes may be as many as 3000 per day, with about 450 strong enough to be felt by people in the immediate vicinity of their origination, referred to as the *epicenter.* About 100 per year are severe enough to cause destruction.

No correlation with other natural phenomena has been determined which may lead to the ability of the scientist to predict when earthquakes will occur.

However, sunspots, tides, planetary positions, and weather are under consideration as possible creators of the internal disturbances and resulting tremors.

Although earthquakes have had a decided effect on the topography of Earth's surface, this catastrophic mechanism is a minor one when it is compared with the slow, gentle processes of erosion and folding which played the major role in sculpturing our lands as we know them.

MINERALS

Elements, the smallest units of matter which are found in nature, were discussed in the chapters related to chemistry. Atoms and their particles were also discussed earlier. In nature, atoms of the various elements are combined to form minerals. A *mineral* is a naturally occurring inorganic substance which has ordered internal structure and a chemical composition that can vary only within defined limits. Each mineral has a set of physical properties that are also fixed within limits. A substance must meet all criteria to be classed as a mineral. Substances such as coal, gas, and oil are classed as mineral fuels rather than as minerals since each contains a dominance of organic minerals.

The geologist has collected and identified over 2600 minerals and discovers about ten new minerals per year. About 50 of the known minerals are elements; the rest are compounds which contain two or more chemically combined elements.

Minerals are crystalline substances, a fact which means they have an ordered internal structure. Substances such as glass do not fit the criteria because of variances in internal structure. Analysis with X-rays which study reflection and refraction produced by the mineral lets the scientist be aware of the substance's internal structure.

Under favorable conditions most common minerals grow as crystals, that is, bodies that have smooth faces which reflect their internal structure. The angle between corresponding faces of any certain mineral is constant regardless of the specimen's size.

The most abundant elements which constitute Earth's crust are presented in Table 22.1. More than 98% (by weight) of the rocks and minerals of the crust are formed from the eight elements listed and the same elements account for all but an insignificant amount of the crust in terms of volume.

According to the abundance of the elements present in Earth's surface, obviously most of the minerals which form rocks are composed predominantly of oxygen and silicon. According to abundance, aluminum and probably one of the other common elements such as sodium or potassium are present. The assumption is quite correct. Rocks are formed from minerals that naturally group themselves and contain over 90% oxygen by volume. This condition exists because of the size of the oxygen atom (ion). Therefore, the possible combinations by which oxygen can arrange itself along with other atoms depend on the relative size of the atoms which attach to the oxygen atom. For instance, the silicon atom is so small that four oxygen atoms can attach to it (see Fig. 22.3).

Plate 3   Olivine, hornblende, quartz, and mica have different internal structures.

Plate 4   The number of cleavage planes differs in each specimen presented. Asbestos (none), obsidian (conchoidal), mica (one), orthoclase (two), and halite (three).

Plate 5 Shown from left to right above are diamonds, topaz, orthoclase, fluorite, and gypsum. The second row shows corundum, quartz, apatite, calcite, and talc.

Plate 6 The rock-forming minerals found in Earth's crust. Orthoclase feldspar, plagioclase feldspar, purple quartz, pyroxene, muscovite, biotite, gypsum, halite, calcite, chlorite, sepentine, dolomite, and kaolinite (Listed from top left to bottom right).

Plate 7 Bowen's reaction series. The degree of lightness or darkness of color is an indication of the relation of preceding specimens, though the observation is not totally reliable because of the potential presence of impurities.

Plate 8 The common sedimentary rocks shown in this illustration are displayed in the order discussed in the chapter.

Plate 9 Some common metamorphic rocks. Anthracite, gneiss, marble, quartzite schist, and slate.

TABLE 22.1

The major elements present in Earth's crust.

Element	Percent Abundance by Weight	Percent Abundance by Volume
Oxygen	46.60	93.77
Silicon	27.72	0.86
Aluminum	8.13	0.47
Iron	5.00	0.43
Calcium	3.63	1.03
Sodium	2.83	1.32
Potassium	2.59	1.83
Magnesium	2.09	0.29
All Others	1.51	Negligible

This unit, SiO_4, is the building block of the vastly abundant silicate minerals. In all the silicate minerals, silicon and oxygen are joined in a tetrahedral (four-sided) arrangement. Oxygen frequently reacts with atoms much smaller than the silicon atom. The ratio by which oxygen and the tiny atom combine and form ions or molecules is often 3 : 1 and thus the two elements form XO_3.

The silicon-oxygen tetrahedrons may exist as single units in some minerals or as multiple units in others. The difference in internal arrangements such as number and pattern causes the minerals involved to have varying physical properties. Some examples of minerals with different internal structures appear on Plate 3.

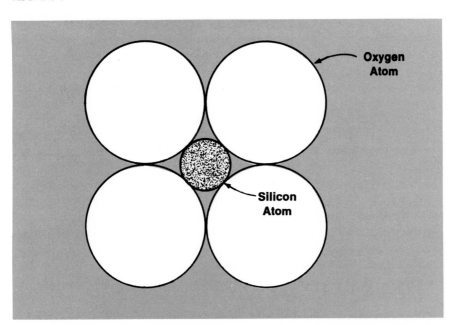

Oxygen
Atom

Silicon
Atom

Fig. 22.3 **The tetrahedral arrangement of silicon and oxygen atoms is a common occurrence in nature.**

The most common minerals composed of the silicates are the feldspars. This class of minerals is closely related to quartz in terms of structure. However, aluminum atoms replace some of the silicon atoms in the tetrahedron, and the elements potassium, sodium, and calcium are also present.

The identification of minerals and the rocks they form is accomplished in several manners. X-ray studies, previously mentioned, are perhaps the most effective, but other methods are more commonly used. One manner of determination is known as petrographic analysis in which thin sections, about 0.3 mm thick, are prepared and viewed under the microscope. Strangely enough, most specimens are transparent at this thickness and a precise study of the crystals present can be accomplished.

The geologist has been able to identify the rocks and minerals by means of correlated physical properties he has noted. A brief discussion of some of the more common properties which are used in identification follows:

1. The internal structure of the rocks and minerals affects the manner in which they tend to break. Some of the specimens break in definite directions along smooth planes of weakness in their internal structure. This property is known as *cleavage*. Plate 4 displays mica, orthoclase, and halite which have one, two, and three cleavage planes, respectively.

2. A mineral which lacks cleavage breaks in some irregular manner. A common type of *fracture,* called *conchoidal,* is the hollow, arc-like break that occurs in glass, quartz, or obsidian. Other specimens fracture into splintery or fibrous faces, such as asbestos (see Plate 4).

3. The *color* of freshly broken surfaces of some minerals may be a diagnostic property. Specimens such as galena and magnetite are reliably determined by color, whereas quartz, calcite, and many others usually contain impurities which produce a wide range of color in the specimen.

4. Many of the minerals when crushed to powder form turn white or yield a paler color than that which is their ordinary appearance. Others produce a color in powdered form which is quite different from the color they appear when viewed as a large fragment. The investigator carries a piece of rough, unglazed porcelain which will powder most rocks and minerals as they are rubbed on its surface. The color of the powder which results is called the *streak*.

5. *Luster* is the manner in which a mineral or rock reflects light. The two main types of luster are metallic and nonmetallic. Among the terms used to identify the luster of a specimen are the following: metallic, bright or dull; nonmetallic, vitreous (glassy), earthy, silky, pearly, waxy, or resinous. Many other terms have been proposed and are used in determinative mineralogy.

6. A mineral resists scratching of its surface in varying degrees. This property of the mineral is called *hardness*. It is a difficult property to use, since many variables exist, including the amount of force used, the shape of the scratches produced, and the variations which might occur on different edges and faces. The mineralogist has established a relative scale of hardness, called *Mohs' scale* after the geologist who proposed the method of measure in the nineteenth century. The scale appears in Table 22.2 and typical specimens are shown on Plate 5.

TABLE 22.2

Mohs' scale of hardness.

1–Talc	3–Calcite	5–Apatite	7–Quartz	9–Corundum
2–Gypsum	4–Fluorite	6–Orthoclase	8–Topaz	10–Diamond

In order to determine the hardness of a mineral, the investigator identifies on the hardness scale the softest mineral that will scratch the specimen. The geologist in the field carries a kit that contains all of the minerals listed on Mohs' scale excluding the diamond for obvious reasons. In addition, he usually has available the common materials with which he can determine the relative hardness of a substance such as his fingernail (2–3), copper penny (3–4), knife blade (5–6), and a piece of quartz (7).

7. Another reliable property of rocks and minerals used in identification is *specific gravity,* the weight (mass) of a given volume of a mineral compared to the weight (mass) of an equal volume of water. Quartz has a value of about 2.6–2.7; gypsum, 2.2–2.4; olivine, 3.2–3.6; and magnetite, 5.0–5.2. The density of each example cited is greater than that of water since each mineral's specific gravity is greater than 1.0.

8. Other useful properties which are commonly tested include electrical conductivity, magnetic tendencies, fusibility, solubility, reactivity, and the tendency of a substance to fluoresce under ultraviolet light. Fine lines, called *striations,* appear on the faces of some minerals and are used in final identification.

All of the minerals which have been identified are present among the rocks. Only a few dominate and are the essential constituents of the rocks. These very common minerals form a group called the *rock-forming minerals.* Some of the more abundant minerals which form the preponderance of the rocks are illustrated on Plate 6.

An *ore* is a mineral or a rock from which various metals and nonmetals can be extracted profitably. The volatiles, including water along with molten quartz that is present in magma, may have caused formation of veins of ore minerals. The outer regions of batholiths are commonly examined for valuable ores since these areas cooled rapidly and solidified first. As the interior of the body cooled, much of the volatiles was trapped, a condition that created a mechanism through which many of the elements could move and concentrate that did not readily react with the rock-forming minerals. This group of elements that did not fit the internal structure of the abundant rock-forming minerals includes gold, copper, lead, zinc, silver, and sulfur. They were carried as suspensions between the large crystals by the fluid quartz and volatiles, and were deposited in the lower extremities of the batholith by gravity. Other small particles of these elements were forced to concentrate along the edges of the great magnetic bodies as the fluids which contained them were squeezed outward by contraction and expansion of the larger minerals.

The miners of the ores search for the large igneous bodies and excavate the masses in search of the valuable veins that formed some of the minor elements that reacted with sulfur and oxygen, both of which were contained in the gases that formed the volatiles. Iron atoms which reacted with oxygen formed magnetite, hematite, and limonite. Iron and sulfur also combined to form pyrite, a common crystal known as fool's gold because of its resemblance to the valuable element. Galena, the major ore of lead, is formed as lead sulfide and often contains silver as an impurity. Cassiterite, tin oxide; sphalerite, zinc sulfide; and bauxite, hydrous aluminum oxide, are the common ores of tin, zinc, and aluminum, respectively. The ores of uranium are the most valuable at the present time. Pitchblende and carnotite are the most yielding and abundant ores of this rare element. They are basically related to magmas and are sought in regions where igneous rocks are abundant.

ROCKS IN GENERAL

Earth is mostly composed of an aggregate of minerals that are massed as a crystalline melt or are cemented together. The wide variations in physical properties and appearance that best describe each specimen depend on the amount and kind of minerals present as well as the manner in which the grains are held together. Rocks are generally composed of two or more minerals which may or may not be associated with a mass of natural glass.

The geologist's chief interest in the rocks is in the information they provide about the physical conditions which existed on Earth when they were formed and laid down. The deciphering of the rock record has yielded fascinating facts about the events that represent the history of Earth. The clues the geologists have fitted together paint a picture of major calamities, such as devastating earthquakes; vast glaciation; great floods; and dry, arid climates which were prevalent in various eras of time. The site where the city of Glasgow, Scotland, is built was once a desert, and areas of Canada were warm enough for tropical plants to flourish in the past. However, the preponderance of rock layers, according to most geologists, was laid down by slow, uniform procedures and not by major catastrophies as the reader might surmise.

The rocks of which the crust of Earth is composed are classified into three types in terms of their origin. Two types of rocks were formed by processes well under the surface of the crust, both of which give valuable information about conditions that exist within Earth. Igneous rocks are composed of solidified magma, and metamorphic rocks are those that have undergone physical changes by virtue of high temperatures and great pressures to which they were exposed under Earth's surface. The third type, sedimentary, is deposited near or upon the surface of Earth by the action of water, wind, ice, and several other mechanisms. Sedimentary rocks are mostly formed on Earth's surface from accumulations of gravel, mud, and sand, all of which are the product of erosion agents from pre-existing rocks. Still other rocks of this type, such as limestone and gypsum, are composed mostly of material deposited from solution. Metamorphic rocks are in effect igneous, sedimentary, or other metamorphic rocks altered by great heat and pressure. The variety of metamorphic rocks that can exist, then, is unlimited. The manner in which the three types of rocks may interact is illustrated in Figure 22.4

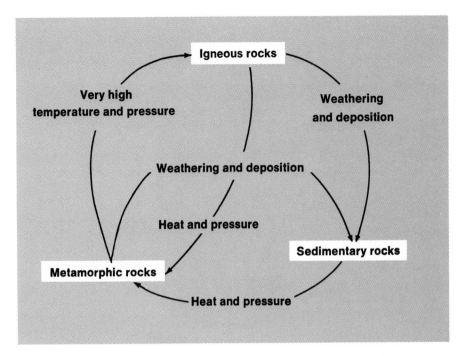

Fig. 22.4
The rock cycle illustrates the manner in which different types of rocks are formed.

IGNEOUS ROCKS

Igneous rocks result from the cooling of hot magma. They are formed from high temperature melts deep in Earth's crust, most of which must be deeply imbedded to crystallize as they generally are found. The classification of igneous rocks is done by means of their textural features such as color, mineral composition, texture, and by their density and strength. Chemical composition is also used to classify the rocks whenever practicable. With the exception of volcanic glass, igneous rocks are crystalline aggregates composed of silicate minerals that have grown together in a melt. A combination of crystalline texture and silicate composition is typical of igneous rocks, since silicates do not crystallize en masse at a very low temperature. The light colored rocks with igneous origin tend to be composed of sialic (silicon-aluminum) minerals including quartz, plagioclase feldspar, and muscovite mica. Others which are dark colors contain simatic (silicon magnesium) minerals of the ferromagnesian group.

Basaltic magma is the parent of all igneous rocks as was surmised through experimentation and observation by N. L. Bowen of the Smithsonian Institute. He originated the concept of a reaction series in which the order of crystallization of igneous rocks is a function of decreasing temperature. A simplified schematic diagram of the formation order appears in Figure 22.5. Note that the feldspars of varying chemical composition were formed at the same temperatures as olivine, pyroxene, amphibole, and biotite. The reaction series on the right is a continuous one; that is, all compositions from calcic to sodic are present. On the left side, the reaction series changes from one mineral to another in discrete steps.

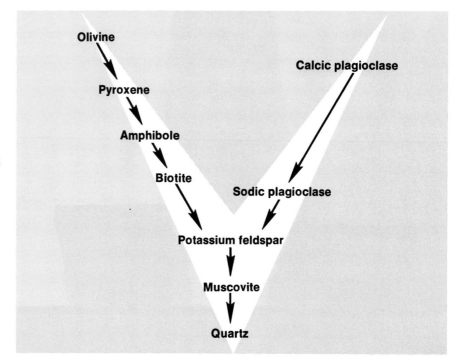

**Fig. 22.5
A schematic representation of Bowen's reaction series.**

The Bowen series includes several of the main igneous rocks. Others found abundantly in Earth are felsite, rhyolite, peridotite, basalt, and obsidian. A brief discussion of each type follows as well as an illustration of each specimen on Plate 7.

1. Basalt is the most abundant of all lavas. It was formed as the lava congealed from its volcanic eruption or as it emerged from fissures. The rock is considered the primary bedrock upon which the continents rest.

2. Diorite is dominant with plagioclase. It contains various dark minerals in abundance.

3. Felsite is extremely fine-grained and is generally banded because of congealing of the lava flow.

4. Gabbro is a granular rock composed chiefly of plagioclase and pyroxene.

5. Granite is another igneous rock of granular texture. Its two most abundant minerals are feldspar and quartz; therefore, it is typically light colored. Some granites are also of metamorphic origin.

6. Obsidian is in reality natural glass. It is characterized by its conchoidal or curved fracture. Most specimens are black with a high luster but they appear transparent as a thin section.

7. Peridotite consists largely of olivine and pyroxene. One variety found in Africa and Arkansas accompanies such metals as chromium, nickel, and platinum as well as diamonds. The rock is characterized by the unusual absence of feldspar.

8. Rhyolite contains fine granules which are predominantly quartz with a small amount of plagioclase and biotite mica also present. The color may be buff, cream, purplish, gray, or green, depending on traces of other minerals which are present.

SEDIMENTARY ROCKS

Sediment is an accumulation of mineral or organic matter that has been transported mechanically to sites of deposition as fragments or that has settled from a liquid as a chemical precipitate while actively carried in solution. Other sediment has been laid down biochemically.

The development of sedimentary rock from lithified sediment occurs in discrete stages of development. First, the sediment is prepared for transport through weathering, chemical precipitation, or biochemical activity. The next stage is its transportation to depositional sites by running water, moving ice, wind, gravity, soil flow, or by organisms. After being transported, the sediment is deposited through loss of velocity of its carrier, loss of volume, gravity, melting, or evaporation. The agents of *lithification* which transform the sediment into rock take over. Some processes are cementation, compaction, chemical changes, organic activity, and recrystallization.

The sites of deposition include the bases of various slopes, glaciers, or streams and at the mouths of rivers where large deltas form.

Some of the most common sedimentary rocks are discussed briefly below and displayed on Plate 8.

1. Sediment from limestone forms a soft and porous rock known as chalk. Large fragments of sea shells also are cemented together to form coquina, generally washed to shore by the oceans.

2. Even though coal, which is about 90% carbon, is organic in origin, it is classed as a sedimentary rock because it is formed in layers. Peat is an aggregate of slightly decomposed plants whereas coal is formed by plant remains that have more thoroughly decomposed and have been compressed into layers. The grade of coal seems to be related to the depth to which it is buried by other sediments and the resultant heat and pressure to which it is exposed.

3. Conglomerates are cemented gravel that is predominant in quartz and chalcedony. The fragments vary in size with the space between larger particles filled with grains of sand.

4. Dolomite resembles the mineral limestone in many ways; however, the metal magnesium has replaced the less active calcium present in limestone, a variation in composition which causes dolomite to be harder and heavier.

5. Gypsum is deposited in thick layers which alternate with other sedimentary rocks that are also formed from evaporation of salt water. Beds of rock salts, accompanied by gypsum, are found in Texas and New Mexico.

6. Limestone can be formed by chemical precipitation but is more often formed by the accumulation of shells and skeletons of organisms which extract calcium carbonate from the oceans. The fragments are surrounded by finely crushed shells which have been ground into powder.

7. Sandstone is composed of grains of sand which contains quartz as the usual mineral. Other sandstone contains largely gypsum or coral. The color is determined by the mineral that cements the grains together, but is generally red, brown, or green.

8. Shale is the most common sedimentary rock that occurs everywhere. Its composition is that of mud, the finest of all sediment composed of silt and clay. Mica and quartz are present in varying quantities, though usually minimal.

METAMORPHIC ROCKS

The third type of rocks is the metamorphic rocks, those that result from any origin but have been visibly changed by great pressure and temperature while the rocks are in a solid state or by fluids that circulate through the solid rocks under these same conditions. Metamorphism involves any process which, as a result of geologic environment, produces changes in original texture or leads to new minerals being developed, or both. The intensity by which metamorphism takes place is usually so great that the nature of the original rock is difficult to determine.

The changes in the rocks include partial or complete recrystallization whereby the grain size of the minerals varies from fine to coarse and the crystalline alignment differs from that of the parent rock. Bending or other parallel arrangements in structure also develop which were not present in the original rock. This property, referred to as *foliation,* causes the rocks to separate along their parallel sides; therefore, they possess cleavage.

The process of recombination is responsible for the creation of new minerals, produced by interactions that occur between rocks of different chemical composition. Most metamorphic rocks, however, are similar in chemical structure to the rocks from which they were transformed.

Several factors determine the character of the rocks exposed to metamorphic processes such as the composition and structure of the original rock, the processes involved, and the degree to which the processes exploit the rock layer. Some typical metamorphic rocks are mentioned in the ensuing discussion and specimens of each are shown on Plate 9.

1. Coal found in abundance in Western Pennsylvania is of a high-grade soft character and is named *bituminous. Anthracite,* or hard coal, is supplied by an area about a hundred miles eastward into the Appalachian Mountain region. This type of coal differs from the bituminous grade in that anthracite practically defies burning. Other characteristics of anthracite are its greater hardness, its

lack of cleavage, its lesser tendency to fracture, and its greater luster. Both types of coal occur in layers between shale and sandstone beds and were laid down in the same geologic period, the Pennsylvanian. The difference in the types of coal is due to the folding and resultant deformations that occur during mountain building, including the increase in temperature and pressure to which the layer is exposed. Also, the gases escaped from this layer more slowly, and the coal was changed in chemical composition and physical properties. Anthracite, then, is a metamorphosed offspring of bituminous coal.

2. Gneiss is the most coarsely banded metamorphic rock and is usually a thing of beauty. Most gneiss contains bands of feldspar and quartz which are composed of larger granular particles than ordinary. The specimen is light in color unless biotite mica, hornblende, or garnet is present in abundance.

3. Marble is formed from the metamorphism of limestone, dolomite, or both. The resultant specimen generally contains streaked light patches alternated with darker ones. It commonly contains pyroxene, amphibole, and serpentine in varying quantities.

4. Quartzite results from the alteration of quartz sandstone. Its appearance is glassy, and it has a sugary-texture. Unlike sandstone, it breaks across the grains rather than around them. The geologist can distinguish quartzite from silica-cemented sand by the rocks which are found in the immediate area.

5. Schist is similar to gneiss except for the dominance of micaceous materials. Mica schist is the most common metamorphic rock. Most of this type of schist is the result of transformed shale and tuffs; therefore, it is similar to slate.

6. Slate is very fine-grained and finely foliated; thus it separates into thin sheets. The grains are visible only under a microscope, and the specimen shows some properties common to the sedimentary rocks in that it is layered and contains fossils. Most slate is the metamorphic form of shale. The black variety contains carbon whereas the red or green types contain ferromagnesian minerals.

THE PHYSICAL LAWS OF GEOLOGY

The early scholars who studied Earth had several purposes in mind: to determine the likely regions where the useful minerals might lie; to uncover the secrets of the history of Earth; and to learn about its origin.

The bits of information gathered over the centuries were eventually set to principle and ultimately led to the development of geology as a science.

One of the earliest observations, uniformitarianism, has been previously mentioned. As a matter of review, the concept points out that geological products visible about us such as minerals, rocks, and physical environment are results of geological processes comparable to those that take place on Earth today.

Nicolaus Steno, one of the earliest esteemed geologists, first stated the *Law of Original Horizontality* in 1669. He declared that rocks formed from sediment and carried by water to depositional sites were laid down in strata in a horizontal fashion that was generally parallel to the surface on which it accumulated. An outstanding illustration of this law is presented in Figure 22.6. Steno also stated the *Law of Superposition* which contends that in a normal marine depositional sequence, the younger strata lie over the older, unless, of course, there have been disturbances at the site in the form of folding or overturning since the sediments accumulated. The law applies to both sedimentary rocks and igneous rocks, particularly successive lava flows. In the illustration, Figure 22.7, the horizontal layers have been exposed to tremendous internal forces which created uplifting and the resultant overturning. Such features as mud cracks, ripple marks, and wave marks offer mute evidence of the stresses to which the surface was exposed. In addition, the escaping of volatiles from molten lava leaves cavities in the top of the individual layers.

Though more of a historical nature than physical, another known relation between various layers of sediment is the difference in assemblage of fossils present in each deposit. The *Law of Faunal Succession* takes into consideration the fact that the plants and animals lived in particular environments and that the various species evolved in a given manner. The younger layers contain fossils which display evolutionary tendencies, including a greater complexity in the

Fig. 22.6 The variation in color and banding provide evidence of sedimentary deposition.

our physical earth

organisms, whereas the older layers of rock contain more primitive organisms. The law designates that each sedimentary rock layer contains characteristic fossils that reveal the time and place the organisms lived. Outcrops at various places on Earth's surface are correlated through the law of faunal succession.

The records available by analysis of the rocks and their contents yield gaps where layers of sediment have been removed by erosional forces or where deposition failed to occur. These gaps of unrecorded time are called *unconformities,* the conspicuous absence of layers of deposition during various environmental conditions. The unconformities make plain that geologic history will never be complete.

Environmental conditions were similar over large areas of Earth's surface at the same geologic time, as is the case today. The geologist relies on the *Law of Continuity,* a generalization which points out that a sedimentary rock unit was originally continuous but parts may be missing because of erosion and faulting. Some formations stretch over hundreds of square miles, others over even larger areas. Both the Law of Superposition and that of Faunal Succession aid in the correlation of identical depositional formations.

Fig. 22.7 **The folding and uplifting of sedimentary and igneous rocks reveal the great turmoil always active near Earth's surface. This formation is undergoing decided distortion which will continue indefinitely.**

our physical earth   473

The last law, the *Law of Cross-cutting Relationship,* concludes that as one geologic body cuts another, the body which is cut is older than the body which does the cutting.

Several of the concepts presented in this immediate discussion are illustrated in Figure 22.8.

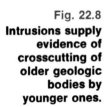

Fig. 22.8
Intrusions supply evidence of crosscutting of older geologic bodies by younger ones.

OUR MINERAL RESOURCES

As vast as is Earth and its seemingly endless supply of minerals, its limitations are now recognized. Earth is in reality a closed system in that no new resources are being added, at least at a rate appreciable to that of consumption. The length of time that our resources last depends on our culture and the way it conserves them. Technology must find a way to remove the minerals from neglected low-yield sources and a way to reprocess industrial and municipal wastes (see Chapter 24).

At the present rate of consumption, by the year 2000 A.D. our available supply of the following minerals will probably be exhausted in the United States: aluminum, copper, gold, lead, nickel, platinum, silver, tin, and uranium.

The minerals which constitute Earth's surface were first used by man of the Paleolithic period (perhaps 500,000 B.C.) as evidenced by the paints he used. Neolithic man (about 5000 B.C.) designed his tools from chert, flint, obsidian, and quartz as did his ancestors, but he also discovered the malleable property of copper and gold. With the Bronze Age (3,000 B.C.), man learned to cut stone for building blocks to build his magnificent temples and pyramids. The proficiency of polishing gemstones to increase their value also was established during the Bronze Age.

Expeditions in quest of various minerals started as early as 2000 B.C. as prospectors ventured to the Red Sea, the Sinai Peninsula, and Africa from their homes in Egypt and Babylonia. They dug to depths approaching 1000 feet into the sides of the mountains for emeralds and turquoise without the aid of machinery.

The mining of gold and silver in Greece was a thriving industry at Cassandra from 2500 B.C. to 360 B.C., and was accomplished by tunneling through hills into the veins of the rare metals. Since the rock strata varied so greatly in different locations, there was undoubtedly an interest in creating a taxonomy to identify the differences. The available records represent the initiation of geology and reveal the classification of minerals into sixteen categories. Aristotle, in his writings, discussed the oil fields and asphalt deposits of Albania and suggested possible utilization of them. De Metallica, 1556, prepared a treatise which concerned the origin of rocks and minerals. The theories he proposed were vague but are considered acceptable by today's scientists.

Nicolas Steno (1631–1687) authored the first discussion of the history of Earth, based on fossils he observed embedded in the rocks of Earth's crust. William Smith (1769–1839) is noted for determining procedures to apply in the search for oil, gas, and water.

From the limited discussion above, the reader should be able to realize that the mineral resources of Earth have been exploited for over 4000 years. There is no second crop of minerals, and yet more metals have been mined in the last fifty years than in all preceding history. At the present rate of consumption, disregarding the increased need for the ores because of inevitable increases in population and additional uses, our next generation will find the supply exhausted. We have no choice but to conserve that amount of natural resources which still remains and to find better methods of recycling them.

CHAPTER SUMMARY

James Hutton (1726–1797) is commonly known as the founder of scientific geology. The principle of *uniformitarianism*, the doctrine that all changes of Earth's topography are the results of the same physical laws that are in operation today, is attributed to him. Uniformitarianism, in general terms means that the present is the key to the past.

Earth's surface is constantly undergoing change as the result of numerous geologic processes. *Gradation* includes the various means by which the surface experiences degradation by erosion and is filled in by sedimentation, a process called aggradation. Gradation is generally attributed to the action of water, ice, and wind. *Diastrophism* includes all movements of the solid parts of Earth in which displacement (faulting) or deformation (folding) occurs. Some of our mountain ranges are produced by diastrophism. Other mountain ranges result from vulcanism, the movement of molten rock and the formation of solid rock from the molten state in Earth's interior or on Earth's surface. Earthquakes are the result of geologic processes, not the cause.

A *mineral* is a naturally occurring inorganic substance which has an ordered internal structure and a chemical composition that can vary only within defined limits. A mineral, thus being homogeneous, exhibits uniformity and constancy in its chemical and physical properties. The most common physical properties of minerals are *cleavage, fracture, color, streak, luster, hardness*, and *specific gravity*. Most minerals are composed of two or more of the ninety-two known elements which have reacted to form a compound. Minerals are an integral part of the rocks; however, only a few of the many minerals are the essential constituents of the rocks. Some mineral deposits are mined for the valuable elements contained in them.

Rocks are generally aggregates of minerals. Their wide variation in physical properties is a result of the amount and kind of minerals they contain and the manner in which the grains are held together. Although the geologist studies the chemical composition and physical properties of rocks, his chief interest is the record that rocks reveal about the environment of Earth at the time minerals aggregated into rock. The reading of the rock record by a trained scientist reveals many spectacular events in Earth's history. Earth's crust is composed of three main types of rocks, classified according to their modes of origin. Certain members of each type of rock grade into another member of the same origin. The three types of rocks are *igneous, sedimentary*, and *metamorphic*.

Igneous rocks are formed from the solidification of cooling molten matter deep in Earth's crust. This type of rock is further classified according to textural features or chemical composition. Density and strength are also used to classify igneous rocks. The texture of both intrusive and igneous rocks usually indicates the condition under which they cooled. The size of the mineral grains in the rock generally reveals the pressure, temperature, rate of cooling, and volatile materials present during the period of solidification. Light-colored igneous rocks usually are composed of *sialic* minerals and dark colored specimens contain *simatic* minerals. The *Bowen reaction series* includes many of the main igneous rocks.

Sedimentary rocks are formed from material ultimately derived from *weathering, chemical precipitation,* or *biochemical activity*. The sediment is transformed into rock by such lithification processes as cementation, compaction, organic activity, and recrystallization. Some common sedimentary rocks are chalk, coal, conglomerates, gypsum, limestone, sandstone, and shale.

Metamorphic rocks are formed from igneous, sedimentary, or other metamorphic rocks. This type of rock is the result of great pressure and temperature that effect material changes without the original material melting. The changes may be in the form of *recrystallization, foliation*, or *recombination*. Some typical metamorphic rocks are anthracite coal, gneiss, marble, schist, and slate.

The Law of Original Horizontality concludes that sedimentary rocks were laid down in a horizontal fashion parallel to the surface on which the sediment accumulated. The Law of Superposition contends that typically the younger strata formed from sedimentary rocks and lava accumulate in such a manner that the younger strata lie over the older; therefore their relative ages are known.

our physical earth

The Law of Faunal Succession, one of the basic laws of geologic history, stresses that each sedimentary formation contains its characteristic assemblage of fossils that reflect the environmental conditions and the time during which the plants and animals lived. The Law of Continuity points out that a sedimentary rock layer was continuous over an area of hundreds of square miles. A break in the rock layers in which various parts are missing constitutes an unconformity. The Law of Cross-cutting Relationship indicates that as one geologic body cuts another, the body which is cut is older than the other.

Earth is a closed system, meaning that no new resources of any significant amount are being added to our supply of minerals that contain aluminum, copper, gold, lead, and other valuable metals. Our mineral resources have been exploited for centuries, but, at the present rate of consumption, our supply will be exhausted in the next fifty years. Our remaining natural resources of minerals must be conserved, at the site of original deposition and in the manner in which they are consumed.

QUESTIONS AND PROBLEMS

1. Physically and chemically, how do minerals differ from rocks? Enumerate your response for the purpose of comparison. (p. 462)
2. The streak produced by a mineral is more reliable for the purposes of identification than is the color of the mineral. Can you conclude why the streak is more reliable? (p. 464)
3. Geologists who apply the various laws of geology infer that vast climatic changes have occurred throughout Earth's history. List various factors on which they might have based this hypothesis. (p. 466)
4. Sandstone, obviously composed of sand, is a substance which is finely grained and difficult to cause to adhere to other bits of sand. How, then, do sandstone fragments hold together? (p. 470)
5. A layer of rock three feet thick lies over a layer of rock one foot thick. From this observation, what hypothesis might be surmised? (p. 471)
6. What conclusions might be reached from the observation that two identical layers of rock are located fifty miles apart but the layers are 150 feet and 200 feet, respectively, above sea level? (p. 457)
7. What laws of geology apply to the conclusion that a fault has occurred; that folding has occurred? (p. 471)
8. What general characteristics would be used to distinguish an igneous rock from a sedimentary rock? a sedimentary rock from a metamorphic rock? (p. 467)
9. What observations can you envision geologists might use to locate new ore deposits? What observations might suggest a potentially new source of water? (p. 456)
10. Table 22.1 lists the most abundant elements in Earth's crust. From this observation what minerals are the most abundant? (p. 463)

sculpturing the land

—

weathering and erosion

23

GRAVITY AND THE HYDROLOGIC CYCLE
WEATHERING
THE ACTION OF RUNNING WATER
GROUNDWATER
WIND EROSION
THE WORK OF GLACIERS

vocabulary

Chemical Weathering	Erosion
Ground Water	Mechanical Weathering
Snow Line	Water Table

The surface of Earth has been boldly sculptured to produce scenes of inspiring beauty and grandeur, and scenes of harsh ugliness. For example, great majestic mountain ranges a thousand miles long reach beyond the clouds like massive pillars supporting the sky, and immense chasms, such as the Grand Canyon, scar Earth's surface. Not all the lands of Earth, however, have been carved into mountains and valleys as is revealed by vast plains that spread in all directions toward distant horizons. It is doubtful that any artist could have conceived on such a grand scale the varied and beautiful forms that nature has carved on the crust of planet Earth.

What causes this sculpturing of the land? Throughout Earth's history portions of the land have been uplifted above the surrounding land masses. Whenever this transformation has occurred, water, wind, and ice have started immediately to erode the uplifted masses. The process of erosion is a very slow process, but Earth has plenty of time. Thus, when a person sees a few rocks fall down the slope of Mount Everest, he is witnessing the destruction of the greatest mountain man has ever known.

GRAVITY AND THE HYDROLOGIC CYCLE

The old adage "Everything that goes up must come down" is beautifully illustrated in the process of erosion. The instant a mass is uplifted, gravity starts its relentless pull on the mass, a pull that never ceases until the mass is once again at the same level as its surroundings. As soon as a small piece of the mass breaks off, it inexorably finds its way downhill because of the pull of gravity, or it is dissolved and carried downhill by water. Because of the pull of gravity, the loosened material produced by weathering is transported to a lower level, a process called *erosion*. What forces are at work that cause the erosion of uplifted masses? Basically, there are three erosional agents: water, wind, and ice. Of the three, the most important is water.

Where does the water come from that falls upon the land? Of course, it comes from rain clouds in the sky, but the real question is, "Where do the clouds come from?" Clouds are actually part of a great cyclic system that operates on our planet—the *hydrologic cycle*. The source of energy for the hydrologic cycle is the sun, and because of the nearly inexhaustible supply of energy that the sun has, the hydrologic cycle has been in perpetual motion since the world began and will continue to operate until our sun burns out.

The hydrologic cycle starts when the energy from the sun causes evaporation of ocean waters. The ocean waters are literally distilled by the sun's heat and converted from a salt solution into gaseous water in the atmosphere. The atmospheric gaseous water collects into small droplets that form clouds. These clouds then release their hoard of water on the sea and on the land in the form of rain or snow. The water that falls upon the land slowly finds its way back into the sea through rivers, and the cycle starts over again.

WEATHERING

The weathering of rocks is accomplished both by chemical means and by mechanical means. We have observed the effects of *chemical weathering* on many of the manufactured articles that we see every day. We have seen objects made of iron slowly rust by the action of oxygen and water, and we have observed the paint on houses slowly fade and peel. In a similar manner, the oxygen and carbon dioxide of the air attack rocks on the surface of Earth.

Carbonic acid, a mixture of carbon dioxide and water, is the chief agent in the chemical weathering of most rocks. In granite, carbonic acid attacks the feldspar minerals and reduces them to soluble carbonates and insoluble clay minerals and silicon dioxide. Similarly, the iron content of various igneous rocks is oxidized in a process akin to that of the rusting of iron to produce iron oxide minerals such as hematite (Fe_2O_3) and limonite ($Fe_2O_3 \cdot H_2O$). Hematite and limonite are responsible for the red and yellow colors of so much of our soil. Of all rocks, marble and limestone are most easily attacked by carbonic acid. The insoluble mineral calcite ($CaCO_3$), the main constituent of limestone, readily reacts with carbonic acid to produce soluble calcium bicarbonate ($Ca(HCO_3)_2$), which is then transported in solution by water to different sites. Some minerals, such as quartz, are almost completely immune to chemical weathering and, therefore, must be removed from the site of weathering by mechanical means.

Mechanical weathering results from the physical breakdown of rocks by such means as the freezing of water in crevices and the expansion of plant roots in rock cracks. Just as an automobile engine block can be split by the force of expansion when water inside the block is changed into ice, the freezing of water that has collected in rock cracks and joints produces sufficient expansion pressure to widen existing cracks and sometimes to split off part of the rock. Likewise, roots of plants growing in rock cracks slowly grow larger and increase the size of the crack until eventually a portion of the rock is broken off. The process of weathering slowly disintegrates small rocks and mighty mountains alike. Although the destruction process is a very slow one, a person can see the result of this destruction is the buildup of loose debris at the foot of cliffs and mountains. These deposits of loose material are called *talus deposits* and are a direct measure of the disintegration that has already taken place.

The process of weathering has been extremely important in the spread of life onto the lands of Earth. The result of weathering is still essential to man's survival. The weathered products of Earth's rocks are the soil that is so necessary for the growth of agricultural products needed to sustain the enormous human population on our planet.

Fig. 23.1 **Talus deposits.**

THE ACTION OF RUNNING WATER

The greatest agent of erosion is running water. When water falls upon the land, a large percentage of the water finds its way back into the air by direct evaporation and by plant transpiration (water given off through a plant's leaves). About 20% of the water that falls upon the land returns to the oceans via streams and rivers. The water that does return to the oceans is called *runoff* water. It is this runoff water that has such a great ability to erode the landscape. During the long, tortuous journey that runoff water makes in its trip to the ocean, the running water carries with it both dissolved chemicals, from the chemical weathering of rocks, and suspended material from the mechanical weathering of rocks. The suspended material carried by a river is used by the river as tools for scouring, abrading, and gouging the bed and banks of the river. Again, it is the force of gravity that is responsible for the erosive "cutting" action of a river's suspended load. Consequently, the greater the difference in elevation of the "head" of a river and the "mouth" of the river, the faster the water flows and the greater the erosional effect of the river's load.

The load of a river or stream may consist of anything from minute clay particles that have very little cutting power to great boulders, which by their sheer massiveness can gouge out huge chunks of a river's bed or banks. The mighty Mississippi River alone carries 1½ million tons of soil and rock each day to the Gulf of Mexico. With that much material the Mississippi River must be doing a lot of erosional work.

Why does a river flow in the lowest part of a valley? At first thought, we might be tempted to reason that since water seeks the lowest level, a river naturally flows in the lowest part of a valley. This explanation is only partly true, however. Actually a river flows at the bottom of a valley because the river has made that valley by erosion of what was probably once a level plain or plateau. At first, land had to be uplifted higher than the surrounding land mass so that a difference in gravitational potential existed. As rain fell upon this elevated plain, the water ran off in whichever direction would most quickly lead it to lower elevations. The runoff water would cut many gullies into the

Fig. 23.2 **Gullies cut by running water.**

(Photograph Courtesy of Mr. C. E. Erdmann, U. S. Geological Survey)

plain. We have seen gullying on a small scale in fields along the highway where loose unconsolidated material has been quickly washed away by a heavy rain. The valleys formed from the initial gullying action on the plain appear in cross section to be v-shaped. V-shaped valleys are formed by the rapid downcutting of rivers that is a direct consequence of the speed of flow of the rivers. V-shaped valleys represent the youthful stage of valley formation and are characterized by rapids and waterfalls.

During the mature stage of valley formation, downcutting is appreciably reduced, and increased widening of the valley is produced by lateral erosion of the river's banks. This increased lateral erosion in a mature valley causes the river's banks to be undercut. The undercut banks then slump into the river and are carried away. Whereas the rivers in youthful valleys flow in relatively straight lines, rivers in mature valleys follow a curving path.

In old-age valleys, lateral erosion is still occurring, but downcutting has almost ceased. Rivers in old-age valleys meander along a sinuous path, and some meanders (loop-like bends in a river's channel) have been cut off to form oxbows. The flow of water in the rivers of old-age valleys moves very sluggishly through the river's floodplain.

sculpturing the land—weathering and erosion   **483**

**Fig. 23.3
A youthful valley.
Note the v-shape
of the valley sides.**

(Photograph Courtesy of Mr. W. T. Lee, U. S. Geological Survey)

Obviously, if the erosional effects of streams and rivers on the continents of Earth were the only geologic forces operating on the continents, all continents would eventually be eroded down to a peneplain—land worn down almost to sea level. At the present rate of erosion and transportation of eroded material, the continents would be reduced to a peneplain in about fifteen million to thirty million years. Since Earth has been around for four and one-half billion years

**Fig. 23.4
A mature valley.
Note the river's
meandering path.**

(Photograph Courtesy of Mr. W. T. Lee, U. S. Geological Survey)

sculpturing the land—weathering and erosion

Fig. 23.5 **An old age valley.**

(Photograph Courtesy of Mr. G. W. Stase, U. S. Geological Survey)

and the continents are not peneplains, it is quite obvious that other geologic forces are working against the erosional effects of streams. The reason that the great continents of Earth have not been eroded to peneplains is that as the erosional work of streams and rivers is constantly proceeding, uplifting of the land in various regions is simultaneously occurring. There is an equilibrium between land uplift and land erosion. It is this equilibrium that is responsible for the beautiful artistry of our sculptured landscape.

Fig. 23.6 **Oxbows in an old age valley.**

(Photograph Courtesy of Mr. J. R. Fakley, U. S. Geological Survey)

sculpturing the land—weathering and erosion 485

GROUNDWATER

As rain falls on the land, it is observed to run downhill into valleys and supply the water for streams and rivers. What is not so easily observed, however, is that much of the rain water soaks into the soil and porous rocks of the ground to become *groundwater*. There is thirty-five times more water existing as groundwater in the United States than there is in all this country's streams, rivers, and lakes. In fact, there is enough water in the form of groundwater to cover the entire United States to a depth of 100 feet.

A lot of groundwater eventually flows through layers of porous rock to emerge as river water, lake water, and natural springs. The weathered topsoil and porous rock layers of the ground act as giant sponges to soak up rain water. The spaces between particles of soil and the pores and small cracks in layers of rock serve the same function as the pores of a sponge. As the water flows down through the pores of Earth, the pores at a certain depth become completely filled with water. The topmost level at which the pores are saturated with water is called the *water table*. At the same time that groundwater moves downward, it also moves laterally through porous rock layers down mountain and hill slopes until eventually it merges with the level of water in lakes and rivers. The groundwater may also appear as springs if the rock layers carrying the groundwater outcrop on the surface.

Because the movement of groundwater is relatively slow, groundwater does not have the mechanical weathering ability that running water has in streams and rivers. On the other hand, since groundwater is in contact with rocks for such extended periods of time, the action of carbonic acid in the groundwater results in considerable erosional work in the form of chemical weathering. In some parts of the United States, such as Kentucky, where beds of limestone (which is easily attacked by carbonic acid) are quite extensive, the erosional work of groundwater can exceed the erosional work of streams and rivers. The chemical dissolution of limestone is responsible for the formation of caves, which are the most easily seen examples of the action of groundwater.

Since groundwater does contain various dissolved chemicals, these chemicals can be used to cement various rock particles together. Rock particles making up a loose sediment can be cemented together to form hard sedimentary rocks such as breccia, sandstone, or shale. Common cementing materials are calcite, silica, and iron oxides.

WIND EROSION

Southern France sometimes receives red rain because the rain drops are colored with minute dust particles of red hematite. This red dust, which originates in the Sahara Desert region of North Africa, is picked up by prevailing winds and carried across the Mediterranean Sea to become the red rains of Southern France. Occasionally, individual dust storms pick up and transport hundreds of millions of tons of material over a thousand miles. It is quite likely

(Photograph Courtesy of Luray Caverns, Va.)

that the dust in the area in which you live contains minute amounts of dust from almost every country in the world.

Wind carrying dust and sand particles can have an appreciable abrading and pitting effect on objects. In sand and dust storms the windshields of cars are sometimes so pitted that they become opaque.

Wind that is carrying sand has a natural sandblast action on objects. As a result of this sandblast action, natural features of rocks of different hardnesses are sometimes carved into rather grotesque shapes as shown in Figure 23.8.

Fig. 23.8 **The result of wind erosion.**

(Photograph   Courtesy of Mr. H. E. Gregory, U. S. Geological Survey)

sculpturing the land—weathering and erosion   **487**

The material carried by the wind, besides being scattered all over the surface of Earth, is sometimes deposited as sand dunes or sediments of silt-and-clay-sized material called *loess* deposits. Loess deposits are sometimes hundreds of feet thick.

Fig. 23.9 **Sand dunes—the result of wind deposition.**

(Photograph Courtesy of Mr. W. C. Mendenhall, U. S. Geological Survey)

THE WORK OF GLACIERS

On the upper slopes of very tall mountains that have a plentiful snowfall, not all the snow that falls during the winter melts or sublimes away during the summer. As a result snow accumulates year after year. The minimum altitude at which snow does not completely melt during the summer is called the *snow line*. Whereas the snow line at the equator is about 20,000 feet above sea level, the snow line at Antarctica and the North Pole region is at sea level.

As snow slowly accumulates year after year in some mountain valleys above the snow line, the snow is compressed by its own weight. The resulting pressure slowly transforms the delicate flakes of snow into granules called *névé*. The névé is then changed into solid ice. Just as gravity is responsible for the movement of running water, gravity causes the huge mass of ice to move slowly downhill. Once the ice has started to move, it is known as a glacier.

There are two main classifications of glaciers: *Valley glaciers* and *ice sheets*. The glacier whose formation was discussed in the above paragraph is a

Fig. 23.10 **This Saskatchewan glacier is a tongue of the Columbia Ice Field, Province of Alberta, Canada.**

(Photograph Courtesy of Mr. H. E. Malde, U. S. Geological Survey)

valley glacier in that it is a stream of ice flowing down a mountain valley. Ice sheets, on the other hand, are huge, broad masses of ice that cover a very large land surface and move outward from the middle of the ice sheet. If an ice sheet is large enough to cover the major portion of a continent, it is called a *continental glacier*. Examples of continental glaciers are the Greenland and Antarctic ice sheets. The Antarctic continental glacier is by far the largest ice sheet on Earth. The Antarctic glacier covers an area approximately one and one-half times greater than the area of the United States and contains 90% of all the world's ice and 75% of all the world's fresh water.

Rocks firmly frozen into ice at the bottom of a glacier are slowly scraped along by the glacier's movement to gouge and polish the bedrock on which the glacier is moving. The glacier produces more spectacular erosional work when it enters a youthful valley and widens the valley until its cross section is changed from a v-shape to a u-shape.

At the front of a glacier that is melting at a speed equal to or faster than the glacier's forward movement, glacial deposits are formed. The deposits contain material of all sizes and shapes from minute clay particles to huge boulders. These deposits made by glaciers are called *moraines*.

The most spectacular example of glacial erosion and deposition in the United States is the formation of the Great Lakes. These lakes were formed by glacial action during the last ice age.

sculpturing the land—weathering and erosion   **489**

Fig. 23.11 A terminal moraine at the margin of a glacier.

(Photograph Courtesy of Mr. I. C. Russell, U. S. Geological Survey)

CHAPTER SUMMARY

Weathering is the chemical and physical disintegration of rocks.

The various methods by which rocks are disintegrated and by which the resulting debris is carried away involve the process of *erosion*. The three most important erosional agents are water, wind, and ice, water being the most important.

Running water erodes the land to produce valleys and hills. During this process of erosion, running water transports the eroded materials in suspension and in solution to lower levels. *Groundwater* erodes mainly by chemical solution of rock layers.

Wind transports small particles that erode rocks by abrading them with a natural sandblast action.

Glaciers erode by a scouring action resulting from rocks firmly frozen into the bottom of the glacier.

The ultimate cause of the transportation of material by erosional agents is the relentless force of gravity. Gravity causes eroded material to be pulled down from a higher level to a lower level where eventually the material is deposited.

QUESTIONS AND PROBLEMS

1. Write an equation to show the action of carbonic acid on limestone. (p. 481)
2. Why does a river make a poor boundary between states? (p. 483)
3. Give an explanation for finding a layer of rock that is fifty million years old resting on top of a layer of rock that is twenty-five million years old.
4. Would chemical weathering or mechanical weathering be more important in very cold regions?
5. Black sand beaches are usually derived from the mechanical weathering of what rock?
6. What is the difference between chemical weathering and mechanical weathering? (p. 481)
7. How do youthful valleys differ from old-age valleys? (p. 483)

(Courtesy of Ryan Photographic Service, Inc.)

chemical
and
physical
pollution
24

PESTICIDES AND HERBICIDES
COMMERCIAL AND DOMESTIC WASTE
POLLUTION
AIR POLLUTION AND THE INTERNAL
COMBUSTION ENGINE
NOISE POLLUTION
THERMAL POLLUTION
THE CHALLENGE IS OURS

vocabulary

Herbicides	Inversion Layer
Noise	Pesticides
Pollution	

The species referred to as man has existed on Earth for a relatively short time. Since the beginning, his life has been a continuous series of potential risks in a world that is constantly undergoing changes. Many of the changes which man faces are the result of his own past ecological practices, and they add to the danger to which he is exposed. Throughout this planet, particularly in America, man finds himself threatened with an environmental crisis. He has tainted the air he breathes, the water he drinks, and the food he eats. The concentration of people in large cities, along with the factories that attract them, results in the accumulation of wastes of all sorts, wastes which produce various health hazards. During the late 1960s, the environmental crises gained prominence in the eyes of the public. Industry became aware of the environmental problems which had been created, and many factories are now gainfully attacking the pollution problems over which they have control. But industry cannot cope with the situation without the aid of each individual who must recognize the urgent need to do his part. Our only hope is to develop an informed and interested society before the problem becomes more severe.

In the past, the concern of some farsighted individuals for our environment and its need to be controlled has been met with indifference. The warnings of many great minds which forecasted deterioration of our surroundings have gone unheeded from the days of Darwin. But today, businessmen, housewives, politicians, scientists, teachers, and youth all realize the importance of rigid control.

The contents of this chapter are presented separately for greater emphasis. The chapter is an attempt to develop the reader's awareness to the broad range of pollution problems that are encountered as various types of energy are consumed. The reader should be able to identify the kinds of energy within the various topics discussed. He will also note that the topics under discussion are commonly present in our daily newspapers. Newspaper articles indicate that all forms of pollution along with their effects on man and his environment are under surveillance. Typically, the articles present the partial result of a debate between a spokesman who defends the major source of pollution and one who attacks the pollution from an ecological point of view or one of potential health hazard. The authors have attempted to present the topics that follow objectively, but empathy is always present.

PESTICIDES AND HERBICIDES

A staggering amount of poisonous substances is added daily to our surroundings in the form of pesticides and herbicides. Pesticides are applied to our soil, water, and atmosphere to destroy unwanted animals and insects. Herbicides are used to retard or destroy unwanted plant growth at home and on the farm. About thirty-five million acres of land are treated annually with pesticides and herbicides in the United States alone.

The application of pesticides has been brought about because insects offer a serious threat to our comfort, food, clothing, and well-being. The insects have decided advantages over man in that they have evolved through hostile environmental changes for 250 million years, whereas man has existed on Earth a scant period of time. The insect has evolved to such a degree that some forty species survive even in the Antarctic region, and countless others thrive in desert areas. Nor does altitude destroy them, for some varieties have been found at 20,000 feet above sea level. Others live in sea water, and some in the hot springs, which reach temperatures of our morning coffee. Still others have been found exposed to −40°F but still alive. For all of man's attempt to annihilate various species of insects with the pesticides, he has yet to meet with total success.

Herbicides, which inhibit the growth of unwanted plants, are important to agriculture. Other chemical compounds that have high nitrogen content stimulate the growth of plants. But both categories of substances have affected our streams, wells, rivers, and lakes as the chemicals are absorbed by the soil and are carried off by the groundwater. Nitrogen run-off into groundwater and its concentration in slow-moving bodies of water have caused an increase in the growth of algae. This intensified growth has caused a corresponding decrease in oxygen content, hence the retardation of marine animal growth and survival.

Trees and plants are susceptible to about 1500 different diseases. Without herbicides and pesticides, such deterrents as insects and blights would undoubtedly destroy many species. However, the direct dangers to mankind from the chemical substances used are evident, as are the dangers of starvation and of the diseases carried by insects if they are not destroyed. Before insecticides, for example, millions of people succumbed to human diseases transmitted by ticks and mosquitoes. Later the pesticides were effective in minimizing the danger from typhus and various other plagues, as the pesticides annihilated the insects which carried the diseases. What is the solution to the problem? At present, the best solution seems to be rigid control of the quantities of the pesticides used and the areas over which they are applied.

But, what are the pesticides? The chemist recognizes five categories classified according to active constituents as follows:

(1) Chlorinated hydrocarbons are the most commonly used class of pesticides. Among the most popular is DDT (dichlorodiphenyltrichloroethane), which has been in use since 1937. The first major application of DDT, however, occurred during World War II as it was applied to stop the spread of typhus among the military and civilians in occupied territories. Other members of this group are benzene hexachloride (BHC), chlordane, and heptachlor. The difficulty with this class of pesticide ensues from its poor solubility in water; therefore, it contaminates the area over which it is spread for years to come.

(2) Another category of pesticides is the inorganic pesticides which contain active metals, such as arsenic, lead, and mercury. These inorganic pesticides are effective in destroying insects only if used rarely, because insects and plant

diseases, at the species level, develop an immunity to a given poison. For example, some strains of insects require doses of DDT fifty times the strength which was required to annihilate their immediate ancestors. If these poisons have been rarely used, some species, such as the beetles, will be readily victimized by the inorganic compounds because the insects lack a natural resistance which would be produced through overexposure to given poisons.

(3) Organophosphate compounds, another category of pesticides, are the by-products of war research. This group contains the most deadly poisons to insects in use. Some examples of this class are diazinon, malathion, and parathion. Like nerve gas, the action of these poisons is to paralyze nervous systems, hence muscular systems. The organophosphates are dangerous to a large spectrum of life among which are many domestic animals. For instance, various flea collars available for dogs and cats contain organophosphates and under confined conditions, are hazardous to animals and to man.

(4) Natural pesticides have been used for many centuries. For example, the powder produced from beautiful chrysanthemums has offered relief from the invasion of insects, such as the ant and the housefly. In addition, rotenone, an extract from various plant roots, and nicotine have also been used as poisons in agricultural sprays with great success.

(5) A fifth class of pesticides includes the petroleum distillates which in the crudest form have destroyed the mosquito in its lair. The larva is smothered by the oils that are less dense than water as they are spread over the swampy regions where mosquitoes breed. Other insects, such as flies, have been controlled in the same manner.

The danger from herbicides at the present time is minimal to the general public because of their limited applications. But man must heed the warnings of nature and use care in their use. Like the pesticides, herbicides have served a valuable purpose in that both types of poisons have helped us maintain a friendly environment. The threat to the general public from both types lies in the unwanted and uncontrollable concentrations which are harmful to all life.

COMMERCIAL AND DOMESTIC WASTE POLLUTION

Sewage is composed of the wastes from homes, industrial plants, hospitals, and similar sources. Not only human wastes are involved, but industrial wastes as well. Everyone is aware of the various refuse which accumulates at home, including waters from the bath and kitchen and the soap, oils, and organic materials they contain. Industry adds to the accumulation of undesirable materials from meat packers, milk processors, canneries, breweries, and drug manufacturers.

The housewife is extremely interested in a nice white laundry, since whiteness represents cleanliness. For this reason she adds to her wash various bleaches, detergents, and soaps. Strictly speaking, a detergent is a substance which breaks the surface tension of water. The soap manufacturers include those chemical combinations which break down grease and place dirt in suspension. The detergent particles attach to the grease and foreign particles in such a manner as to make them readily loosened by agitation. However, water with a high mineral content loses its ability to cause the dirt to dissolve. This problem is solved through the addition of phosphates, additives which render the minerals inactive so that the detergent can perform its intended function. However, as the phosphates find their way to rivers, lakes, and streams, they accelerate the growth of algae. The algae in time perish and decompose. But the decomposition process requires oxygen and in turn produces hydrogen sulfide, a condition that spells doom to most aquatic life. Eventually the lake or stream fills with decayed matter and is left as an unproductive marsh area.

Again, science must find a solution by replacing the detergent and phosphate compounds with those that do not affect the normal growth rate of algae or otherwise upset environmental balance. Enzymes and biodegradable compounds are currently under investigation, for both have serious side effects which create other unwanted environmental conditions.

The problems in handling sewage also derive from the presence of organic solids, such as carbohydrates, fats, and proteins. A city the size of New York City accumulates over 500 tons of solid waste per day which, regardless of the method of removal, accumulates to form staggering quantities. The problem lies in disposition, for if the sewage is dumped into lakes and oceans, the solids provide abundant food for algae and other microbes. Thus the propagation of

Fig. 24.1 **Gigantic plants such as the one illustrated are designed to treat commercial and domestic wastes before the wastes are returned to our environment.**

(Courtesy of Ryan Photographic Service, Inc.)

chemical and physical pollution

viruses, protozoa, and pathogenic microbes can cause the spread of disease through groundwater as it penetrates the storage area and enters into the streams and rivers. Over twenty diseases are spread from solid wastes to unsuspecting consumers of contaminated water.

One manner of disposing of the solid wastes is to process it in an oxidation pond. The solids are suspended in solution as they enter the pond. As the water velocity decreases, the heavy solids precipitate; then bacteria and microbes thrive on the organic matter which remains suspended in the water. Through this process the organic matter is consumed; then the remaining heavy solids can be removed from the bottom of the pond and buried.

The burning of solid wastes also spreads disease as well as pollutes the air. This process has been discontinued in many localities because of public reaction, and probably will never be permitted again.

Some European countries are converting solid waste materials into cinders by burning the wastes under controlled conditions. The cinders are then used to pave roads and to extend land areas into the seas. Other projects are underway in the United States in which compressed solid wastes are treated with adhering plastics, molded into bricks, and used to construct buildings. Still others are investigating the feasibility of building islands around populated areas in the shallow, calm inlets that encompass our eastern seacoast.

Attempts to reclaim solid wastes are under serious study because of the immense storage problems and because of the rapid rate at which our natural resources are being consumed. At the present time, few, if any, efficient methods of reclamation of the constituents in solid wastes have been discovered. However, a solution to the problem must be determined before our resources are completely depleted. Future generations may consider our landfills, dumps, and auto graveyards as a major source of mineral deposits. In the future, our standard of living will be determined by our ability to salvage and recycle our waste products. The national cost of sewage collection, transportation, and treatment amounts to over four billion dollars annually. Three-fourths of this amount is spent on collection and transportation alone; still many cities have insufficient systems. More money must be allotted also for the treatment and possible recycling of the wastes to preserve our overall economy and our remaining resources.

The increase in the number of nonreturnable containers used, although they are more expensive than those which the consumer returns to the store, has not aided the waste product situation. The seriousness of waste disposal greatly outweighs the convenience afforded the consumer. Biodegradable containers have been initiated, but have not proved satisfactory at the present time.

Although such waste treatment techniques as electrodialysis, reverse osmosis, compaction, distillation, and dissolution are under various stages of development and consideration, the problem of treatment is severe and immediate. Our generation must assume the problem through combined efforts of the interdisciplinary sciences, technology, and society if civilization as we know it is to continue.

AIR POLLUTION AND THE
INTERNAL COMBUSTION ENGINE

Earth is surrounded by vast amounts of air, enough to last man another three million years without regeneration. Of the six quadrillion tons of air that enclose Earth, 50% is within 18,000 feet of the surface. The cause for concern develops in that the lower level of our atmosphere is used, reused, and used again, particularly air in the lower 100 feet. Disregarding the water vapor and traces of various gases, the main elements which comprise our atmosphere are oxygen (about 20%) and nitrogen (approximately 80%).

An unwanted ingredient of the air is the particulate matter and gases that man has added to the lower levels. Soot and solid unburnable particles form the preponderance of the particulate matter. Also included are dust, paint, pigments, and countless other contaminants which remain suspended in the air by virtue of the prevailing winds and other air currents. Air pollution is primarily retained, however, by an *inversion layer* that forms when the air is practically motionless. As a warm air mass slowly approaches cooler air, the warm air is turned upward and over the cooler air mass. The moderately warm air that surrounds Earth rises until it comes in contact with the progressively warmer air mass and becomes stationary. As this phenomenon occurs, wastes from home and industry add to the amount of pollution previously trapped in the lower levels, and the total amount of particulate matter and gases accumulates rapidly (see Fig. 24.2).

**Fig. 24.2
Inversion layers create hazardous conditions around large cities and the countryside as gaseous industrial wastes and auto exhausts concentrate.**

(Courtesy of Ryan Photographic Service, Inc.)

chemical and physical pollution   **499**

Also, an inversion layer forms every evening after the sun sets. The ground cools off quickly as does the air immediately above it. Air farther from the surface of Earth retains its warmth; therefore an inversion forms at the level of the contact region of the two air masses. The height of this inversion may range from a hundred to several thousand feet.

Frequently, the inversion layer does not break apart, but remains intact for days. Then areas which may cover thousands of square miles experience an accumulation of polluted air, a condition which becomes hazardous to health. The layer subsides somewhat daily and constantly moves closer to Earth; thus the situation becomes much more serious.

Inversions are quite common over the central Plains, and only slightly less common over the Atlantic Coast area. Various investigations underway hint that air pollution is increasing the average temperature of Earth. An overall increase equivalent of 4°C (about 7°F) is contended to be enough to cause polar ice caps to melt appreciably, a condition which would flood numerous cities. Other studies suggest that the average temperature is decreasing. An overall decrease of 4°C possibly would cause the climate of Miami to be similar to that of the city of Boston, and Seattle to be similar in climate to the coldest regions of Alaska. Both changes would occur in less than four centuries, and the United States would experience a new ice age.

Among the gases added to the atmosphere by industry are nitrogen and sulfur oxides. Nitrogen oxides are unwanted by-products that accompany combustion. Their odor is obnoxious; they irritate eyes, noses, and throats; and they absorb sunlight. Sulfur oxides accumulate in areas where fossil fuels and wood are burned such as over power plants, coke-producing plants, and large apartment buildings. Some estimated twenty-three million tons of irritating gases composed chiefly of sulfur oxides are added to the atmosphere annually.

In the United States, the worst air polluters of all are the 100 million automobiles and their internal combustion engines which add some 200,000 tons of waste into the atmosphere each day. The auto industry now spends millions of dollars annually to find methods whereby exhausted gases and particulate matter can be made less offensive. Devices to cut pollution such as afterburners and catalytic mufflers are now in use and have been installed on most new automobiles. Carbon monoxide, a by-product of incomplete combustion, is one of the most common dangerous pollutants produced by the automobile. About sixty-six million tons of the deadly gas are produced and released each year from exhaust systems.

Two other contaminants that result from the internal combustion engines are the nitrogen oxides and metallic toxins, mostly lead. The nitrogen oxides, formed by most combustion processes, require expensive and complicated devices to absorb them. Lead pollution results from tetraethyl lead, an ingredient of gasoline used to prevent "knock" by low octane fuels and to provide additional power. Nickel is also placed in gasoline as a beneficial additive. Both metals are thrust into the air as combustion of gasoline occurs or as a result of evaporation, the manner in which an estimated 15% of all our gasoline is lost. A higher percentage of lead pollutants is added to the air by the high-compression engines that were placed in autos from 1966 to 1969, but protective devices have been installed in recent models to combat the problem. In addition, many petroleum companies are marketing grades of gasoline which are free from lead and other polluting additives.

Another type of atmospheric contaminant, known as photochemical air pollution or photochemical smog, has become a major concern for various geographic areas. Smog is a mixture of gaseous and particulate matter which results from reactions that occur in the atmosphere between substances placed there by incomplete combustion from motor vehicles and industry. Ultraviolet radiation from the sun initiates chemical reactions between nitrogen oxides and organic residues which include aromatic hydrocarbons and aldehydes, both common categories of organic compounds. The accumulation rate is sometimes great enough to cause a dense haze to form, the density of which reduces visibility much the same as does fog.

In order to counter the automobile's unwanted contribution to our health hazards, electrically powered automobiles are reappearing in urban areas. The electric car does not emit carbon monoxide, sulfur and nitrogen oxides, or hydrocarbon compounds. It does, however, produce ozone, a molecule composed of three oxygen atoms that attack with zest such substances as rubber and asphalt. Ozone is also harmful to life when the gas is present in concentrated quantities. The electric auto operates on batteries which must be charged electrically. The electricity must be produced by some means, be it the burning of coal, nuclear power, or some other energy source. The problem is a cyclic one, indeed, since the source of electricity creates pollution problems.

A propane-powered vehicle may well be the least polluting type of engine available at the present time. The combustion of propane offers little waste since propane burns practically to completion and thus forms primarily carbon dioxide and water.

Most knowledge about the effects of air pollution on health has been accumulated during periods of acute inversion formations or after industrial accidents. Scientists have determined that air pollution is a complicated process whose severity depends on the climate, the density of traffic, the heating methods, and the topographic surroundings. There is no doubt that air pollution is the most serious environmental deterrent to our health and that it must be attacked with enthusiasm.

NOISE POLLUTION

Another serious environmental problem is excessive noise. Within limits, noise is both unpreventable and necessary if our society is to function effectively. However, noise produced commercially is suspected of causing unfavorable conditions in our cities. The general public is aware of the annoyance created by power home vacuum sweepers, loud pneumatic hammers used on our streets, or noisy automobiles equipped with muffler systems altered from the manufacturer's design.

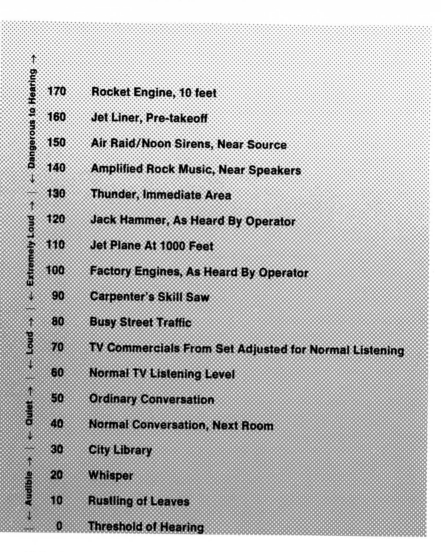

Fig. 24.3 **Noise levels are measured in units called decibels. Note the number of intense sounds produced in our homes. The United States is actively attacking this problem to lower the intensity of many sources of noise.**

dB	Source
170	Rocket Engine, 10 feet
160	Jet Liner, Pre-takeoff
150	Air Raid/Noon Sirens, Near Source
140	Amplified Rock Music, Near Speakers
130	Thunder, Immediate Area
120	Jack Hammer, As Heard By Operator
110	Jet Plane At 1000 Feet
100	Factory Engines, As Heard By Operator
90	Carpenter's Skill Saw
80	Busy Street Traffic
70	TV Commercials From Set Adjusted for Normal Listening
60	Normal TV Listening Level
50	Ordinary Conversation
40	Normal Conversation, Next Room
30	City Library
20	Whisper
10	Rustling of Leaves
0	Threshold of Hearing

(left axis labels, bottom to top): Audible ↑ — Quiet ↑↓ — Loud ↑↓ — Extremely Loud ↑↓ — Dangerous to Hearing ↑)

Noise is purported to cause inefficiency, loss of sleep, nervous disorders, and heart disease, but all conditions supposedly caused by noise lack reliable and sufficient proof of cause. However, science has undertaken the task to determine whether sounds of various frequencies and intensities are in any sense deterrent to good health. The public will be advised of research results, and action to diminish the noises will follow if the findings reveal noise as a definite hazard.

Various professions in the past produced isolated cases of hearing loss such as that found among blacksmiths, boilermakers, and other metal workers who used an anvil and a hammer or other noisy devices extensively. During the last several decades, the loudness and frequency of sounds have been under study as a potential cause of hearing loss by the medical and other professions.

More sophisticated testing devices which measure characteristics of sound and their effects on hearing have opened the way to concentrated research on the suspected correlation. The seriousness of the problem of hearing loss is revealed by studies that found about ten million people in the United States who wear hearing aids. Surveys made by industry show that 20% of those workers tested have impaired hearing. Those people who work in factories have twice the frequency of hearing loss as those in similar age groups who work in offices.

Often hearing loss is caused by a sudden loud noise, but there are many cases of deafness among those people who are exposed to low-level noises over long periods of time. The constant exposure to sounds of low intensity, when used as a device for torture, caused many prisoners of war to reveal vital information to the enemy when other methods of interrogation failed.

Industrial and home appliances operate by the conversion of energy from one form to another. Sound, as well as heat and light is often produced as wasted energy since no device is perfectly efficient. Each of the three forms of energy create serious concern for industry since the three forms cause discomfort and inefficiency among the workers.

Research reveals that sounds of equal frequencies can vary in annoyance according to loudness. In addition, sounds in the frequency range between 200 and 10,000 hz can be heard at lower intensities than can sounds above 10,000 hz or below 200 hz. The greatest range of sensitivity lies between 500 and 4,000 hz.

Noise is considered an unwanted sound and, therefore, is subjective indeed. Sounds that annoy some people cause no discomfort to others. Specifically, noise is a form of energy that is transmitted through some medium such as air, water, or solids and then is converted to sound by the mechanisms of the ear it reaches.

One of our most disturbing sources of noise is the jet engine. The engine, at advanced throttle, creates noise that exceeds 130 db, almost the limit of human endurance. Technical advances have found means to muffle the sounds significantly, but the devices required to do so are expensive and heavy. A plane equipped with the suppressive devices is forced to carry less cargo or else risk overburdening the engine. Therefore, costs to the consumer increase due to smaller payloads and to the costly installation of the muffler systems. Under present circumstances, the roar of jets and the impending potential of the sonic booms they produce annoy many people who live on the routes over which the planes fly. Those people who live in the immediate regions of commercial airports are even more affected by the voluminous noises.

A governmental study released in 1966 by the Office of Science and Technology revealed that the sonic boom is not the result of the plane's breaking the sound barrier (traveling at velocities greater than the velocity of sound), a misconception held since World War II. In reality, the boom is produced by a tremendous pressure wave generated by the nose section and tail assembly of planes that exceed about 660 miles per hour. The pressure wave eventually is transmitted through the air to Earth's surface and is interpreted as a thunderous boom by all those who hear.

THERMAL POLLUTION

In the past decade, electric utilities have felt the public demand rising for cooler and cleaner air, a point discussed earlier in the chapter. The burning of coal, the major source of electrical energy since its development as a source of commercial power, creates vast pollution problems and undoubtedly will be replaced by another energy source. One source of power presently being initiated is nuclear energy. However, both the typical electric and the thermonuclear generating plants may create a serious hazard to aquatic life, since both release warm water that has served as a coolant during their operation. Thermal pollution, a problem only in the immediate area of the generating plant, may well spread to all heavily populated and industrialized areas as electric demands increase. For example, projections in regard to electrical consumption point to an increase of over 200 times the present need by the year 2000. Unless technology can increase the efficiency of thermonuclear power and of traditional power plants by the turn of the century, an equivalent amount of the entire runoff of fresh water in the United States would be used each week by the projected 100 nuclear power plants required to furnish electrical power and heat for industrial and commercial consumers. At the present time, the Commonwealth Edison Company of Chicago is constructing a "fast breeder" nuclear power plant which will furnish 500,000 kilowatts of electricity. This prototype reactor may prove to be the answer to our problem of thermal pollution, since its efficiency is such that it will minimize the number of generating plants required to furnish our electrical power required in the future.

Heated water that is discharged back into the stream or river may be used by several industrial plants before it has reached ambient temperatures. The effect upon ecological relationships is therefore increased. An increase of water temperature of only 10°F can be lethal to various species of plants and animals that serve as a food supply for still other species. The result is the death of numerous species. Also, warmer water contains less oxygen than cooler water; therefore, oxygen-breathing fishes and other aquatic animal life perish or move to regions which contain more natural waters. The heat also serves as a catalyst to expedite the rate of chemical reactions and biological activity, both of which have a retarding effect on the ability of the water to purify itself. The addition of heat, then, magnifies other water pollution problems as well as directly endangers man's food supply.

Industry has undertaken several unique methods to overcome the threat of thermal pollution. The warm water is often pumped into a cooling lake where it cools by evaporation and by heat loss to the atmosphere. Another system makes use of a cooling tower where the heated water is sprayed into the air and cooled by the atmosphere or by giant fans. But both systems create fog under certain atmospheric conditions and add to the expense of electrical power to the consumer.

Not all studies are concerned with cooling the water from thermonuclear power plants or other industries. Instead, the warm water is being put to beneficial use. For instance, the warm water is added directly to exceedingly cold water, an action which increases fish production in regions where low temperatures inhibit aquatic growth. Another innovative use of the warm water from generating plants is that of plant irrigation. Plants, such as corn and beans, grow much faster with warm water irrigation than with water of ambient temperatures. Also, warm water sprayed over strawberry plants and orange groves minimizes frost damage and produces a hothouse effect about the plants. Other crops are being exposed to warm water irrigation at the present time, and attempts to extend the length of growing seasons are also under consideration.

Although thermal pollution is no widespread hazard at the present, all branches of science must become involved in the concern and pool their knowledge to determine locations and designs of future power plants whereby mankind can be most beneficially served.

THE CHALLENGE IS OURS

The task of protecting our environment belongs to each individual, now and in the future. The nature of the task is simple: (1) We must control and lower pollution levels in all of our environment; (2) We must restore the natural resources as much as possible and maintain that beauty of our environment which remains; (3) We must use our natural resources wisely and learn to reuse those which can be separated from total waste.

Pollution results mainly from the use of natural resources in a manner by which they are not recoverable. The total waste products remain in our environment for long periods of time once they are placed there. For civilization to continue and at least to maintain its present standard of living, the industrial and domestic wastes must be handled properly. The solution seems straightforward:

1. Care must be exercised to see that solid wastes such as garbage, junk, and trash are disposed of in a manner that will not contaminate our environment —land or water.

2. Smoke and dust particles must be removed at the source prior to their release into the atmosphere.

3. Commercial and domestic wastes that serve as a haven for bacteria must be converted into substances which inhibit unwanted bacterial growth. Poisonous wastes must be rendered harmless before they are released into our environment.

4. Proper control must be maintained over the chemicals that are used as insecticides, herbicides, cleansing agents, and food preservatives to provide protection for both domestic animals and plants as well as for all species of wildlife.

All of the above answers can be made available only through extensive research. As new techniques for treatment of waste products are discovered, new regulations must be provided by society. A great expense to each of us lies in the necessary educational program that must be initiated to make all citizens aware of the seriousness of environmental pollution. The individual must realize his need to share the expense and the burden of treating wastes properly. He must also develop an awareness to the demands of all society which has as an ultimate aim the rebeautification of cities, parks, and rural areas. All of our problems which concern pollution are serious, even the potential ones. But progress is being made and will certainly flourish when society looks at our environment as our greatest resource of all.

CHAPTER SUMMARY

Pesticides are chemical compounds that are applied to our immediate environment to destroy unwanted animals and insects. Chemical compounds designed to control or to destroy undesirable plant growth are called *herbicides*. Through research, herbicides and pesticides have been developed that attack only specific forms of life. However, numerous chemical compounds disintegrate into other compounds that are harmful to man and to various species of animals, insects, and plants which are advantageous to mankind. Other pesticides and herbicides in original form but in concentrated amounts are also harmful. A consideration of the effects, from the salutary to the adverse, must be maintained to guide users of these chemicals to employ them in a manner most beneficial to mankind.

Wastes from homes and industry pose serious problems to our environment. The most critical problems from domestic wastes are the discharges of soaps and organic solids into our waters and the accumulation of nonreturnable containers in our solid waste disposal sites.

Solid wastes are often combustible, yet the burning of solid wastes in turn adds unwanted gases and particulate matter to our atmosphere. *Inversion layers,* suspended layers of air caused by unusual temperature variations, concentrate the gases and particulate matter and hold them over many densely populated areas. Industry adds many other contaminants to our atmosphere including nitrogen and sulfur compounds. However, most of the pollution of our air is attributed to the automobile and its internal combustion engine. Automobiles with electric motors or with engines that consume propane are more than just speculative replacements for the internal combustion engine. Air pollution is not only a health hazard, but it also reportedly affects our climate in a manner that is not conducive to mankind and his environment.

Various studies indicate that noise, unwanted sounds of all sorts, affect our health and comfort in numerous ways. Sound, heat, and light energies are often produced during the conversion of one form of energy to another. Public reactions to noise have not gone unheeded. The thunderous sounds from jet engines, faulty mufflers, and pneumatic hammers are among those sources of greatest concern.

Industrial areas and the accompanying large population demand that sources of electrical energy be developed nearby. Any generating system of electric power for large communities releases energy in the form of heat directly into the atmosphere and into the water. The heated water retards and destroys most ordinary aquatic life, yet it is conducive to unwanted plant and animal growth, including algae. Industry is overcoming this problem by cooling the heated water before its release into streams and lakes as well as by using it in the irrigation of various crops.

To protect our environment, we must employ measures to lower pollution levels and exercise extreme care in the consumption of our remaining natural resources. We must also resort to better means for solid, liquid, and gaseous waste disposal if future generations are to survive and to prosper.

QUESTIONS AND PROBLEMS

1. Do you think that science eventually will develop pesticides and herbicides which will not be toxic to mammals and to man? Elaborate on your response. (p. 495)
2. Recent research indicates that there is little worldwide carbon monoxide pollution from man-made sources other than in urban areas. Does this conclusion mean that man does not have to concern himself with controlling the addition of carbon monoxide to our atmosphere? Justify your answer. (p. 500)
3. What widespread attempts to conserve our available mineral resources have you observed in literature that have been initiated in the decade of the 1970s? (p. 494)
4. What innovations have you read about recently which indicate an honest attempt to retard or remove our problems of air pollution?
5. Noise, as has been discussed in this chapter, can be minimized but at a loss of efficiency and a waste of energy. What ideas do you have for establishing optimum permissive noise levels?
6. The pollution of our streams, rivers, and oceans appears to be both irrevocable and unpreventable. Which types of water pollution do you deem most hazardous to our environment? Justify your answer. (p. 495)
7. Cite the advantages and disadvantages which result from the application of a pesticide that is soluble in water. (p. 495)
8. Suppose science and technical advancement produced a small, inexpensive power reactor which could be placed in each home. Assuming this unit required a water-cooling system, do you think that such a unit would be preferable in terms of thermal pollution potential over the typical large power reactor? Discuss your response. (p. 504)
9. Landfills which cover old garbage dumps are declared unsafe by many health experts as building sites. One of the hazards for people who reside over the area is the presence of poisonous gases, particularly methane. What is the probable source of this methane? (See Chapter 14)
10. One solution to the growing solid waste problem, including radioactive wastes, is to put the material in orbit about Earth or to propel it into outer space. Defend and refute this solution. (p. 505)

chemical and physical pollution

appendices

APPENDIX I
Mathematics Refresher

The use of systems of measurement makes science a precise study. The manipulations of measurements further develops the student's understanding of science. In order to solve problems in science, the student should be familiar with several basic mathematical operations. The following manipulations and examples are typical of those encountered in this textbook.

Addition

Numbers and quantities can be added in any order as long as the units of measurement are identical. For example, the sum of 3 ft, 4 ft, and 18 inches = 3 ft + 4 ft + 1.5 ft = 8.5 ft. Also, $X + Y = Y + X$, if X and Y are real numbers.

Subtraction

The operation of deducting one number or quantity from another is called subtraction. If the positive number 8 is deducted from the positive number 12, the difference is positive number 4. That is, $12 - 8 = 4$. To subtract a negative number from a positive number, change the sign of the negative number and add. For instance, $13 - (-6) = 13 + 6 = 19$. Algebraically, $X - (-Y) = X + Y$. Also, $-X - (+Y) = -X - Y$.

Multiplication

Multiplication is a mathematical operation by which a number is added to itself a specified number of times. As to quantities, other mathematical laws are involved. The following examples are used to point out manipulations involved in multiplication of numbers and of quantities. Note the manner in which positive numbers and negative numbers are treated.

$$(5) \times (4) = (4) \times (5) = 20$$
$$(X) \times (Y) = (Y) \times (X) = XY \text{ or } YX$$
$$(-5) \times (-4) = (-4) \times (-5) = +20$$
$$(-X) \times (-Y) = (-Y) \times (-X) = XY = YX$$
$$(-5) \times (+4) = (+4) \times (-5) = -20$$
$$(-X) \times (+Y) = (+Y) \times (-X) = -XY = -YX$$

Also:

$$(4 \text{ ft}) \times (3 \text{ ft}) = (4) \times (3) \times (\text{ft}) \times (\text{ft})$$
$$= 12 \text{ ft}^2$$
$$(3 \text{ in}) \times (2 \text{ in}) \times (4 \text{ in}) = (3) \times (2) \times (4) \times (\text{in}) \times (\text{in}) \times (\text{in})$$
$$= 24 \text{ in}^3$$
$$(5 \text{ ft}) \times (2 \text{ lbs}) = (5) \times (2) \times (\text{ft}) \times (\text{lbs})$$
$$= 10 \text{ ft-lbs} = 10 \text{ lbs-ft.}$$

Division

The process by which one determines how many times one number or quantity is contained in another is called division. Division is typically indicated by a horizontal or diagonal line between numerator and denominator. For example, 3 divided by 4 is $\frac{3}{4}$ or 3/4. The number 3 is the numerator and the number 4 is the denominator. F divided by M is $\frac{F}{M}$ or F/M. If the numerator, denominator, or both are fractions, division is accomplished by inverting the denominator and multipling. For instance, 4/5 divided by 2/3 is written $\frac{\frac{4}{5}}{\frac{2}{3}}$ and simplified to $4/5 \times 3/2 = 12/10 = 1 \ 2/10 = 1 \ 1/5$. Furthermore, 10 ft divided by 2 sec $= 10 \ \text{ft}/2 \ \text{sec} = \frac{5 \ \text{ft}}{\text{sec}} = 5 \ \text{ft/sec}$. Also, 6 ft divided by 2 ft $= 6 \ \text{ft}/2 \ \text{ft} = 3$.

Constants and Variables

A constant is a quantity whose value does not change in a given problem. A variable is a quantity that may assume any one of a set of values in a given problem. For example, if you travel by a train that runs at a uniform velocity of 60 miles per hour, the velocity of the train is constant, that is, it does not change. The distance you go will depend on the time you ride the train, and thus is a variable. The following table points out how the distance varies with time; velocity $(v) = 60$ mph, a constant.

t (hours)	½	1	2	3	4	5	10
d (miles)	30	60	120	180	240	300	600

Notice also that any two values of t with the corresponding values of d form a proportion. For example, $3:4 = 180:240$. Also, $5:10 = 300:600$. The first proportion is read: 3 is to 4 as 180 is to 240.

Direct Proportion: Direct Variation

As one might surmise from the previous example, the distance is directly proportional to the time, or otherwise stated, the distance varies directly as the time. In specific terms, one variable quantity is directly proportional to another when the ratio of any two values of the one is the same as the ratio of the corresponding values of the other. Again, from the previous example, every value of d equals 60 times the corresponding value of t. Hence, $d = 60 \ t$, or $\frac{d}{t} = 60$. As a principle, one variable quantity varies directly as another value when one equals a constant times the other, or when their quotient is constant. Hence, every direct variation can be expressed as an equation as illustrated in the present discussion.

Inverse Proportion: Inverse Variation

Suppose you were asked: "How long does it take you to drive 60 miles?" Undoubtedly you would want a better qualification of the overall situation before you would respond including direction, time of day, weather, and numerous other details, all of which govern the velocity with which you would drive. The following table illustrates the relation between velocity and time, v and t. The constant involved is distance, $d = 60$ miles.

v (mph)	10	20	30	60	90	120
t (hours)	6	3	2	1	0.67	0.50

The table indicates that if the velocity is 10 mph, the time required for the trip is 6 hours; if the rate is 20 mph, the time required is 3 hours, and so on. As a proportion, the two measurements would be stated: $10:20 = 6:3$. The ratio of the two values of v equals the ratio of the corresponding values of t *inverted*. Similarly $1:2 = 60:30$, etc. In this case one would cite that "t is inversely proportional to v," or "t varies inversely with v." In specific terms, two variable quantities are inversely proportional if their product is constant.

Equations

The solution of problems in science involves the collection of relevant data and an algebraic interpretation as to the manner the data are related. Generally, one quantity cannot be measured directly; however, logic is used to determine its value from known or measurable quantities. Numerous general rules aid in the determination of unknown quantities present in algebraic expressions of scientific principles called equations. In equations (or formulas), single letters are used to represent words or numbers. Some of the rules for solving equations are illustrated below.

a. Any quantity can be added or subtracted from both sides of an equation.

$$A - 10 = 30 \qquad\qquad 6B + 20 = 7B$$
$$A - 10 + 10 = 30 + 10 \quad \text{and} \quad 6B - 6B + 20 = 7B - 6B$$
$$A = 40 \qquad\qquad 20 = B$$

b. Both sides of an equation may be multiplied or divided by any quantity.

$$\frac{1}{5}A = 6 \qquad\qquad 8A = 56 \qquad\qquad 5ab = 15a$$
$$\frac{1}{5}A \times 5 = 6 \times 5 \quad \text{or} \quad \frac{8A}{8} = \frac{56}{8} \quad \text{and} \quad \frac{5ab}{5a} = \frac{15a}{5a}$$
$$A = 30 \qquad\qquad A = 7 \qquad\qquad b = 3$$

c. A fractional equation which states that two ratios are equal is called a proportion. Equations expressed as proportions are simplified by cross-multiplying.

$$\frac{a}{b} = \frac{c}{d} \qquad\qquad \frac{a}{4} = \frac{2}{8}$$

$$a \times d = b \times c \quad \text{and} \quad a \times 8 = 4 \times 2$$

$$ad = bc \qquad\qquad 8a = 8$$

$$\frac{8a}{8} = \frac{8}{8}$$

$$a = 1$$

d. Both sides of an equation may be raised to the same power or have the same root taken.

$$\sqrt{X} = 10 \qquad\qquad\qquad X^2 = 25$$

$$(\sqrt{X}) \times (\sqrt{X}) = (10) \times (10) \qquad \text{and} \qquad \sqrt{X^2} = \sqrt{25}$$

$$X = 100 \qquad\qquad\qquad X = 5$$

Square roots that are not "perfect squares" may be readily determined by calculation. However, the method of approximation is noteworthy, also. For example, the square root of eighty-five is not a perfect square. Note the approximation procedure and method of calculation that follows.

$$X^2 = 85$$

$$X = \sqrt{85}$$

$$(9)^2 = 81$$

$$(10)^2 = 100$$

The square root of 85 lies between 9 and 10.

$85 - 81 = 4$ and $100 - 85 = 15$, so the square root of 85 is approximately 9 and 4/15 or $X = 9.266$.

By Calculation:

```
                9 . 2  1  9
         √85.00′00′00
               81
            |04 00
       182  | 3 64
            |0 36 00
      1841  | 18 41
            |17 59 00
     18429  |16 58 61
            |01 00 39
```

To Check:

$$(9.219) \times (9.219) + 0.010039 = 85.000$$

Scientific Notation (Powers of Ten)

Very large numbers and very small numbers are more readily manipulated when they are expressed as powers of ten. Any number may be written as a number between 1 and 10 (the coefficient) multiplied by a number written as a power of 10 (the exponent). For example, 2800 may be written as 2.8×1000 and as 2.8×10^3. Also, 5,400,000 may be written as $5.4 \times 1,000,000$ or as 5.4×10^6. Numbers less than one are written using negative powers of 10. A decimal fraction such as 0.0000453 may be written $\frac{453}{10,000,000}$ or as 4.53×10^{-5}. The multiplication of 5,400,000 and 0.0000453 becomes much less formidable when written as powers of ten.

$$(5,400,000) \times (0.0000453) =$$
$$(5.4 \times 10^6) \times (4.53 \times 10^{-5}) = 24.462 \times 10^1$$
$$= 244.62$$

The rules of combining numbers written in scientific notation are easily comprehended. In addition or subtraction, the exponents of all numbers written as powers of 10 to be added or subtracted must be the same. For example,

$$(5.6 \times 10^3) + (4.2 \times 10^3) \quad = 9.8 \times 10^3$$

and

$$(3.6 \times 10^4) + (4.3 \times 10^2) \quad = (3.6 \times 10^4) + (0.043 \times 10^4)$$
$$= 3.643 \times 10^4$$

also

$$(4.6 \times 10^2) - (2.4 \times 10^{-2}) = (4.6 \times 10^2) - (0.00024 \times 10^2)$$
$$= 4.59976 \times 10^2$$

To multiply or to divide numbers written as powers of ten, the coefficients, numbers written between 1 and 10, are multiplied or divided. In multiplication, the exponents are combined (added or subtracted) according to sign. In division, the sign of the latter exponent is changed, then combined according to sign.

$$(5 \times 10^3) \times (1.5 \times 10^4) = 7.5 \times 10^7$$
$$(2 \times 10^{-3}) \times (6 \times 10^5) = 12 \times 10^2 = 1.2 \times 10^3$$
$$(6 \times 10^5) \div (2 \times 10^2) = 3 \times 10^3$$
$$(8 \times 10^6) \div (4 \times 10^{-3}) = 2 \times 10^9$$

Significant Figures

A major part of the scientist's time is spent in making measurements. The accuracy of his measurements is indicated by the number of digits recorded. (Digits are 0, 1, 2, 3, 4, ...) A linear measure of 56 cm indicates that distance was measured to the nearest cm. A recording of 56.3 cm is ten times as accurate as one of 56 cm. A recording of 56.0 cm would indicate the same accuracy. If the number were recorded as 56.00 cm, the accuracy is again increased by a factor of ten and so on. No calculation can be made that is more accurate than the data recorded. The answer to calculations is therefore never expressed more accurately than the least accurate recorded data. For example, if a series of measurements were to be added: 13.232 cm + 11.26 cm + 18.2943 cm = 42.7863 cm. In the second measurement in the series the 6 is a doubtful figure, therefore the second digit to the right of the decimal point in the answer is also a doubtful figure. The answer, then, in terms of significant figures, should be expressed as 42.78 cm. Also, the density of a substance whose measured mass is 18.6427 gm and whose volume is 5.62 cm^3 may be calculated to be 3.3172 gm/cm^3. However, the 2 in the divisor is a doubtful figure. In order to adjust for significance, the mass would be rounded off to as many significant figures as the volume, namely, three. The density of the substance, then, should be calculated: $18.6 \div 5.62 = 3.31$ gm/cm^3.

APPENDIX II
Abbreviations and Symbols

acceleration = a
acceleration due to gravity = g
alpha particle = α
alternating current = AC
amplitude = h
amplitude modulation = AM
astronomical unit(s) = A.U.
atomic mass = A
atomic mass unit = a.m.u.
atomic number = Z

beta particle = β

calorie = cal
centi = c
centigrade = C
coefficient of linear expansion = α
cubic centimeter = cc = cm^3
cubic foot = ft^3
current = I

decibel = db
density = D
direct current = DC
distance = d

effort = E
electron = $\bar{e}$
energy = E

fahrenheit = F
feet per second per second = ft/sec^2
focal length = F
foot, feet = ft
force = F
frequency = f
frequency modulation = FM
fulcrum = F

gallon = gal
gamma ray = γ
gram(s) = gm

half-life = $t_{1/2}$
heat = J
height = h
hertz = hz
horsepower = H.P.
hour = hr

intensity = I

joule(s) = j

kelvin = K
kilo = k
kilowatt-hour = kwh
kinetic energy = KE

length = l
light year = LY
liter = l

mass = m
mass number = A
mechanical advantage = m. a.
micro = μ or u
mile = mi, m
milli = m

neutron = n
neutron number = N
number of whole number multiples = n

ohm = Ω
ounce = oz

potential energy = PE
pound(s) = lb.
pounds/square inch = lb/in^2
power = P
proton = p^+

quart = qt

radius = r
resistance = R

speed = s

time = t

velocity = v
velocity of light = c
volt = V

wave length = λ
weight = w
work = w

yard = yd

APPENDIX III

Temperature Equivalent Table

Centigrade (C°)	Fahrenheit (F°)
300	572
250	482
200	392
150	302
120	248
100	212
95	203
90	194
85	185
80	176
75	167
70	158
65	149
60	140
55	131
50	122
45	113
40	104
35	95
30	86
25	77
20	68
15	59
10	50
5	41
0	32

Centigrade (C°)	Fahrenheit (F°)
− 20	− 4
− 40	− 40
− 50	− 58
−100	−148
−150	−238
−200	−328
−250	−418
−273	−459.4

glossary

Absolute Zero—The temperature at which atoms or molecules have zero kinetic energy ($-459.7°F$ or $-273.2°C$).

Acceleration—The rate at which velocity changes with time. Acceleration may involve a change in magnitude, direction, or both.

Acceleration due to Gravity—The rate of change of velocity of a free falling object in the absence of air resistance or other source of friction.

Acid—A chemical compound that produces hydrogen ions in water solution, turns blue litmus pink, and has a sour taste.

Activation—The process of producing artificial radioisotopes by bombarding stable elements with charged particles or neutrons.

Activation Analysis—A procedure that permits the detection and measurement of artificially produced radioisotopes for the purpose of identification of trace elements present.

Alcohols—Organic compounds that have an — OH group attached to a carbon atom. For example methyl alcohol (CH_3OH) and ethyl alcohol (CH_3CH_2OH).

Aldehydes—Organic compounds containing the $-\overset{\overset{\textstyle O}{\|}}{C}-H$ group. Thus, acetaldehyde is $CH_3-\overset{\overset{\textstyle O}{\|}}{C}-H$.

Alpha Particle—The nucleus of a helium atom, a particle ejected by many heavy radioisotopes. The alpha particle is composed of two protons and two neutrons.

Alpha Rays—A stream of alpha particles.

Alternating Current—An electric current that reverses direction per unit time. A complete reversal is generally known as a cycle. Typically, alternating current is produced at fifty cycles/sec or sixty cycles/sec.

Amber—A brown translucent resin that accumulates an electric charge readily.

Amines—Organic compounds containing a carbon to nitrogen single bond. Thus, ethyl amine is $CH_3CH_2NH_2$ and methyl ethyl amine is $CH_3CH_2\overset{\overset{\textstyle H}{|}}{N}CH_3$.

Amino Acids—Organic compounds containing an amino group and a carboxylic acid group. Thus, the amino acid glycine is $H_2N — CH_2 — CO_2H$.

Ampere—The rate of flow of electric charge equal to one coulomb per second.

Amplitude—The maximum distance through which vibrating particles in a wave disturbance move from a rest position. See Figure 6.5.

Amplitude Modulation (AM)—Modulation of a radio carrier wave by varying its amplitude.

Anode—A positively charged terminal in an electrical device.

Antarctic Circle—Parallel of latitude 66½° S.

Aphelion—That point in the orbit of an object revolving around the sun in which it is farthest from the sun.

Armature—The rotating coils of wire in a motor or a generator.

Aromatic Compounds—Compounds similar to benzene that have a cyclic structure containing alternating double and single bonds. This peculiar structure results in considerable electron delocalization with the result that the double bonds present do not normally undergo addition reactions.

Arctic Circle—Parallel of latitude 66½° N.

Asteroid—One of tens of thousands of planetoids that orbit the sun between the orbits of Earth and Mars.

Atom—The smallest part of an element that can exist and still have all the properties of an element.

Atomic Mass—The mass of a neutral atom expressed in atomic mass units (amu). The amu is defined as exactly equal to one-twelfth the mass of $^{12}_{6}C$.

Atomic Number—The identifying number of an element which corresponds to the element's positive charge. The symbol for atomic number is Z.

Atomic Weight—The average relative weight of the atoms of a given element compared with those of $^{12}_{6}C$, the accepted standard.

Autumnal Equinox—The time during autumn (northern hemisphere) when the sun crosses the equator. Occurs about September 23.

Background Radiation—That detected activity in the absence of a radioactive sample which is attributed to cosmic radiation, electronic noise, and natural-occurring radioactivity present in building materials.

Bacteriophage—A microscopic organism that destroys disease-producing bacteria.

Barycenter—The center of mass of two bodies that mutually revolve due to gravitational attraction.

Basalt—A fine-grained igneous rock consisting of feldspar and ferromagnesian materials.

Base—A chemical compound that produces hydroxyl ions in water solution, turns pink litmus blue, and has a bitter taste.

Beat—A series of outbursts of sound followed by intervals of comparative silence.

Beta Particle—An electron (or positron) emitted during the disintegration of a radioactive atom.

Beta Rays—A stream of beta particles.

"Big Bang Theory"—A theory of cosmogony in which it is believed that the universe started with a great explosion and the universe has been expanding ever since.

Binary System—Two stars mutually revolving about each other.

Binding Energy—The energy which holds the particles in an atom's nucleus.

Black Dwarf—The final stage in the evolution of a star. All the sources of energy are exhausted and the star no longer emits light.

Black Hole—The result of uninhibited gravitational collapse of a star. Since the escape velocity of a black hole exceeds the speed of light, neither matter nor electromagnetic radiation can escape from a black hole.

Boiling Point—That point (temperature and pressure) at which the vapor pressure of the liquid slightly exceeds the pressure of the atmosphere upon the liquid.

Brownian Motion—The random motion of small particles caused by collisions with molecules.

Calorie—The amount of heat required to raise the temperature of one gram of water one degree C.

Carbohydrates—Organic compounds containing many hydroxyl $(-OH)$ groups and usually an aldehydo $(-\overset{\overset{\displaystyle O}{\|}}{C}-H)$ or a keto $(-\overset{\overset{\displaystyle O}{\|}}{C}-)$ group. Sugar and starch are carbohydrates.

Carbon Cycle—The process whereby carbon nuclei fuse with hydrogen nuclei to eventually produce helium nuclei and regenerated carbon nuclei. Much energy is released in the process.

Carboxylic Acid—Organic compounds that have a $-CO_2H$ group. Thus, acetic acid is CH_3CO_2H.

Catalysts—A substance that speeds up or slows down the rate of a chemical reaction, yet undergoes no permanent chemical change itself.

Cathode—A negatively charged terminal in an electrical device.

Cenozoic Era—From seventy million years ago to the present.

Centrifugal Force—The outward reaction to a centripetal force. The centrifugal force is equal in magnitude but opposite in direction to the centripetal force.

Centripetal Acceleration—The acceleration produced by a center-seeking force impressed on a body. The resultant change in velocity may be in direction only.

Centripetal Force—The center-seeking force which causes a body to assume a circular path.

Chemical Change—A change which alters the chemical composition of a substance, such as cooking or burning.

Chemical Properties—Those characteristics of a substance that display the manner in which the substance reacts chemically. For example, a chemical property of water is that water can be decomposed to form two new substances.

Chemical Weathering—The breakdown of rock by various chemicals in water solution.

Chill Factor—An expression of equivalent temperature when both present air temperature and wind velocity are considered.

Chromosphere—The part of the sun's atmosphere directly above the photosphere. The color (red) of the chromosphere gives it its name.

Comet—A collection of material revolving around the sun in highly elliptical orbits. Comets are best known for the spectacular tails they often form when near the sun.

Compression—In sound, that part of a wave form in which the density of the vibrating medium is more than normal.

Conduction—The transmission of thermal energy from molecule to molecule.

Conductor—Any substance through which electric charge (or heat) readily flows. Conductors are generally metallic in nature.

Convection—The transfer of thermal energy in a gas or liquid by moving currents in the fluid.

Core (of Earth)—The center portion of Earth. The core is liquid, contains molten iron and nickel, and is about 4,000 miles thick.

Corona—The outermost portion of the sun's atmosphere. The ivory-white corona is seen as a halo around the totally eclipsed sun.

Coulomb—A measure of electric charge that equals the total electric charge carried by 6¼ billion billion electrons.

Covalent Bond—A chemical bond formed by two atoms mutually sharing a pair of electrons.

Critical Angle—That angle of incidence at which the refracted ray produces a right angle with the normal. See Figure 9.10.

Crust (of Earth)—The outer layer of Earth. The crust varies in thickness from three miles to about twenty-five miles.

Curie (Ci)—The total number of disintegrations per second from approximately one gram of radium. The accepted value is 3.7×10^{10} dis/sec. Also, a curie refers to that quantity of any radioisotope which is decaying at the rate of 3.7×10^{10} dis/sec.

Density—The mass of a substance per unit volume. Commonly, density is a measure of weight per unit volume. The density of water, for example is 62.4 lbs/ft^3.

Diastrophism—The processes of deformation in Earth's crust that produce the continents and ocean basins.

Diffraction—The spreading or bending of a light wave around an obstacle or through a narrow opening in such a manner that parallel fringes of light and dark bands are produced.

Diffusion—The reflection of light from an irregular reflecting surface.

Diode—A vacuum tube that contains only two electrodes, a plate and a filament. The tube rectifies alternating current.

Direct Current—An electric current, pulsating or steady, which flows in one constant direction.

Dispersion—The separating of a beam of light into its component frequencies by passing the beam through a glass or plastic prism.

DNA (deoxyribonucleic acid)—A large molecule present in cells that contains the genetic code and transmits the hereditary pattern.

Domain—A region of aligned magnetic atoms.

Doppler Effect—The variation in frequency or pitch of sound waves created by the relative motion of the listener and/or the source. The principle also applies to electromagnetic waves emitted from a source moving relative to an observer.

Dyne—The force which causes a mass of one gram to accelerate 1 cm/sec^2.

Eclipsing Binary—Two stars mutually revolving about each other in a manner so that one star fully or partially eclipses the other during each full revolution.

Efficiency—The ratio of useful work to total work.

Electric Current—The flow of electric charge in which energy is transported from one point in an electric circuit to another. The unit by which current is measured is the ampere.

Electric Potential—The potential energy per unit electric charge. Electric potential is measured in volts and is often referred to as voltage.

Electricity—The flow of electric charge through a conductor. Static electricity is an accumulation of ions (including electrons) on the surface of a poor conductor.

Electrode—A positively or negatively charged wire or other terminal in an electrical device such as a vacuum tube or battery.

Electromagnet—An electrically produced magnet generally made of a soft iron core and a coil of insulated wire through which a current flows.

Electron—The negatively charged elementary particle which orbits the nucleus of an atom.

Electron Volt (eV)—The unit used to measure the energy released during radioactive decay.

Electronegativity—A measure of the attraction an atom has for an "extra" electron.

Electroscope—A device used to detect the presence of electric charge.

Elliptical Galaxy—A galaxy that has a shape that varies from spherical to near lens shaped.

Energy—The ability or capacity to do work. The unit of energy is identical to the unit of work within a system of measure.

Enzymes—Proteinlike substances formed in plants and animals that act as organic catalysts.

Erg—The work done by a force of one dyne acting through a distance of one centimeter.

Erosion—The transportation of loosened material from a higher to a lower level mainly by the action of water, wind, and ice.

Esters—Organic compounds that have a — CO_2R group. Thus, methyl acetate is $CH_3CO_2CH_3$.

Ethers—Organic compounds that have a C — O — C bond. For example, ethyl ether is $CH_3CH_2OCH_2CH_3$.

Fault—A fracture in rock layers along which there has been obvious movement.

Fauna—Animal life.

Filament—A wire used in a vacuum tube that emits electrons when it is heated.

Fission—The splitting of a nucleus into fragments of approximately equal mass. As a result, several lighter atoms are formed along with the release of energy and neutrons.

Flora—Plant life.

Fluorescence—The process by which electromagnetic radiation of one frequency is absorbed and reemitted at lower frequency.

Folding—The bending, flexing, or wrinkling of rock layers during the time the layers were in a plastic state.

Fossil—Any remains, traces, or impressions of prehistoric life.

Frequency—The number of waves, complete or fragmented cycles, which arrive at a given point per unit time, usually the second. See Figure 6.4.

Frequency Modulation (FM)—Modulation of a radio wave by varying its frequency rather than the carrier wave's amplitude.

Friction—The force that resists the rolling or sliding of one object on another. Air friction retards the acceleration of a freely falling body.

Fusion—The combining of light nuclei to form a heavier nucleus. Energy is generally released during fusion, a process that takes place only at extremely high temperatures.

Galaxy—A large (millions to hundreds of billions) collection of stars bound together by gravitation.

Gamma Ray—Electromagnetic radiation of high frequencies emitted from the nucleus of an atom.

Geocentric (concept)—The idea that Earth is the center of the solar system.

Geosyncline—A basin in which sediments accumulate. As sediments fill the basin, the basin floor sinks, a feature which permits geosynclines to reach depths of thousands of feet. The folded mountain ranges were formed from geosynclines.

Globular Cluster—A spherical collection of stars. Globular clusters are associated with our Milky Way Galaxy.

Gradation—The lowering or smoothing of Earth's surface by processes which produce sediment. The wearing down of rock layers is degradation and the building up of low areas is aggradation.

Granite—A coarse-grained igneous rock consisting of the minerals, feldspar, quartz, and mica.

Gravitation—The force that every body in the universe exerts on every other body because of its mass.

Gravity—The force of attraction which causes all freely falling bodies at Earth's surface to accelerate about 32 ft/sec². Gravity varies among heavenly bodies.

Groundwater—Water that is present in porous rock strata and soils.

Half-life—The time required for one-half of a given number of atoms of a radioisotope to undergo decay.

Health Physics—The science which deals with the task of protecting man and his environment from unnecessary exposure to ionizing radiation.

Plasma—Extremely hot gases that are composed of electrically charged particles, in reality fragments of atoms and nuclei. Plasma is sometimes considered the fourth state in which matter can exist.

Plate—The positively charged terminal (anode) in a vacuum tube.

Polarization—The alignment of electromagnetic waves in such a manner that all light waves are parallel to each other.

Poles (magnetic)—The end of a magnet where a magnetic field appears to concentrate.

Pollution—The unnatural addition of contaminants to our air, land, and water.

Polymer—A large molecule produced by the chemical combination of many small molecules. Thus, the polymer, polyethylene, is made by the reaction of thousands of individual ethylene ($CH_2 = CH_2$) molecules.

Potential Energy—The stored energy an object possesses because of its position relative to other bodies.

Power—The rate of doing work. Power is commonly measured in watts and horsepower.

Precambrian Era—From the time of formation of Earth to 600 million years ago.

Precession (of Earth)—The conical motion of Earth's spin axis. It takes Earth about 26,000 years to precess one time.

Pressure—The force exerted per unit area. Pressure may be measured in such units as lbs/in^2 and gm/cm^2.

Principle of Equivalence—A statement of fact that indicates that the amount of inertial mass a body contains is exactly equal to the body's gravitational mass.

Proteins—Organic polymers consisting of polymerized amino acids.

Proton—The positively charged elementary particle found in the nucleus of an atom.

Proton-Proton Reaction—The fusion process whereby four hydrogen nuclei are fused at very high temperatures into a single helium nucleus with the liberation of considerable energy.

Pulsars—Rapidly rotating and highly magnetic neutron stars that give off light and radio waves of very rapid pulses.

Pulsating Variables—Stars that continually expand and contract. This pulsation of size produces a pulsation of the star's luminosity.

Quality—A property of sound waves that is dependent on the number and prominence of overtones present.

Quasars—A stellar-like object that exhibits a very large redshift and is amazingly luminous. Quasar is the popular contraction for a quasi-stellar radio source.

Radiation—The transmission of energy by electromagnetic waves.

Radioactivity—The spontaneous release of matter and/or energy from the nucleus of an atom:

Radioisotope—A radioactive isotope of an element. All elements have at least one isotope that is unstable and that undergoes radioactive decay.

Rarefaction—In sound, that part of a wave form in which the density of the vibrating medium is less than normal.

Real Image—An image that is actually formed by rays of light and can be projected on a screen.

Rectifier—A device which permits electric current to flow in only one direction. Alternating current is converted by the device to direct current; thus the current is said to be rectified.

Red Giant—A large, relatively cool star of high luminosity.

Redshift—The wave length of light from stars and galaxies has been shifted toward longer wave lengths. The redshift is produced by the Doppler effect.

Reduction—The loss of oxygen by an element or the gain of electrons by an element.

Reflection—The rebounding of energy, such as heat, light, or sound from a surface.

Refraction—The bending or abrupt change in direction of energy such as light as it passes from one medium into another of different optical density.

Relativity—A proposed theory which asserts the equivalence of mass and energy.

Reluctance—A measure of magnetic resistance. The reluctance of an air core is more than one of iron.

Rem—The generally accepted unit measure of human exposure to ionizing radiation.

Resonance—The inducing of vibrations of an inherent rate in matter by a vibrating source that has the same or a simple multiple frequency.

Retrograde Rotation—The rotation of a planet on its axis that is the reverse of the other planets. Only Venus exhibits retrograde rotation.

Reverberation—Multiple echoes created in a confined space.

Roche's Limit—The minimum distance from a primary body that a satellite can approach the body without being disintegrated by tidal forces caused by the primary body.

Rock—An aggregate of minerals of different kinds and that lack a definite chemical composition of given properties.

Roentgen—That quantity of ionizing radiation, generally gamma or X-radiation, which produces 1.6×10^{12} ion pairs in one gram of living tissue.

Scalar Quantity—A physical quantity that does not have a direction associated with it. Examples of scalar quantities are candlepower, speed, and density. Temperature also is considered a scalar quantity.

Scattering—A process in which a molecule absorbs electromagnetic waves and releases them in a different direction.

Science—The general areas of nature which incorporate matter and energy.

Scintillation—A burst of electromagnetic energy that occurs in some crystals and liquids upon their interaction with gamma or X-radiation.

Sedimentary Rock—Rock formed from the accumulation of sediment that has undergone cementation, compaction, chemical change, or other processes of lithification.

Sidereal Day—The length of time it takes Earth to make one rotation on its axis. The sidereal day is twenty-three hours and fifty-six minutes.

Slug—The mass of a body, measured in the English system, which is accelerated 1 ft/sec^2 when a force of one pound acts on it.

Snow Line—The altitude above which snow does not completely melt during the summer.

Solar Day—The apparent period of rotation of Earth as measured in relation to the sun. The solar day is twenty-four hours.

Sound—Successive compressions and rarefactions that represent longitudinal waves which travel through a medium such as air. Sound is considered a form of energy.

Specific Gravity—The ratio of the density of a substance to the density of a standard. Air is the standard for gases, and water is the standard for liquids and solids.

Spectrum—A beam of electromagnetic waves that has been separated into components according to wave lengths. Often a continuous sequence or range of energy is referred to as a spectrum because the array can be readily separated according to some varying characteristic.

Spiral Galaxy—A galaxy flattened by rotation to produce pinwheel-like arms.

Spring Tide—The highest tide of the month.

Stellar Parallax—The apparent displacement of a star when viewed from different positions of Earth's orbit.

Stratosphere—The layer of Earth's atmosphere between the ionosphere and the troposphere.

Summer Solstice—The time during the summer (northern hemisphere) when the sun is farthest north of the equator. Occurs on June 21 or 22.

Sunspots—Dark spots on the photosphere of the sun. The spots appear dark because they are cooler than the surrounding photosphere.

Superior Planet—A planet whose orbit is farther from the sun than Earth's orbit.

Supernovae—Stars that explode and increase their luminosity hundreds of thousands to millions of times.

Synchrotron Radiation—Radiation emitted by charged particles when they are accelerated near the speed of light in a strong magnetic field.

Talus Deposits—Deposits of loose rocks at the base of cliffs and mountains.

Temperature—A measure of the average kinetic energy among molecules in a body. Temperature is measured in degrees Celsius, Fahrenheit, and Kelvin.

Terminal Velocity—The maximum velocity a moving object attains because of the combined forces in action on it. For a falling object, terminal velocity is reached when the downward force of gravity is precisely balanced by the counteracting upward forces exerted on the object by the gas or liquid through which it is falling.

Thermostat—A temperature-operated, heat-regulating device used to control mechanical or electrical equipment automatically.

Transformer—A device for increasing or decreasing the voltage of alternating current.

Transistor—A device that has commonly replaced the vacuum tube in electronic circuits

Transmutation—The process by which one element is changed into another. The emission of any charged particle by an atom's nucleus will produce instant transmutation.

Transverse Waves—A wave motion in which the disturbance is at right angles to the direction the wave travels. Electromagnetic waves are considered transverse waves.

Triode—A vacuum tube that contains a filament, a plate, and a grid. The tube is used in an electronic device to regulate the flow of electrical energy through a circuit.

Triple-alpha Process—The fusion of three helium nuclei at extremely high temperatures into one carbon nucleus with the liberation of considerable energy.

Tropic of Cancer—Parallel of latitude 23½ ° N.

Tropic of Capricorn—Parallel of latitude 23½ ° S.

Troposphere—The layer of Earth's atmosphere from Earth's surface to about ten miles up where most weather occurs.

Umbra—The completely dark shadow of a body.

Unconformity—The conspicuous absence of layers of sedimentary deposition that have been removed by erosional processes.

Uniformitarianism—An observation that the present is a key to the past.

Vacuum—A space completely void of matter.

Valence—The degree of combining power of an atom with other atoms. The valence is equal to the electrostatic charge which ions exhibit in chemical reactions.

Van Allen Belts—Doughnut-shaped areas surrounding Earth that contain charged particles, trapped from the solar wind by Earth's magnetic field.

Vector Quantity—A physical quantity that possesses both magnitude (size) and direction. Examples of vector quantities are force, velocity, and weight.

Vernal Equinox—The time during spring (northern hemisphere) when the sun crosses the equator. Occurs about March 21.

Vibration—Back-and-forth motion of a body. Vibrations are generally measured in cycles.

Virtual Image—An image that only appears when the eye follows diverging rays back to their apparent source. The illusionary image cannot be projected on a screen.

Volcano—A land form which develops during the accumulation of molten rock (lava) around a central vent.

Volt—The unit of measure of electrical potential.

Vulcanism—The processes by which molten rock moves and forms solid layers with Earth's crust.

Water Table—The level below which the ground is saturated with water.

Wave length—The distance from any given point on a wave to the corresponding point on the successive wave. See Figure 6.6.

Weight—A measure of the gravitational attractive force Earth or other heavenly body exerts on an object.

White Dwarf—A small star (about the size of Earth) that has exhausted most of its nuclear fuel and has collapsed to a small size.

Winter Solstice—The time during the winter (northern hemisphere) when the sun is farthest south of the equator. Occurs on December 21 or 22.

Work—The product of the force exerted on a body and the distance the body is moved in the direction of the force.

Zenith—The point in the sky directly overhead.

index

DNA (deoxyribonucleic acid), 63, **312**, 314, 424, 426
domains, **151**
Domain theory, **152**
Doppler redshift, **127,** 339
double helix, 312, **315**
dynamite, **299**
dyne, **33**

Earth,
  circumference, **18**
  core, **405**
  chemical composition of, **405,** 406
  crust, **406**
  decreasing spin, **386**
  earthquakes, **457,** 461
  future of, **389**
  orbital speed, **22**
  mantle, **405**
  primitive atmosphere of, **418**
  reversal of polarity, **402**
  shape, **383**
  surface features, **456**
Earthlight, **386**
echo, **128**
eclipses, **386**
eclipsing binary, **323**
effective value of AC, **156**
efficiency, **75**
effort, **75**
Einstein, Albert, **217**
element, electric, **176**
electric charge, **140,** 162, 163, 275
electric current, 135, **145,** 162
electric field, **156**
electric motor, **178**
electric potential, **143**
electric potential energy, **142**
electricity, **134,** 137, 156
electrodes, **136**
electromagnet, **152**
electromagnetic waves, **188,** 189, 190
electron, **137**
electron volt, **229**
electronics, **178**
electronegativity, **256**
electrons,
  properties, **60**
electroscope, **142**
elementary particles, **224,** 226
emitter, **181**
energy, **83,** 84, 217, 229
English system of measure, **7**
enzymes, **307,** 309, 311, 312, 315
epicenter, **460,** 462
epochs, **438**
Eratosthenes, **18**
erg, **75**
erosion, **480**
escape velocity, **50,** 327, 420

esters, **301**
ether, ethyl, **299**
ethyl acetate, **301**
ethylene, **295**
ethylene glycol, **298**
extrusive rock, **459**

Fahrenheit temperature scale, **67**
Faraday, Michael, **136**
farsightedness, **205**
fats, **302,** 303
faults, **460**
fauna, **437**
filament, **179**
fire damp, **293**
fission, **232,** 233, 237, 238, 242
fluorescent lamp, **174,** 175
focal length, **201**
focal point, **201**
folding, **457**
foliation, **470**
force, 26, 28, 30, 31, **52**
force and weight, **31**
formaldehyde, **423**
fossils, **436**
fossil fuels, **235**
Foucault pendulum experiment, **376**
Fox, Sidney, **425**
Franklin, Benjamin, **135**
frequency, **110,** 121, 154, 190
frequency modulation, **182**
friction, **90**
fructose, **308,** 309
fruit flavors, **301,** 302
fulcrum, **75**
fundamental tone, **116**
fungicide, **301**
fusion reactions, 246, **320,** 321, 322

galaxies,
  Andromeda, **336**
  barred spiral, **335**
  cluster of, **338**
  Coma Berenices, **333**
  elliptical, **336**
  irregular, **336**
  local group, **339**
  Messier 74, **333**
  Milky Way, **332**
  spiral, **335**
Galilei, Galileo, **17,** 26, 350, 359
gamma radiation, 225, **228,** 241, 244
Ganymede, 21, **372**
gases,
  kinetic molecular theory of, **65**
  properties, **64**
gasoline, **294**
Geiger-Muller system, **243**
generator, electric, **169,** 178
genes, **312**